THE
HUMAN EVOLUTION
COLORING BOOK

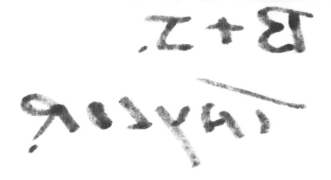

Also available in the Coloring Concepts series

Zoology Coloring Book by L. M. Elson

Human Brain Coloring Book by M. C. Diamond,
    A. B. Scheibel, and L. M. Elson

Marine Biology Coloring Book by T. M. Niesen

Botany Coloring Book by P. Young

Computer Concepts Coloring Book by B. M. Glotzer

Biology Coloring Book by R. D. Griffin

The
HUMAN
EVOLUTION
COLORING
BOOK

by Adrienne L. Zihlman

Illustrations by
Carla Simmons, Wynn Kapit,
Fran Milner, and Cyndie Clark-Huegel

 HarperPerennial

A Division of HarperCollins*Publishers*

Adrienne L. Zihlman is Professor of Anthropology at the University of California, Santa Cruz. She has done research on human origins, human locomotion and the role of women in evolution, and published numerous scientific articles. Her proposal that pygmy chimpanzees are the best living prototype of the common ancestor of humans and African apes was featured in the Science section of *Time* magazine. Dr. Zihlman's work has taken her several times to eastern and southern Africa, where she has studied the fossil evidence of human evolution. A regular participant in international conferences, her biography appears in *World Who's Who of Women, International Who's Who in Education,* and *American Men and Women of Science.*

This book was produced by Coloring
    Concepts Inc.
P.O. Box 324, Oakville, Ca. 94562

The book editors were Carol Denison
    and Joan Elson
Layout was by Wynn Kapit
The copy editor was Sylvia Stein
Type was set by Ampersand Design,
    San Francisco
Page makeup and production coordination
    was by Donna Davis
The proofreader was Sue Gamlen

ISBN: 0-06-460304-0

94 95 18 17 16 15

DEDICATION

This book is dedicated to my teacher
and colleague Sherwood Washburn,
University of California, Berkeley and to
my students at the University of
California at Santa Cruz.

ACKNOWLEDGMENTS

In the preparation of this book, I thank Larry Elson who encouraged me to take on the task and Wynn Kapit who worked closely with me as artist and provided continuity and encouragement throughout the project.

Catherine Borchert served as research assistant and generously contributed her vast knowledge, imagination, enthusiasm and hard work to this project and helped me survive many discouraging moments! Karen Wcislo and Lynda Brunker offered invaluable help on research and editing. Carla Simmons as artist helped make the book a reality.

Jerold Lowenstein as colleague, critic and companion sustained me throughout the project with both practical and emotional support. I cannot sufficiently express my appreciation.

Finally, I give love and thanks to Carl Zihlman, my father, who provided the long-time support and encouragement that led to this project, and to Ada Bales Zihlman, my grandmother and first teacher of anatomy, animal behavior, and ecology, who is responsible for shaping my subsequent interest in the evolution of it all.

TABLE OF
CONTENTS

Part II.   Genetics and Evolution

Part III. Characteristics of Living Primates

Part IV: Primate Evolution

Part V.   Fossil Evidence for Human Evolution

PRODUCER'S ACKNOWLEDGMENT

The Producer acknowledges, with gratitude, the reviews and constructive criticisms of Dr. Mary Ellen Morbeck of the University of Arizona, Dr. Mark L. Weiss of Wayne State University, Dr. Kathleen K. Smith of Duke University Medical Center, Dr. Clark Howell of University of California, Berkeley, Michael Guthrie of City College of San Francisco, and Betty Goerke of Mill Valley, California during the preparation of this work.

Coloring Concepts had tremendous support during the development of this and the other three books of this initial series (Botany, Human Evolution, Marine Biology, and Zoology). The people at Harper and Row, especially Irv Levey and Tom Dorsaneo, have encouraged us, responded to our needs, and demonstrated patience and understanding in our struggle to make a deadline. The following friends have directly aided our endeavor: Howard Nemerovski, Tom Larsen, Stu Boynton, Kathy Dahl, Bill and Danielle Brown, Don Jones, Terry Anderlini, John Moran, and Dr. Jack Lange. Their support and confidence in us is greatly appreciated. Dr. Lange (Lange Medical Publications) has given us advice and direction as we develop as publishers. His philosophy of publishing has been a source of inspiration to us. We are much indebted to Donna Davis who coordinated our book productions and the fine people who worked with her: Libbie Schock and Nancy Steele at Ampersand Design, Sue Gamlen, Bob Nemerovski, and our copy editors, Sylvia Stein and David Cross. The concern and reassurance of our friends Gene Mattingly, Regan Anderlini, Ken and Ute Christensen, Ellis and Merilyn Bowman, Maurice and Ann MacColl, Bob and Nancy Teasdale, Sharon Boldt, and Julie and Nancy at Schoonmaker provided the background so necessary for us to achieve our goal. We are very grateful.

Joan Warrington Elson
Lawrence M. Elson

July 1981

PREFACE

Four years ago, Larry Elson and I were independently asked to be expert witnesses involving evidence of footprints in a court case. We had been graduate students together at University of California at Berkeley (in different departments) but had not seen each other for ten years. Meanwhile, Larry had co-authored the hugely popular *Anatomy Coloring Book.* When we renewed our acquaintance, as witnesses for the defense, he asked me to consider writing a *Human Evolution Coloring Book.* Anatomy, of course, is a natural for this format, but since human evolution includes so much more than anatomy, I wondered for some time how this format might apply.

As I thought about it, I realized that many of the human characteristics I would be writing about are perfectly adapted for this kind of book—namely, color vision, hand-eye coordination, manual dexterity, and a brain especially evolved for tool-using!

As a student of Sherwood Washburn in the early 1960s, I learned the importance of short clear verbal presentations and good visual aids in effective teaching—lessons which I have applied to my own teaching of anthropology at the University of California at Santa Cruz during the past fifteen years. A fellow graduate student at Berkeley with whom I have remained friend and co-worker, Ted Grand, has also strongly influenced this book. An original and creative anthropologist/anatomist, Ted has always stressed the value of good illustrations for clearly communicating anatomical concepts. Doug Cramer, a fine and recognized illustrator as well as anthropologist, has shown me through our many collaborations how interrelated the two fields are. Sherwood Washburn, Ted Grand, and Doug Cramer have all, without knowing it, had an important role

in shaping this book, though I do not hold any of them responsible for its contents.

Several years of collaborating with Jerold Lowenstein on popular as well as technical articles have further taught me the skills and pleasures of communicating clearly and with humor to a wide audience. Recent joint work with paleontologist colleague Léo Laporte has further brought home to me the rewards of taking a broad view of one's own subject.

Ultimately, I undertook this book with my own students in mind, particularly those in my Anthropology I class, "Introduction to Human Evolution." Having experienced the difficulties of getting across certain technical concepts, like the structure and function of DNA, I could immediately see the didactic value of a coloring book based on these concepts. Many of my students have already benefited from using the *Anatomy Coloring Book.*

Naturally, this book has a point of view, one not always shared by other texts on human evolution. For example, I consider molecules to be facts about evolution as real as fossils and comparative anatomy, and so molecular data are given equal emphasis here with the more familiar evidence. Females, and young, so often invisible in works on the evolution of "man," appear here as equal participants. I have presented much of my own original research on comparative anatomy, fossils and pygmy chimpanzees but have tried to balance my own views with those of others. This is still a very controversial subject, human evolution. Many still deny that there is such a subject.

For me, it has been a long, hard path between those footprints Larry Elson and I examined over four years ago and the 3 million year old fossil human footprints on the cover of this book that mark the end of my effort—though only the beginning of the human journey. Each of these 111 plates is a condensate of dozens of scientific articles and book chapters and years of thought. Some of the plates may appear complex, but remember that they have been designed so that you can learn their contents through the coloring process, and they have been tested by both experts and novices to this end.

So prepare to take the human evolutionary journey from molecules to monkeys to modern people. All you need are some coloring implements and the primate brain-hand combination that has made us and our ancestors such successful survivors for the past 60 million years.

Adrienne L. Zihlman

July 1981
Santa Cruz, Ca.

INTRODUCTION

When embarking on a study of human evolution, one may be confused by the variety of seemingly unrelated topics (fossils, biochemistry, comparative anatomy, genetics, ecology, and others) to be explored. What do all of these have to do with the story of our ancestors?

Reconstructing human prehistory requires many kinds of knowledge. The evolutionary scientist must examine the biochemical processes of life, principles of physiology, behavior, population genetics and ecology, as well as the particular sequence of events which gave rise to the first vertebrates, the mammals, primates, and only recently, human beings. These are among the topics to be explored in this book.

The plate-by-plate visual presentation will teach you about the information available for the study of human origins. More importantly, you will learn how to organize the scattered pieces of this intricate puzzle. Illuminating lifeways no longer available for direct scrutiny entails *asking the right kinds of questions.* The 111 plates in this book are designed to illustrate how modern scientists ask questions by developing and testing hypotheses about human evolution. You will learn to evaluate for yourself the controversies which surround the story of our ancestry.

The coloring book is necessarily limited in scope. However, you are invited to use the bibliography at the end, which includes both popular and technical literature to get you started on investigations of your own.

Part I begins with a rapid excursion through earth history, a journey which presents a somewhat startling perspective on time and the human condition. During what is merely an instant in the eons of evolutionary time, humans *(Homo sapiens)* have populated almost every inhabitable region of the globe. Tragically, our species has managed to disrupt ecosystems to the degree that our own food and energy resources are now in danger of depletion. This brief overview of earth history reveals that although our biological evolution is an

extremely recent event, the rapidity of our recent *cultural evolution* (rise of industrial nations, technological control of resources, and the tremendous human population explosion during the past 10,000 years) is even more extraordinary. In view of the pace of change previously experienced on earth, the impact of human beings on this rather small planet is staggering indeed.

Part I illustrates our place in the animal kingdom, for in order to comprehend what it means to be human, it is essential that we understand our evolutionary heritage. We are vertebrates—meaning that we belong to the group of animals with backbones—of the class Mammalia which evolved from reptiles over 150 million years ago (mya). Thus, a number of illustrations concentrate on distinguishing mammals from reptiles. In coloring these plates, you will encounter two concepts which will emerge again and again in your study of human evolution. First, *an animal's lifestyle is largely determined by the kind of food it eats and the manner in which that food is acquired.* Second, *the anatomy, or form, of an organism reflects the way it behaves, or functions, in the particular environment it inhabits.*

Part I stresses that human evolution must be examined within the context of evolutionary principles that apply to all species, living and extinct. Such principles include *conservatism, adaptive radiation, convergence,* and *natural selection.*

Evolution is *conservative* in that *in the emergence of new life forms, new structures are modified from the old.* Coloring your way through the evolution of the mammalian ear and brain, for example, will reveal the continuity between diverse life forms, as well as the "ingenuity" of nature for finding new uses for inherited, but "outdated" characteristics.

Evolutionary theory emphasizes the role of the environment in shaping body form. Millions of years of evolution have modified the mammalian forelimb, for example, to enable some mammals to inhabit the skies, others the depths of the seas, and still others the slender tree limbs of the forest canopy. Nature has "experimented" with teeth to yield varieties of dentition, each adapted for very different styles of using the mouth. *The diversification of form to meet the demands of the environment and lifestyle is called adaptive radiation.*

The phenomenon of convergence further clarifies the role of the environment in evolutionary change. Plates 16 and 17 illustrate how animals which are only very distantly related—but which share similar habitats and lifestyles—may be almost identical in body build. Why should this be so?

The modern scientific theory of evolution by natural selection, first proposed by Charles Darwin in 1859, enables us to understand why each species is adapted to its particular environment. In fact, it allows us to *predict* what kinds of characteristics we might expect to find in animals which exploit certain resources. Darwin first noticed that within each species, *individuals vary* in their anatomical and behavioral characteristics, and furthermore, that these *variations are inheritable,* that is, passed on from parent to offspring. He found too, that *some individuals, by nature of their particular characteristics, are more suited for survival than are others.* "Suitability," of course, depends on the nature of the environment (food sources, predators, climate, etc.). The young Darwin also observed that *all species over-reproduce,* creating more progeny than will survive to reproductive maturity. From these observations he concluded that *those individuals whose characters best adapt them to the environment are more likely to survive and produce offspring.* In this way, some characteristics are promoted within populations, while others are "weeded out." Darwin called this evolutionary mechanism *natural selection.*

Thus, the amphibian forelimb was modified from the lateral fins of ancestral fish because natural selection operated over many generations to favor individuals whose limb structure was suited to life on dry land. Natural selection is also responsible for convergence. So strong is the role of the environment in determining what is adaptive—in determining what varieties of organisms will be favored—that even distantly related life forms like the North American placental mammals and the pouched Australian marsupials have nearly identical body forms when they share similar strategies for making a living.

In *Part II* you will learn about evolutionary process at the level of molecules. One hundred years ago, Darwin could discuss evolution without mentioning genetics, because "genes" as basic units of heredity had not yet been discovered. Mendel, in the 1860's, had set down some basic principles of inheritance which Darwin would have found compatible with his theory of natural selection had he been acquainted with them. However, Mendel's laws remained largely unknown in the scientific community until they were "rediscovered" at the turn of the century. Still, it was not until the years following 1953, after James Watson and Francis Crick discovered the structure of DNA, that anthropologists began to consider the role which genes play in evolutionary process. The modern synthetic theory of evolution emerged with the realization that evolution must be investigated simultaneously at the levels of genes, organisms, and populations.

Once the structure of DNA—the molecule which carries the hereditary information from one generation to the next—was known, the biochemical study of genes burgeoned forth with a whole new array of data for how evolution occurs. Genetics became established as a scientific discipline, and the definition of evolution became "changes in the gene frequencies of populations through time." Simultaneously, a background in genetics became essential for thinking about human origins. The biochemical behavior of genes revealed the source of variation within populations, variation which Darwin had observed but not been able to explain. Modern genetics also revolutionized Mendel's laws of inheritance, because now the units of inheritance can be identified as sequences of nucleotides along chromosomes, rather than as hereditary "particles," as Mendel had referred to them.

Today, the synthesis between the microscopic and macroscopic aspects of evolutionary process is by no means complete; for example, it remains unclear how changes in the genes are related to changes in anatomy and behavior. But it *is* clear that a fundamental component of evolutionary process takes place at the molecular level, and physical anthropologists can no longer content themselves with a specialty in comparative anatomy, paleontology, or primatology without also understanding the molecular basis of evolution.

Towards this end, Part II will introduce you to the basic principles of genetics, including the biochemistry of DNA, the genetic code, and the cellular processes by which proteins—the building blocks of bodies—are manufactured. You will learn about the latest biochemical evidence for evolution, the techniques by which scientists measure the rates at which genes change through time, and how cross-species comparisons of molecules can help us to construct family trees. It is important to understand the propensity of DNA for self-duplication, and for packaging itself in the sex cells for its passage to the next generation. Plates 32–36 cover Mendelian inheritance, which you—with your newly acquired understanding of the genes and DNA—will understand with greater sophistication than did even Mendel himself.

Part II also presents selected topics on the genetics of human adaptation: the major blood groups, lactose tolerance, and the sickle cell trait. These provide instructive examples of how chemical, biological, ecological, and cultural processes are all intertwined in the evolution of human populations.

Our place within the animal kingdom is within the group of mammals called primates. Appropriately, the theme of *Part III* is "What is a primate?" How do primates gather information about their environment? What do they eat? How do they move around and acquire food? You will learn about new methods for comparing the body builds of different species, and how to interpret what anatomical differences imply about behavior. *Primates are social animals,* with complex communication systems, strong ties between individuals, and an extensive socialization period during which youngsters must learn the communication and technical skills they need to be successful adults. Examining the central features of the primate adaptation sets the stage for evaluating what makes us human.

As you color the characteristics of the more primitive primates, the prosimians, and then those of the "higher" primates—the monkeys and apes—you will discover distinctive trends in primate evolution. Many human traits are further elaborations of these same trends, but others seem to represent qualitative "leaps": bipedalism and spoken language, for example, bestow on us a distinctive way of life not shared with other primates. However, due to the conservative nature of evolution, even bipedalism and human language have their precursors in primate ancestors; these are the kinds of preadaptations you want to be looking for as you work your way through this section.

Throughout Part III, I have emphasized *development*—of the locomotor system and of the face and dentition, for example—because it is important to realize that *natural selection modifies adult body build and behavior by operating on the growth patterns of immature individuals.* Restricting our comparisons to adult animals obscures what two species may have in common. Moreover, the routes by which each kind of primate becomes adapted to its particular lifestyle only become apparent when we examine the entire life cycle.

An important trend in the evolution of primate life cycles is illustrated on Plate 61. Monkeys and apes have extended periods of infancy and childhood, and reach reproductive maturity at a much later age than do prosimians. This trend attests to *the significance among higher primates of learning within the context of play.* Anthropologists emphasize the "plasticity" of the primate intellect, meaning that because primates, especially humans, are born in such immature stages of development, the role of social interactions in shaping adult behavior has become indispensable.

You will learn about the reproductive cycle of rhesus monkeys, which serves as a brief introduction to the relationship between hormones and social behavior. As you color, you will see that human beings share only some aspects of this cycle with nonhuman primates. By studying how other primates do things, you can formulate questions relevant to the study of human evolution. For example, human females have retained a monthly cycle of ovulation and menstruation, but unlike monkeys and apes, we do not have sexual swellings and behavioral estrus signaling fertility.

The last three plates in Part III deal with communication. As you color these plates, look for trends in communication which correlate with trends in the evolution of the primate senses. You will discover that primate vocalizations are quite a different phenomenon from human language. So let this be a clue to you as to how anthropologists might most successfully investigate what makes one certain primate such a highly cultural and symbol-dependent creature. This section of the book is designed to help you understand what characteristics we share with other primates, while simultaneously leading you to see the differences by which we might consider ourselves unique.

In *Part IV* we turn to the fragmentary fossil record, adding a time dimension to our characterizations of the primate adaptive pattern. You'll be introduced to a paleontological vocabulary, to maps of the earth as it existed millions of years in the past, and to the delightful world of living primates. As you study the primates of the past sixty million years, you will find that the anatomical specializations of our arboreal ancestors did not all evolve simultaneously. When the earliest primates took to the trees, they were equipped with typical mammalian senses, limbs, and claws, and were identifiable as primates only on the basis of their teeth and minute details of their skulls. Modern primate characters, such as depth perception, grasping hands with nails on padded fingertips, and a large brain evolved later, as did new locomotor and dietary patterns. The appearance of each of these higher primate characteristics at different periods of their history illustrates *the mosaic nature of primate evolution.* As you color, you will discover the order in which primate characters evolved, and the behavioral correlates which apparently accompanied each change in anatomy.

As you travel through the fossil record, you will encounter dozens of extinct and living primates. I have interspersed the family trees of living primates with representations of the fossil record. The family tree of living prosimians, for example, follows the plates on Paleocene and Eocene fossils. The rationale for this is two-fold: First, you will notice that the family trees are all based on what the molecules tell us about degrees of relatedness between different species of living primates. I want you to see how the molecular phylogenies are complementary to those based on dated fossil remains. *When fossils and molecules are considered together, they can, and do, create a coherent picture of primate evolution.* Second, the comparison of living and extinct species should help you to understand how primatologists make informed guesses about eating habits, locomotor patterns, and social behavior of extinct fossil species. Again, the emphasis is on the relationship between *form and function.*

Whereas in Part III the focus is on characteristics shared by primates, Part IV stresses the diversity of primate patterns. You will learn what makes a tree shrew more primitive than other prosimians and how to distinguish the two kinds of New World monkeys, the two Old World monkey groups, and the greater and lesser apes. By studying the life-styles of many different primates you can discover the range of strategies adopted for finding and utilizing food sources, for avoiding predators, for socializing youngsters, for allocating space among groups, and for getting along with group members. The species-specific relationships between anatomy, environment, and behavior are the analogies used to bring the primate fossils to life.

In light of the conservative pattern of evolutionary process, it is appropriate to look to our nearest living relatives, the gorillas and chimpanzees, for clues about early humans. I suggest that the anatomy of the pygmy chimpanzee can help us to imagine what the common ancestor of apes and humans looked like. Present-day chimpanzees can also help us to imagine the *behavior* of the earliest humans, behaviors which would have enabled them to leave the forest they had occupied for millions of years and stake out a new lifestyle and a new direction of evolution on the African savanna.

As you turn to the human fossil record in *Part V,* you will have a basic background in evolutionary principles, genetics, and primate behavior, anatomy, and paleontology. But there are still a few more technical and interpretive skills that you will need to make sense of the brittle bones and teeth which fossil hunters dig up.

What do these fragments tell us about the early humans? How were they meeting nutritional requirements, avoiding predators, and organizing their social groups? How has the human lifestyle changed during the past five million years? To answer these questions, you will need to consider the time period and ecological zone in which the fossils were deposited, and if possible, reconstruct the particular events by which fossil specimens enter the paleontological record. Analogies must be drawn from the *form–function* relationships in living species in order to infer the behavior of humans long extinct. Our material evidence for these scenarios is mainly human teeth and bones, the plant and animal remains from nonhuman species, and stone tools in deposits younger than two million years of age. Clearly, the analysis of human fossils requires a team of experts, for no one scholar has such a range of needed scientific specialties. I have drawn from several disciplines in this final section to give you a brief overview of the many kinds of data which one must bring to bear on interpreting human fossils.

As you color your way through Part V, you will tour four major African hominid sites, and examine in detail the major fossils at each. Analyses of the dental, cranial, and postcranial characteristics of these early human fossils reveal that there was more than one species of hominids inhabiting Africa between two and one million years ago: one robust and one gracile, each perhaps occupying slightly different eco-niches on the savanna or woodland fringes. The teeth of these two australopithecine groups suggest that they had different diets. You will examine the pelvis and the earliest footprints which document that humans were indeed walking upright before they acquired their large brains.

Tools were probably already being used by the human-like ancestor of 5 mya, if living chimpanzees are representative of how these proto-humans might have behaved. However, the archeological record of tools—of simple stone flakes and choppers probably used to prepare vegetable foods, to skin carcasses of scavenged animals, and perhaps to sharpen wooden digging sticks—begins only about 2 mya. We need to imagine that before that date, early humans used organic tools which would have decomposed and left no trace. Like gathering-hunting peoples today, the australopithecines were probably using digging sticks to get at roots and tubers beneath the ground. It seems likely that they invented containers to carry both small food items and their increasingly immaturely born infants.

You will follow the early humans as they leave Africa after 1½ mya, migrating first to the Middle East, China, and Indonesia, and later to Europe. The illustrations to be colored present anatomy, site distribution, and reconstructed behavior patterns of *Homo erectus* and early *Homo sapiens,* including the famous Neanderthals.

The last two plates focus on the evolution of the human brain, an evolution of unparalleled speed and magnitude. Though the complexities of the human brain are far too vast to detail here, I conclude with a discussion of brain size and brain reorganization because herein lies the key to human adaptation, that is, human culture. By nature of our huge primate brain, we possess the capacity for symbolizing, for acquiring language, for complex thought about the distant past, the future, the abstract, without mention of which the human story is radically incomplete. From here, the physical anthropologist must turn her ear and her attention to cultural anthropologists for the lessons of how our species has adapted in different parts of the world through a diversity of lifestyles, each requiring a subtle understanding of rules, norms, and codes which vary from culture to culture.

The study of our origins is a perspective we must now use in reflecting upon our potentials and our limitations for the future. Social anthropologist Peter Wilson has recently dubbed us the "promising primate"; the promise lies in our ability to responsibly and collectively manage the social and physical milieu which evolution has fitted us to inhabit.

HOW TO USE THIS BOOK
COLORING INSTRUCTIONS

1. This is a book of illustrations (plates) and related text pages in which you (the colorer) color each structure indicated the same color as its name (title), both of which are linked by identical numbers (subscripts). In the doing of this, you will be able to relate identically colored name and structure at a glance. Structural relationships become apparent as visual orientation is developed. These insights, plus the opportunity to display a number of colors in a visually pleasing pattern, provide a rewarding learning experience.

2. You will need coloring instruments. Colored pencils or colored felt-tip pens are recommended. Laundry markers (with waterproof colors) and crayons are not recommended: the former because they stain through the paper, and the latter because they are coarse, messy and produce unnatural colors.

3. The organization of illustrations and text is based on the author's overall perspective of the subject and may follow, in some instances, the order of presentation of a formal course of instruction on the subject. To achieve maximum benefit of instruction, you should color the plates in the order presented, at least within each group or section. Some plates may seem intimidating at first glance, even after reviewing the coloring notes and instructions. However, once you begin coloring the plate in order of presentation of titles and reading the text, the illustrations will begin to have meaning and relationships of different parts will become clear.

4. As you come to each plate, look over the entire illustration(s) and note the arrangement and order of titles. Count the number of subscripts to find the number of colors you will need. Then scan the coloring instructions (printed in bold face type) for further guidance. Be sure to color in the order given by the instructions. Most of the time this means starting at the top of the plate with (A) and coloring in alphabetical order. Contemplate a number of color arrangements before starting. In some cases,

you may want to color related forms with different shades of the same color; in other cases, contrast is desirable. In cases where a natural appearance is desirable, the coloring instructions may guide you or you may choose colors based on you own knowledge and observations. One of the most important considerations is to link the structure and its title (printed in large outline or blank letters) with the same color. If the structure to be colored has parts taking several colors, you might color its title as a mosaic of the same colors. It is recommended that you color the title first and then its related structure. If the identifying subscript lies within the structure to be colored and is obscured by the color used, you may have trouble finding its related title unless you colored it first.

5. In some cases, a plate of illustrations will require more colors than you have in your possession. Forced to use a color twice or thrice on the same plate, you must take care to prevent confusion in identification and review by employing them on separate areas well away from one another. On occasion, you may be asked to use colors on a plate that were used for the same structure on a previous related plate. In this case, color their titles first regardless of where they appear on the plate. Then go back to the top of the title list and begin coloring in the usual sequence. In this way, you will be prevented from using a color already specified for another structure.

6. Symbols used throughout the book are explained below. Once you understand and master the mechanics, you will find room for considerable creativity in coloring each plate. Now turn to any plate and note:

　a. Areas to be colored are separated from adjacent areas by heavy outlines. Lighter lines represent background, suggest texture, or define form and (in the absence of "don't color" symbols) should be colored over. If the colors you used are light enough, these texture lines may show through, in which case you may wish to draw darker or heavier over these lines to add a three-dimensional effect. Some boundaries between coloring zones may be represented by a dot or two or dotted lines. These represent a division of names or titles and indicate that an actual structural boundary may not exist or, at best, is not clearly visible.

b. As a general rule, large areas should be colored with light colors and dark colors should be used for small areas. Take care with very dark colors: they obscure detail, identifying subscripts, and texture lines or stippling. In some cases, a structure will be identified by two subscripts (e.g., A + D). This indicates you are looking at one structure overlying another. In this case, two light colors are recommended for coloring the two overlapping structures.

c. Any outline-lettered word followed by a small capitalized letter (subscript) should be colored. In most cases, there will be a related structure or area to color. If not, the symbol N.S. (not shown) will follow the word; or, the word functions as a heading or subheading and is colored black (●) or gray (★). Outline titles (headings) with no subscript following are to be left uncolored.

d. In the event structures are duplicated on a plate, as in left and right parts, branches, or serial (segmented) parts, only one may be labeled with a subscript. Without boundary restrictions or instructions to the contrary, these like structures should all be given the same color.

e. In looking over a number of plates, you will see some of the following symbols:

●	= color black; generally reserved for headings/subheadings
★	= color gray; generally reserved for headings/subheadings
-¦-	= do not color
A( )	= set next to titles subscript; signals this structure composed of parts listed below with same letter but different exponents; receives same color; only its parts are labeled in illustration
A^1, A^2, etc.	= identical letter with different exponents implies parts so labeled are sufficiently related to receive same color
N.S.	= not shown
BRS/TRIB	= branches/tributary(-ies)
A/As	= artery/ateries
V/Vs	= vein/veins
M/Ms	= muscle/muscles
N/Ns	= nerve/nerves

7. In the text, certain words are set in *italics*. According to convention, the generic name and species of an animal or plant are set this way (e.g. *Homo sapiens*). In addition, the title of any structure to be colored on the related (facing) plate is set in italics (except for headings and subheadings). This is to enable you to quickly spot in the text the title of a structure to be colored.

1
EARTH HISTORY. PRECAMBRIAN: EVOLUTION OF EARLY LIFE

The geological timetable divides earth history into five eras. Choose five colors for these eras and continue to use them for the next three plates. This plate focuses on the first two eras: the Archean and the Proterozoic, together known as the Precambrian, which account for over 90 percent of all earth history. Color the title for each era as discussed in the text. For each era, color the corresponding section of the time scale and the date to the left of the scale. Color the conditions of the atmosphere and the important events for each time period to the right of the time scale. Use a light color for G.

Approximately *4,500 million years ago (mya),* at the beginning of the *Archean* era, the *earth* began to *form* by condensation of dust particles into a molten mass. As this mass began to cool, a rock *crust* formed on the surface. The oldest rocks known on earth were formed about *3,800 mya.* A primitive atmosphere was formed by gases escaping from the molten rock, composed primarily of hydrogen, carbon dioxide, ammonia, and water vapor. There was *no free oxygen* (O_2) for the first third of earth history.

As the earth continued to cool, the dense layer of water vapor began to condense, flooding the planet with rains, which formed the oceans. In these waters, *biochemical evolution* began as organic molecules (carbon-rich acids, alcohols, and simple carbohydrates) were created by chemical reactions between inorganic molecules. These organic molecules were the essential building blocks needed for the emergence of *earliest life,* about *3,500 mya.*

Early life forms were single-celled *prokaryotes,* cells without a nucleus, similar to modern-day *bacteria.* The earliest *prokaryotes* fed on the "organic soup" of the ancient seas.

The early *prokaryotes* eventually began to deplete the supply of organic molecules upon which they fed. Those organisms that could synthesize their own energy source survived. About *3,000 mya* autotrophic (self-feeding) *prokaryotes,* like *blue-green algae,*

evolved. These single-celled organisms evolved *photosynthesis,* a metabolic process using the energy of the sun to combine carbon dioxide and water to form simple sugars for energy storage. The early atmosphere was rich in carbon dioxide, but as a by-product of *photosynthesis, free oxygen* began to accumulate in the atmosphere.

Early in the *Proterozoic,* the era of "proto-life," the *prokaryotes diversified* into many types, shown here by a variety of single-celled life forms. Although it is difficult to tell from the fossil record, *eukaryotic organisms* may have first appeared as early as *1,200 mya. Eukaryotes,* which package their hereditary material in a cell nucleus (see Plate 19), evolved new methods for cellular division and the passage of hereditary information from one generation to the next. Soon after the evolution of *eukaryotes,* the *plant and animal* phyla diverged. Although this divergence cannot be seen in the fossil record, evidence from the study of proteins (see Plate 28) points to a split in the *Proterozoic.*

The next major evolutionary event was the development of *sexual reproduction.* Up to this point, *prokaryotes* and *eukaryotes* had reproduced asexually, with each cell duplicating its genetic material and passing on a complete copy to two identical daughter cells. *Sexual reproduction* probably evolved around *1,000 mya.* In this new reproductive process, two individuals rather than one contribute hereditary material to form a new organism. The combination of hereditary information from two different parent organisms created more diversity and enhanced the rate of *plant and animal* differentiation in the late *Proterozoic.*

Multicellular plants and animals first appear in the fossil record about *750 mya.* These organisms developed tissue systems of specialized cells and structurally complex body forms. The early soft-bodied animals did not preserve well in the fossil record. *Plant and animal expansion* continues into the *Mesozoic* and *Cenozoic.*

EVOLUTION OF EARLY LIFE.★

ERAS.●
ARCHEAN$_A$
PROTEROZOIC$_B$
PALEOZOIC$_C$
MESOZOIC$_D$
CENOZOIC$_E$

ATMOSPHERE.●
NO FREE OXYGEN (O$_2$)$_F$
FREE OXYGEN
   ACCUMULATES$_G$

IMPORTANT EVENTS.●
EARTH FORMATION$_H$
BIOCHEMICAL EVOLUTION$_I$
EARTH'S CRUST$_J$
EARLIEST LIFE (PROKARYOTE)$_{K(\,)}$
   BACTERIA$_{K1}$
   STROMATOLITES$_{K2}$
PHOTOSYNTHESIS$_L$
   BLUE-GREEN ALGAE$_{L1}$
PROKARYOTES DIVERSIFY$_M$
EUKARYOTES APPEAR$_N$
SEXUAL REPRODUCTION$_O$
MULTICELLULAR PLANTS$_P$
MULTICELLULAR ANIMALS (FIRST
   INVERTEBRATES)$_{P1}$
PLANT AND ANIMAL
   EXPANSION$_Q$

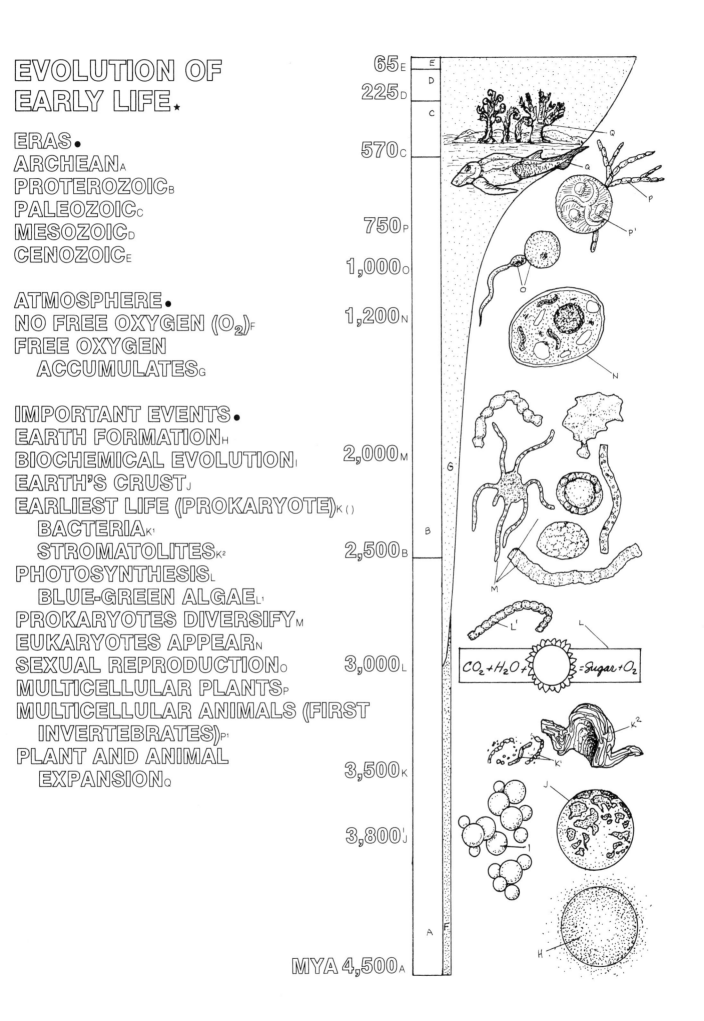

MYA	
65$_E$	E
225$_D$	D
570$_C$	C
750$_P$	
1,000$_O$	
1,200$_N$	
2,000$_M$	
2,500$_B$	B
3,000$_L$	
3,500$_K$	
3,800$_J$	
4,500$_A$	A

$$CO_2 + H_2O + \bigcirc = Sugar + O_2$$

2
EARTH HISTORY. PALEOZOIC: LIFE LEAVES THE SEA

The accumulation of free oxygen in the atmosphere had two important consequences. First, an abundant supply of oxygen created the conditions necessary for the development of multicellular animals, as their metabolism requires oxygen for the efficient extraction of energy from carbohydrates. Second, oxygen accumulating in the upper reaches of the atmosphere created an ozone layer, which filters out the radiation destructive to DNA, the hereditary material in every life form.

Continue using the same five colors for the geological eras. Color the era titles and the corresponding section of the time scale on the bottom left of the page. Choose six light colors for the periods of the Paleozoic. Color the Cambrian section of the large Paleozoic time scale, then color over the Cambrian vignette at the bottom of the page. Work upward, reading about events occurring in each period as you color.

During the *Cambrian,* life remained confined to the seas. Green algae — the probable ancestor of the land plants — appeared, and many kinds of soft-bodied invertebrates flourished. This period marks the appearance of the shelled arthropods, ancestral to modern insects, spiders, lobsters, and crabs. The most common arthropods during the *Paleozoic* were the trilobites, a diverse group, now extinct.

Numerous predators including the cephalopods, represented here by the ancient squid, evolved during the *Ordovician.* The first vertebrates appeared. These were jawless, armored fish whose internal skeleton provided support for their nervous systems, musculature, and gut organs without sacrificing flexibility and mobility. Their external armor protected them from predators.

The *Silurian* period provides our earliest evidence for life on land. The fossil record contains impressions of vascular plants with an internal transport system for moving nutrients between roots and leaves and a waxy cuticle layer to protect leaves and stems from drying out. These early plants were confined to swampy areas so that they could reproduce in water.

In the seas, armored placoderms (fish) evolved an upper and lower jaw from the first gill arch of the jawless ancestors, setting an important precedent for new herbivorous and carnivorous lifestyles. Placoderms had paired fins that set the stage for vertebrate invasion of the land.

All forms of fish proliferated during the *Devonian.* The most important event in the seas was the evolution of the bony fish. True bone provided greater rigidity in the internal skeleton, and descendants of the bony fish were soon to begin moving around on dry land. Taking advantage of new food sources provided by abundant land plants, the earliest amphibians were able to utilize two further adaptations for land dwelling. First, lobed fins with bony internal skeletons allowed the earliest amphibians to move from pond to pond. Second, some bony fish had adapted to life in stagnant ponds, where oxygen levels in the water were insufficient. Many fish had an internal air balloon, known as a swim bladder, that evolved into a primitive lung and allowed some fish to come to the surface to gulp air. Amphibians evolved from these "lunged fish."

The *Carboniferous* was a period of increasing dryness. Gymnosperms ("naked seed"), such as conifers (nonflowering, seed-bearing plants), evolved and replaced seedless plants. The first reptiles evolved from amphibian ancestors. Their innovations for land dwelling included a hard-shelled egg and dry scaly skin, which retains the body's moisture. Thus, both seed plants and reptiles evolved mechanisms that alleviated their restriction to water environments for reproduction.

During the *Permian,* the swampy primitive forests of giant ferns were replaced by gymnosperm forests. While the amphibians and insects flourished, the reptiles diversified, and from this geological period comes the oldest reptilian fossil egg. The emerging reptiles include *Dimetrodon,* the sail-backed lizard (shown here), one of the mammallike reptiles that marks the branch of the evolutionary tree leading to early mammals, primates, and, still millions of years in the future, humankind.

LIFE LEAVES THE SEA.★

ERAS.●
PRECAMBRIAN.★
  ARCHEAN_A
    PROTEROZOIC_B
PALEOZOIC_C
MESOZOIC_D
CENOZOIC_E

PALEOZOIC_C
  CAMBRIAN_F
  ORDOVICIAN_G
  SILURIAN_H
  DEVONIAN_I
  CARBONIFEROUS_J
  PERMIAN_K

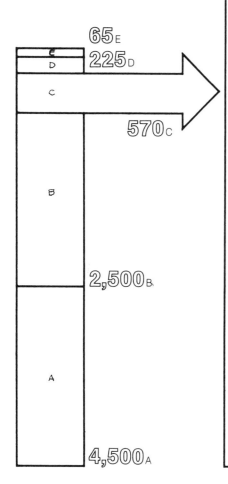

65_E
225_D
570_C
2,500_B
4,500_A

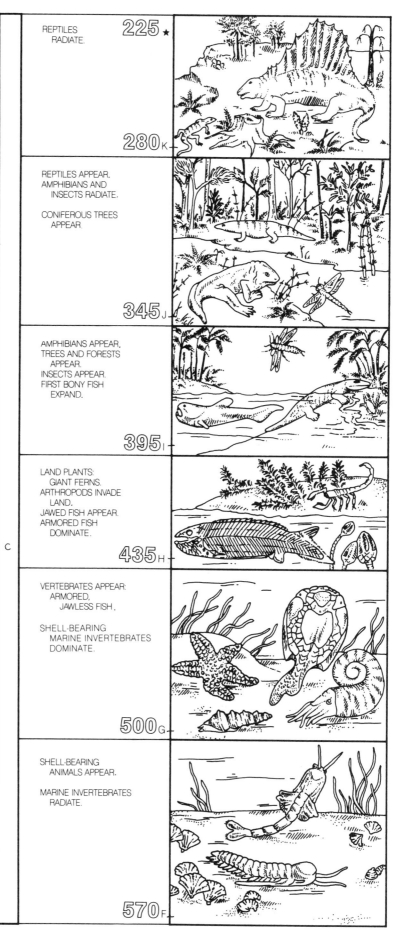

REPTILES RADIATE.
225 ★
280_K

REPTILES APPEAR.
AMPHIBIANS AND INSECTS RADIATE.

CONIFEROUS TREES APPEAR.
345_J

AMPHIBIANS APPEAR.
TREES AND FORESTS APPEAR.
INSECTS APPEAR.
FIRST BONY FISH EXPAND.
395_I

LAND PLANTS:
  GIANT FERNS.
ARTHROPODS INVADE LAND.
JAWED FISH APPEAR.
ARMORED FISH DOMINATE.
435_H

VERTEBRATES APPEAR:
  ARMORED, JAWLESS FISH.

SHELL-BEARING MARINE INVERTEBRATES DOMINATE.
500_G

SHELL-BEARING ANIMALS APPEAR.

MARINE INVERTEBRATES RADIATE.
570_F

3
EARTH HISTORY. MESOZOIC: MAMMALS EMERGE, REPTILES REIGN

The *Mesozoic* ("middle life") era spans 160 million years and is divided into three periods: *Triassic, Jurassic,* and *Cretaceous.* This era is called "the age of the reptiles" because they were the dominant land vertebrates. Birds, mammals, flowering plants, and many modern insects appeared for the first time. Broadleaf flowering trees like elm, oak, and maple became common. Ammonites came to dominate the seas but were extinct by the end of the era. Lands and seas changed; new mountain ranges developed. The flowering plants spread, pollinated by insects, providing mammals with plant and insect food.

Continue to use your five colors for the eras and color those first. Then proceed to the Triassic, Jurassic, and Cretaceous periods, using light colors and the coloring plan of Plate 2.

During the *Triassic,* reptiles became more successful due to biological innovations such as the protected eggshell and advanced body structure. Mammallike reptiles were diverse and abundant until the end of the *Triassic,* when dinosaurs first appeared. The alligatorlike phytosaur pictured here is a relative of the dinosaurs. Pine, fir, and cedar arose in the *Triassic,* forming great forests. Today's redwood forest is the closest approximation of the great conifer-cycad forests of the *Triassic* and *Jurassic.* At the *Triassic-Jurassic* border, mammals began to appear. They were quite small, with shrew-sized jaws and teeth. Their fossil remains are found in the western United States, Europe, and South Africa.

During the *Jurassic,* which began *180 mya,* reptiles ruled the air, land, and sea. Winged reptiles ranged in size from those smaller than a sparrow to those with 1.25 meter wingspans. The brontosaurus shown here is one of three groups of dinosaurs, the land reptiles. Streamlined marine carnivores, including ichthyosaurs ("fish reptiles") and plesiosaurs ("near reptiles") dominated the seas. The oldest known bird, *Archeopteryx,* had feathers like modern birds but had reptilian teeth and a reptilian skeleton and short wings relative to body length.

Conifer-cycad-gingko forests were widespread at that time and ferns were common. There were several kinds of primitive mammals, known as *Mesozoic* mammals, the largest the size of a cat. They fed on plants and insects, and they did not constitute an abundant or important part of animal life on land.

The *Cretaceous* ("chalky," for its characteristic deposits), began *135 mya* and lasted for seventy million years. During this period, the Andes and Rocky Mountains were uplifted. Dinosaurs continued to be widespread on every continent, and there were many varieties, like the horned dinosaur pictured here. The carnivorous *Tyrannosaurus rex* stood 6 meters high, with a skull 1 meter long. There were giant turtles and sea-going mosasaurs. Large flying reptiles such as the *Pteranodon* still abounded, and there were diving birds and strong fliers.

Mammals were still small and did not dominate the land. Two distinct groups emerged, the marsupials (opossumlike pouched creatures), and the placentals (mostly insectivores). Living representatives of these groups are distinguished from each other by their type of reproduction. Marsupials lack a placenta, are born very immature, and develop in a pouch. Placentals (named for the cake-shaped uterine structure by which a fetus is nourished) have a longer period of development in the uterus and so emerge more mature. In fossils of this age, the two groups can only be distinguished by their teeth.

The most important new arrivals in this era were the flowering (seed-bearing) plants, the angiosperms ("covered seeds"). (Today 96 percent of vascular plants are angiosperms.) The enclosed seed permits the development of fleshy, edible fruit, sought after by birds and animals, which ate the fruit and dispersed the seeds. Earlier plants, the gymnosperms, had relied on the wind to carry pollen from plant to plant. As birds and mammals guaranteed dispersal of plant seeds, the nutrition provided by angiosperm fruit, seeds, flowers, shoots, and leaves, made it possible for birds and mammals to become increasingly successful. Later, primates exploited these foods, too.

At the close of the *Cretaceous,* there was widespread extinction of many reptiles and of all the ammonites and dinosaurs.

MAMMALS EMERGE, REPTILES REIGN ★

ERAS ■
  PRECAMBRIAN ★
    ARCHEAN_A
    PROTEROZOIC_B
  PALEOZOIC_C
  MESOZOIC_D
  CENOZOIC_E

MESOZOIC_D
  TRIASSIC_F
  JURASSIC_G
  CRETACEOUS_H

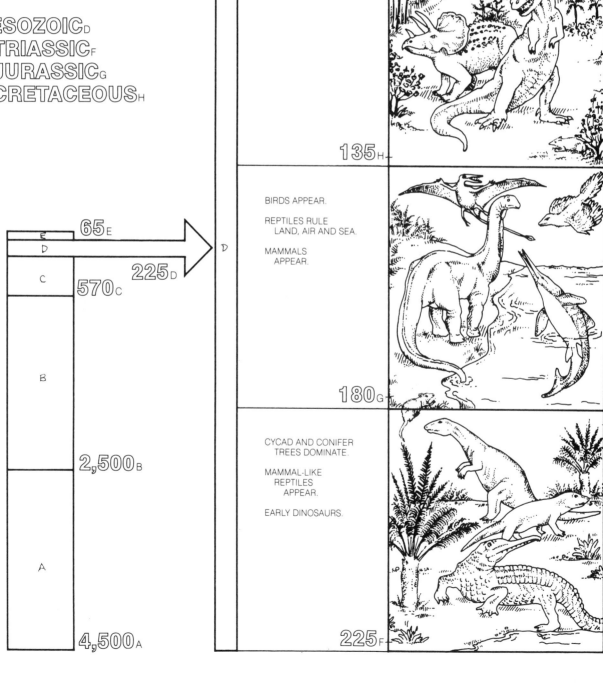

EXTINCTION OF LARGE REPTILES.

MAMMAL RADIATION BEGINS.

ANGIOSPERM PLANTS DOMINATE.

65 ★

135_H

BIRDS APPEAR.

REPTILES RULE LAND, AIR AND SEA.

MAMMALS APPEAR.

180_G

CYCAD AND CONIFER TREES DOMINATE.

MAMMAL-LIKE REPTILES APPEAR.

EARLY DINOSAURS.

225_F

65_E
D
225_D
C
570_C
B
2,500_B
A
4,500_A

D

4
EARTH HISTORY. CENOZOIC: RADIATION OF THE PRIMATES

The extinction of the ruling reptiles and the expansion of the angiosperms at the close of the *Mesozoic* left a variety of habitats available for exploitation by mammals. Even before the last of the dinosaurs had disappeared, several orders of mammals, including the primates, had appeared.

The early mammals were generally small in body size and had relatively small brains. Their longish snouts attest to the importance of smell in finding food and mates. Radiating during the *Cenozoic* from a small Cretaceous ancestor, the mammals quickly took to the sea and to the air, and dominated the land as well, like the reptiles before them. Ancestral whales, bats, elephants, hoofed animals, and carnivores originated and spread during this era, "the age of mammals." This plate focuses on the radiation of primates, the order of mammals to which the species *Homo sapiens* belongs.

Again, use your five original colors for the five eras and color those first. Then color over the events of the Tertiary period, which includes the Paleocene, Eocene, Oligocene, Miocene, and Pliocene epochs, as they are discussed in the text.

The oldest evidence of fossil primates consists of jaws and teeth from the very late Cretaceous in the Rocky Mountains and the early *Paleocene* of the western United States. These early primates were small, insect eating, and possibly ground living. In the *Paleocene,* we begin to see a diversity of primates, including some that took to the trees and supplemented their diet of insects with seeds, nuts, fruits, and leaves (Plate 69).

More solid evidence appears during the *Eocene* that primates were tree dwelling (arboreal; arbor = tree) animals. The *Eocene* primates had evolved grasping hands and feet, with opposable thumbs and great toes. Claws evolved into nails with fleshy tactile pads beneath them, which allowed better grasping of branches and gathering of small food items. Early primates had slightly longer lower limbs for power in jumping from branch to branch. Their eye sockets were directed toward the front of the skull, indicating development of stereoscopic vision and depth perception which contributed to skill in leaping among tree branches.

Prosimians were abundant in the tropical *Eocene* forests, but decreased in number during the *Oligocene*. At this time the primates developed completely enclosed bony eye sockets and a shorter snout, suggesting changes in the senses, with less reliance on smell and greater reliance on eyesight.

During the *Miocene,* there was increasing seasonality of climates and development of savannas around the world. In Africa, we find the earliest evidence of apes, the dryopithecines, and the closely related ramapithecines, which, by the middle of late *Miocene,* are found over Europe, Africa, and Asia (Plates 83 and 84). These early apes had teeth similar to modern apes, but their limbs resembled living Old World monkeys.

Around 5 mya, at the beginning of the *Pliocene,* the earliest hominids, direct human ancestors, stepped into the picture. They were adapted for bipedal, erect walking, but with small brains, similar to their ape ancestors, and obtained their food using tools made of wood on the African savannas. Bipedal locomotion for walking and carrying, tool using and regular food sharing are the main ingredients of this early savanna way of life.

Complete the plate by coloring the events of the Quaternary period, which includes the Pleistocene and Holocene epochs.

In the *Pleistocene,* the human fossil record is supplemented with evidence of human activities: first with the appearance of stone tools, found later with food remains.

The beginning of the *Holocene,* or recent epoch, marks the first human "revolution" with the domestication of plants and animals around 10,000 years ago. Cultivation of plants and the keeping of animal herds allowed humans to establish a greater degree of control over food sources. Though our species continues to evolve biologically (Plates 41 and 42), the remainder of our story of human evolution is primarily a cultural one, told by archeologists, historians, and storytellers through the ages.

Turn your attention again to the bar graph of all earth history on the bottom left of the plate. You can now see what an incredibly small proportion (five million years) of our evolutionary heritage is "uniquely human." For almost 3,500 million years, other life forms preceded us on this earth.

RADIATION OF THE PRIMATES.

ERAS.
PRECAMBRIAN.
ARCHEAN_A
PROTEROZOIC_B
PALEOZOIC_C
MESOZOIC_D
CENOZOIC_E
TERTIARY_F
PALEOCENE_G
EOCENE_H
OLIGOCENE_I
MIOCENE_J
PLIOCENE_K
QUATERNARY_L
PLEISTOCENE_M
HOLOCENE_N

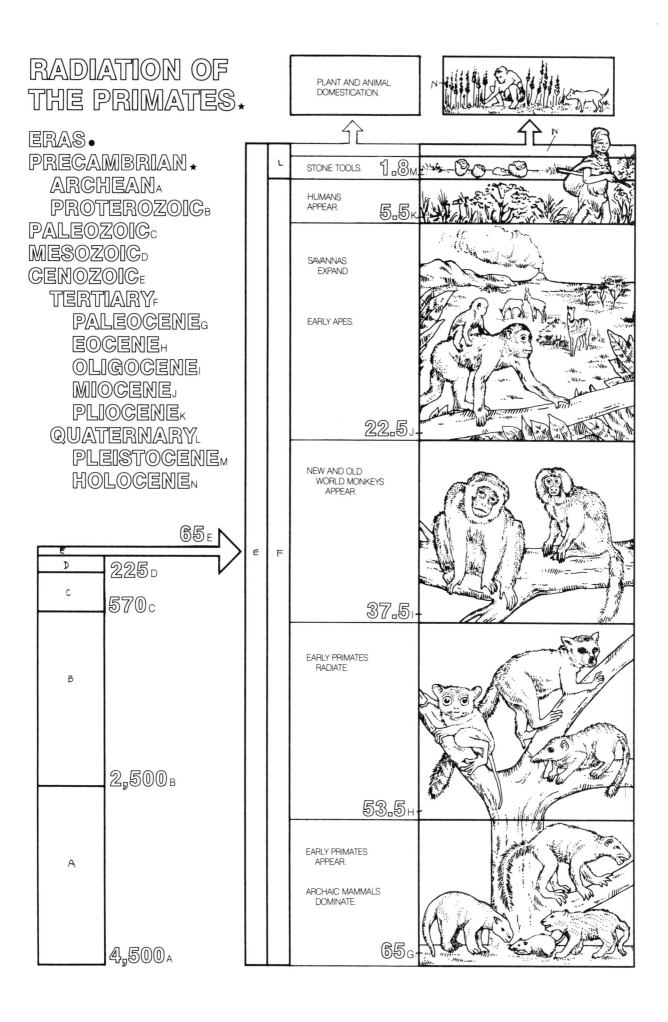

PLANT AND ANIMAL DOMESTICATION.

L — STONE TOOLS. 1.8_M

HUMANS APPEAR. 5.5_K

SAVANNAS EXPAND

EARLY APES.

22.5_J

NEW AND OLD WORLD MONKEYS APPEAR.

37.5_I

EARLY PRIMATES RADIATE.

53.5_H

EARLY PRIMATES APPEAR.

ARCHAIC MAMMALS DOMINATE.

65_G

E
D — 225_D
C — 570_C
B — 2,500_B
A — 4,500_A

65_E

E   F

5
THE KINDS OF PRIMATES

Primates are one of eighteen orders of living mammals, adapted for tree living through prehensile (grasping) hands and feet for climbing. Primates rely a great deal on their excellent vision and less on the sense of smell than do other mammals. The primate diet is omnivorous, consisting of combinations of fruits, leaves, insects, and small prey. Primate mothers carry their young on their bodies until weaned, and strong mother-infant ties develop. Social groups and a relatively large brain are other primate features.

Within the order primates there are many specialized adaptations in the various groups. The living primates will be studied in detail in Parts III and IV. This plate introduces the kinds of primates living today and their evolutionary relationships to each other.

Primates consist of three major groups: *tree shrews*, prosimians, and anthropoids.

Color the tree shrew, the most distantly related of the primates. Color the tree shrew branch of the graph. The graph illustrates the evolutionary relationships and approximate separation times of each primate group.

The numerous species of *tree shrew* range in size from a mouse to a rat and do not look much like other primates. However, they seem to be more closely related to the primates than to any other mammal; so we include them here. Living *tree shrews* serve as a model to help us understand primate beginnings.

Color the prosimians, the tarsier, lemur, and loris and galago. A loris represents the lorises and galagos here.

Tarsiers are small, insect eating, and nocturnal and live in Asian tropical forests. The *lemurs* live on Mada-

gascar (the only nonhuman primates living there). They are a diverse group, ranging from the mouse-sized dwarf *lemur* to the cat-sized ring-tailed *lemur* (pictured here), to the large indri. Notice that the *lorises and galagos,* a third kind of prosimian, separated from the *lemurs* about 55 mya. They are small, nocturnal, tree-living primates, and we explore their ecology in Plate 72.

Color the anthropoids, which diversified later but originated from some ancient prosimianlike ancestor.

The radiation of the anthropoids began 35–40 mya when the *New World monkeys* separated from other anthropoids. In Central and South America, the only nonhuman primates are monkeys, including the tiny marmosets and the larger forms with prehensile tails, like the woolly monkey pictured here.

After the separation of the *New World monkeys,* the *Old World monkeys* and the line leading to *apes* and *humans* (this group is referred to as the hominoids) shared a common evolutionary history until their separation about 20 mya.

As their name implies, *Old World monkeys* live in a variety of habitats in Africa and Asia; a number of species feed and move on the ground as well as in trees.

Humans separated from the other *apes* about 5 mya. The *apes* include chimpanzees, pictured here, and gorillas; both live in central Africa. The gibbons, siamangs, and orangutans live in tropical Asia.

Our closest living relatives are the African *apes,* and we share many anatomical and behavioral features with them. These similarities and differences will be explored in Parts III and IV.

A detailed classification of the kinds of living primates appears in the Appendix.

KINDS OF PRIMATES.★

TREE SHREWA

PROSIMIAN.●
TARSIERB
LEMURC
LORIS AND GALAGOD

ANTHROPOID.●
NEW WORLD MONKEYE
OLD WORLD MONKEYF
APEG
HUMANH

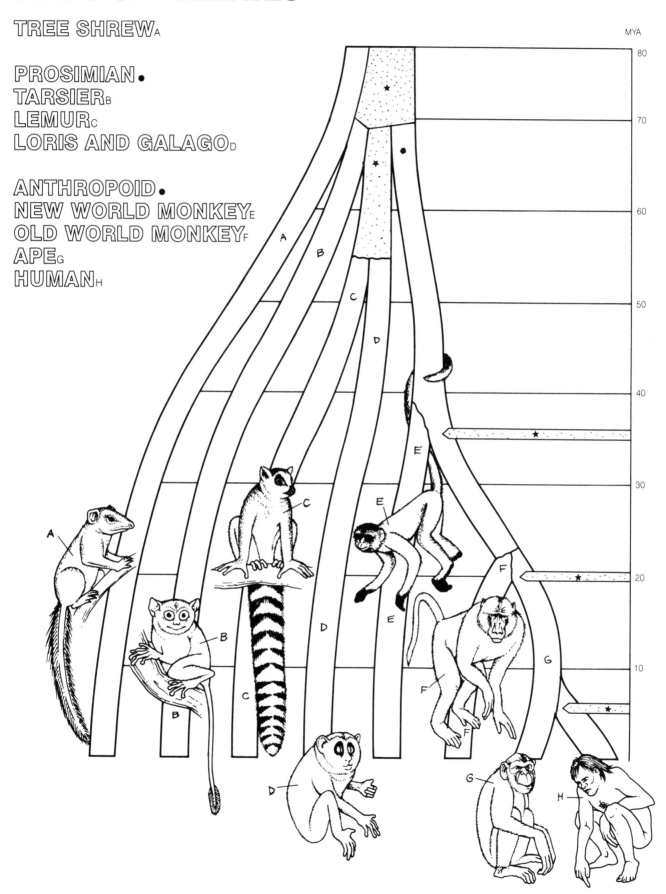

Living primates are found most often in tropical regions. Turn your attention to the illustration and notice that the equator passes through areas where primates live: the northern part of *South America,* central *Africa,* and through the *islands of Southeast Asia.* Three main geographical radiations of non-human primates exist: *New World monkeys* in *Central* and *South America, lemurs* in *Madagascar,* and *tree shrews* and prosimians, *Old World monkeys,* and apes in *Africa* and *South Asia.*

Color Central and South America and the name of the primates living there.

In these areas live the *New World monkeys* as far north as Mexico and as far south as Argentina. The Andes Mountains, which run along the western side of *South America,* act as a barrier, and no monkeys live on that coast. Besides humans, who came to the Americas about 15,000 years ago, monkeys are the only primates found in the New World and have been geographically separated from other monkeys for 35–40 million years.

Color Madagascar, a large island off the coast of Africa, and the name of the primates living there.

Madagascar has been isolated from *Africa* for over 60 million years (as we see on Plate 68). A diversity of *lemurs* inhabits the island, from solitary, nocturnal species to others living in large social groups active in the day. Because there are no anthropoids on *Madagascar,* it is the only region where prosimians have adopted day-living activity patterns.

Color the continent of Africa and the names of the primates living there.

South of the Sahara Desert we find a rich primate population. *Lorises and galagos,* small, nocturnal prosimians, inhabit the tropical forests in western and central *Africa. Galagos* also live in southern *Africa.*

Old World monkeys are found all over *Africa,* in the tropical forests in western and central *Africa* and also in the savanna regions in the east and south. Apes, which include *chimpanzees* and *gorillas,* live in a smaller area of western and central *Africa,* extending into western Tanzania. (Their distribution is indicated by the dots and the c¹ label.)

The areas labeled c² indicate the home of *Old World monkeys* (macaques), which live in and around the Atlas Mountains of Morocco and Algeria. A few baboons are found in the small area on the Arabian peninsula.

Color the rest of the plate.

Many kinds of primates live in *Asia,* which, together with *Africa,* forms the third area of primate radiation. As indicated by the macaques still living in northern *Africa* and the baboons on the Arabian peninsula, they were once more widely distributed.

In *India, Sri Lanka,* and *South Asia,* there are *tree shrews, lorises,* and *Old World monkeys. Gibbons,* the smallest of the apes, live on the Malay peninsula. The northernmost living primate, besides humans, are the *Old World monkeys,* the macaques, which live in *Japan.* Some of these *Old World monkey* populations live in the snow and are called "snow monkeys."

Island Southeast Asia, which includes the Philippines, Indonesia, and the islands of Sumatra, Java, Borneo, and Suluwasi, has a large variety of all kinds of primates. *Tree shrews* are found on several islands, as are *tarsiers* and *lorises,* as well as *Old World monkeys. Gibbons* and *orangutans* are found on Sumatra, Java, and Borneo. Primates were more widespread in the past, when the continents had a different configuration and climates were warmer.

The human primate inhabits all parts of the globe in the greatest diversity of habitats. Part V demonstrates that humans originated in the warm savannas of eastern *Africa,* about 5 mya.

WORLD DISTRIBUTION
OF LIVING PRIMATES ★

**CENTRAL AND
SOUTH AMERICA**ₐ
NEW WORLD MONKEYSₐ

MADAGASCARᵦ
LEMURSᵦ

AFRICAc
LORISES AND
GALAGOSc
OLD WORLD MONKEYSc, c²
APES ★
CHIMPANZEESc¹
GORILLASc¹

**INDIA, SRI LANKA, SOUTH ASIA,
JAPAN**ᴅ
TREE SHREWS, LORISES,
OLD WORLD MONKEYSᴅ
APES ★
GIBBONSᴅ

ISLAND SOUTHEAST ASIAₑ
TREE SHREWS, TARSIERSₑ
LORISES, OLD WORLD
MONKEYSₑ
GIBBONS, ORANGUTANSₑ

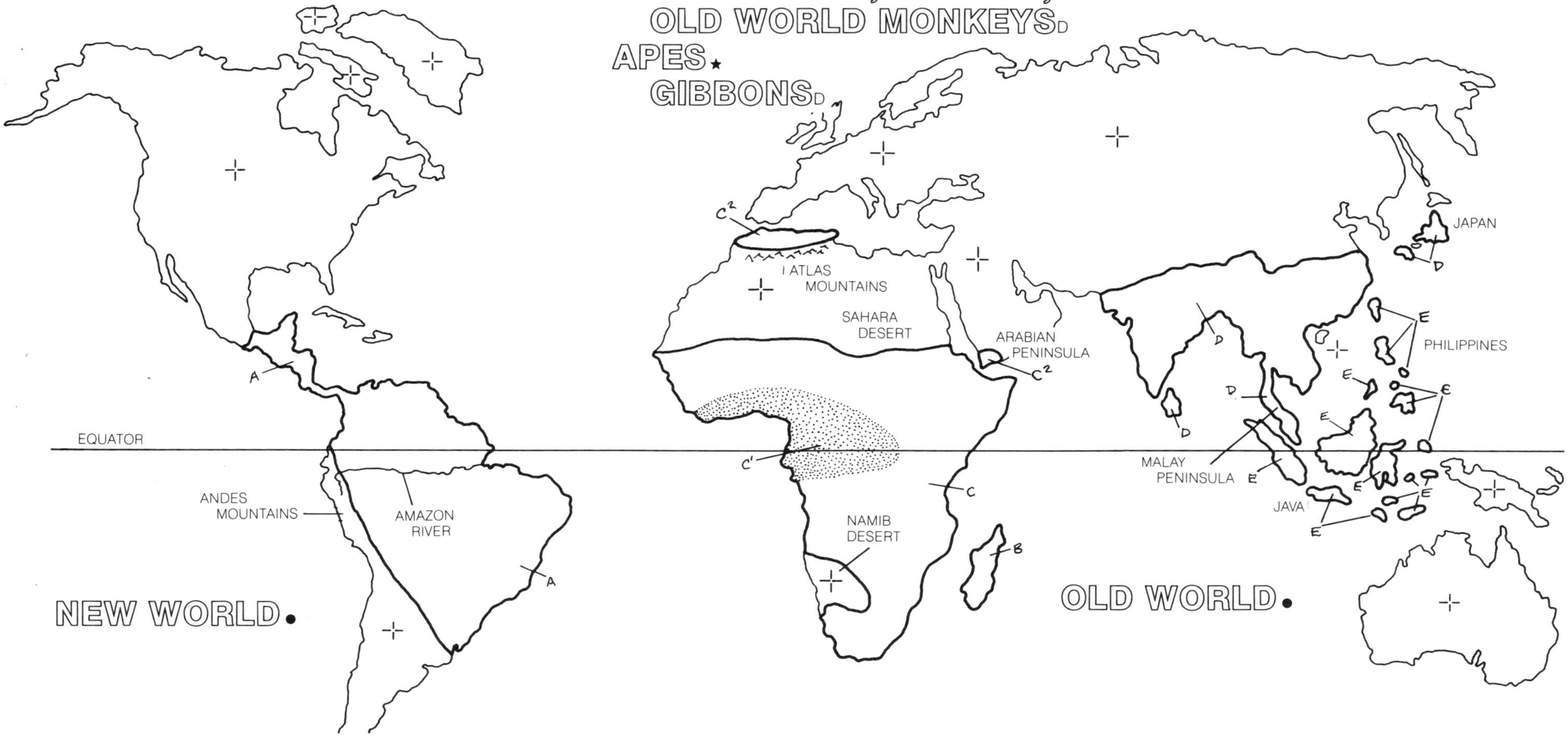

OLD WORLD ●

NEW WORLD ●

JAPAN

PHILIPPINES

JAVA

MALAY
PENINSULA

ARABIAN
PENINSULA

ATLAS
MOUNTAINS

SAHARA
DESERT

NAMIB
DESERT

AMAZON
RIVER

ANDES
MOUNTAINS

EQUATOR

7
DISTINGUISHING REPTILES FROM MAMMALS: BODY TEMPERATURE AND REPRODUCTION

Primates belong to the class Mammalia, named for their characteristic *mammary glands,* by which the young are fed. In addition to *mammary glands* and intrauterine fetal development, mammals have evolved the capacity to maintain a high internal body temperature. This plate and the three that follow examine major differences between mammals and reptiles and discuss how mammalian adaptations have led to their success in the Cenozoic.

Much of the success of the mammalian line of evolution is due to homeothermy or "warm-bloodedness." In contrast to fish, amphibians, and reptiles, mammals maintain a high and constant body temperature. Metabolic processes occur rapidly at this temperature and provide quick energy for locomotion and other activities independent of the external environmental temperatures. Mammals can maintain normal activity during changes in environmental temperatures in which amphibians or reptiles might freeze or fry.

Color the temperature diagram at the top of the plate.

In most habitats, *air temperature* fluctuates cyclically, rising during the day and falling rapidly at night. In some environments, the air may be as cold as 0 °C in the early morning yet rise to a high of 43 °C in midday. The *reptile body,* represented here by a lizard, has a temperature of about 22 °C in early morning. As the *air temperature* rises and the lizard basks in the sun, its body temperature gradually rises, then stabilizes for a time in midday at 35 °C. The stable level varies from species to species, but all depend upon external heat sources. Fluctuating *reptilian body temperature* causes metabolic processes, such as food digestion, to proceed slowly when the *body temperature* is low, thus accounting for their "sluggish" behavior in the early morning and late evening.

By contrast, the *mammalian body,* represented here by a cow, has a constant temperature of about 37 °C, which does not fluctuate significantly with *air temperature.* Mammals have several unique adaptations to maintain constant temperature: hair and fur keep out the cold and retain warmth, shivering produces internal heat through rapid muscular contractions, and sweat glands and panting dispel excess heat after physical activity or in hot weather.

The center of the plate illustrates differences in reproduction. Begin with the embryo and color this and subsequent structures in both the reptile and mammal.

Reptiles were the first true land vertebrates; unlike amphibians, they do not return to the water to reproduce. The *amniotic* egg of the reptiles evolved structures that replaced the protective, nutritional, and waste-disposal functions of a watery milieu. Notice the structures in the reptilian egg and mammalian uterus: Much of the architecture of the *amniotic* egg has been retained and modified in the embryonic membranes of viviparous ("live birth") mammals.

A yolk enclosed in a *yolk sac* provides all the nutrients needed for reptilian development in the egg. The membranous *chorion* lies below the porous *eggshell* and acts as a primitive lung for the *embryo,* mediating the exchange of gaseous oxygen and carbon dioxide. Another membrane, the amnion, encloses a large cavity called the *amniotic sac,* which substitutes for the watery environment of fish and amphibian eggs. The *allantois* stores waste products from the *embryo* until it hatches.

In the mammalian *embryo,* notice that the *eggshell* is replaced by the mother's thick muscular *uterus.* The *amniotic sac* persists, but the *yolk sac* and *allantois* have diminished and play lesser roles due to the evolution of the *placenta.* The *placenta* provides nutrients to the mammalian *embryo* and removes wastes via the mother's blood supply. It is connected to the *embryo* by the umbilicus.

Color the parental care illustration.

The intrauterine development of mammals is more extended than that of unhatched reptiles. Even so, reptiles are ready to fend for themselves as soon as they hatch, and by and large there is *no care by mother* (or father) of the young. By contrast, mammalian offspring are fairly helpless and immature at birth and depend upon their mothers for care and for nutritious *milk* from *mammary glands.*

Maternal care is a primary feature of the mammalian adaptive pattern, and it is the basis for adult social behavior.

BODY TEMPERATURE AND REPRODUCTION. ★

TEMPERATURE.
AIR~A~
REPTILE BODY~B~
MAMMAL BODY~C~

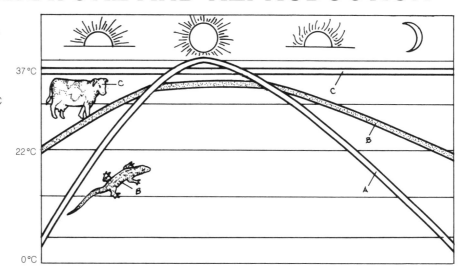

REPRODUCTION.
EMBRYO~D~
AMNIOTIC SAC~E~
ALLANTOIS~F~
YOLK SAC~G~
CHORION~H~
EGG SHELL~B¹~
PLACENTA~I~
UTERUS~C¹~

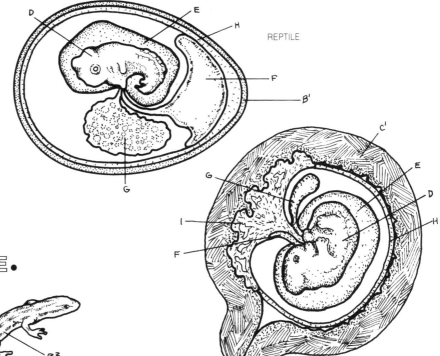

REPTILE

MAMMAL

PARENTAL CARE.
NO CARE BY
MOTHER~B²~

MAMMARY GLAND,
MILK, MATERNAL
CARE~C²~

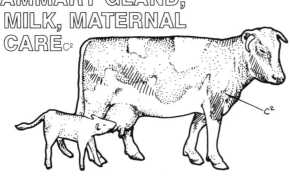

A critical factor in the higher activity level of mammals lies in the construction of their skeleton and musculature—the "machinery" for moving an animal around its environment to find food, avoid predators, and meet mates. This plate examines the locomotor systems of reptiles and mammals. The reptile depicted is an advanced Triassic mammallike reptile from which all mammals arose. The mammal, a tree shrew, resembles a small Mesozoic mammal that had a flexible locomotor system for scampering around low shrubs and bushes.

Color the main motion of the reptile and mammal, illustrated on the left.

In most reptiles, the limb bones are held out horizontally from the body; mammals stand with the limb bones held directly beneath the body. When reptiles move, they do not flex and extend the back, but swing it laterally from side to side. Their legs *rotate* outward in a sideways fashion. By contrast, when a mammal runs, it moves the limbs backward and forward under its body, with the upper and lower parts of each limb moving in the same plane. Mammals *flex* and *extend* their backbones, which helps increase the length of each stride.

The skeleton has two parts: the axial skeleton, which includes the head (or cranium) and the backbone (or vertebral column, which extends from the neck to the tip of the tail), and the appendicular skeleton, which includes the *pectoral* and *pelvic girdles* and the *limb bones.*

The vertebrae, the individual bones of the spine, are fairly uniform in the reptile, and there are ribs from the neck to the tail. In the mammal, there are five different kinds of vertebrae, each specialized for a different function. The cervical or neck vertebrae lack ribs, allowing greater ability to move the head around. The thoracic vertebrae bear ribs and enclose and protect the heart and lungs. The lumbar vertebrae lack ribs. At the juncture of the thoracic and lumbar region, there is a center of motion—of *flexion* and *extension*—of the back, which confers to mammals a great deal of flexibility and speed. The sacrum and tail, or caudal vertebrae, are the last two regions of the spinal column.

Color the pectoral girdle and its parts in the reptile and mammal. Also color gray the bones of the forelimbs.

The *pectoral girdle* connects the *forelimbs* to the trunk. In the reptile, the *pectoral girdle* is solid: the right and left *girdles* are bound in the front of the neck and remain relatively stationary when the reptile moves its *forelimbs.* In the reptile, the *pectoral girdle* is composed of several bones: the *scapula, cleithrum, coracoid, procoracoid, interclavicle,* and *clavicle.* In mammals, however, there are only two bony elements: the *scapula* (or shoulder blade) and the *clavicle* (or collar bone). In most mammals, the *pectoral girdle* is thus lightened and not fused solidly to the other side. This allows the *scapula* to move back and forth on the rib cage, adding to stride length. The socket for the humerus points outward in a reptile but downward in a mammal. This allows the *forelimb* to be drawn up under the mammalian body; in the reptile, the *forelimbs* extend more sideways out from the trunk.

Color the pelvic girdle and its parts. The pelvic girdle binds the hind limb to the trunk. Color the hind limbs gray.

The *pelvic girdle* consists of a pair of *innominate* bones, connected beneath the animal's tail by a sacrum (not seen in this view). In both reptiles and mammals, three bones: the *ilium, ischium,* and *pubis* comprise each *innominate* and develop from three separate ossification centers. However, in mammals they fuse into a single bone at adulthood. The reptilian pelvis has a nearly vertical orientation; the *ilium* is short and equal in size to the *pubis* and *ischium.* In mammals, the *ilium* is about twice as long as the *pubis* or *ischium* and is orientated more horizontally, running parallel to the spinal column.

In the reptilian *pelvic girdle,* the *ischium* is relatively large. This large surface area below the socket for the hip joint (acetabulum) indicates that the muscles that rotate the lower limb around the body are well developed. In mammals, the extension of the *ilium* and reduction of the *pubis* and *ischium* correlate with a development of the muscles that flex and extend the lower limb.

Thus, the locomotor system of many mammals (mobile yet supportive limbs, flexible back, and so on) makes possible movement around their environment with considerable agility, grace, and speed.

THE LOCOMOTOR SKELETON.★

MAIN MOTION.
ROTATION_A
FLEXION_B/EXTENSION_{B'}

APPENDICULAR SKELETON.

PECTORAL GIRDLE_C
SCAPULA_D
CLEITHRUM_E
CORACOID_F
PROCORACOID_G
CLAVICLE_H
INTERCLAVICLE_I

BONES OF THE
FORELIMB_J★

PELVIC GIRDLE_K
INNOMINATE★
ILIUM_L
ISCHIUM_M
PUBIS_N

BONES OF THE
HIND LIMB_O★

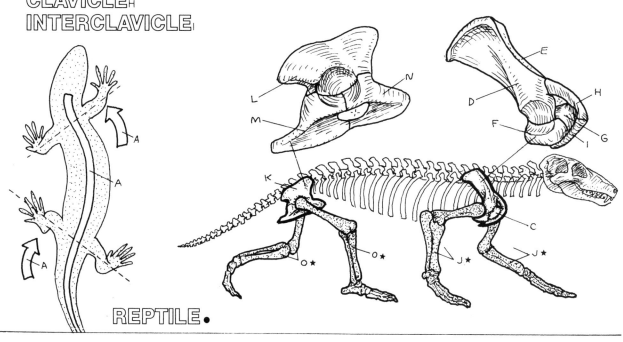

REPTILE.●

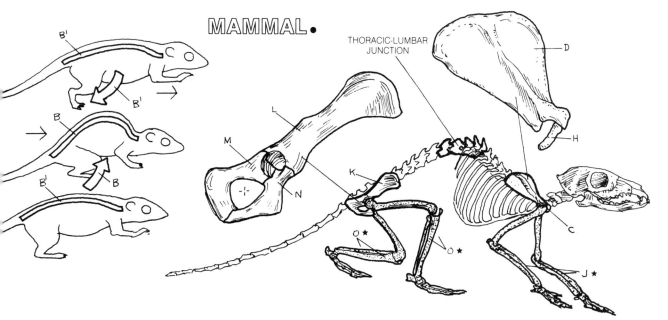

MAMMAL.●

THORACIC-LUMBAR
JUNCTION

9
DISTINGUISHING REPTILES FROM MAMMALS: BODY AND BONE GROWTH

Reptiles and mammals have distinctly different growth patterns: Mammals go through definite stages, whereas reptiles grow continuously.

Color the infant and adult human figures, representing mammals, and the young and adult alligators, representing reptiles.

Human babies have a different body shape than their adult counterparts: most conspicuously, their heads are large in proportion to the rest of their body. Newborn alligators, on the other hand, are essentially miniature adults.

Color the chart on the right, which illustrates bone growth.

Mammals grow rapidly during infancy and youth and do not grow at all in adulthood. (In humans, various parts of the body and specific bones grow at different rates; by age six, the skull and brain are nearly as large as they will ever be. The long bones, like those shown here, do not complete their growth until the teen years or even the early twenties.) In contrast, reptilian growth is *continuous throughout life*. As you can see on the graph, reptilian bones (and body size) enlarge throughout an individual's lifetime.

The skeleton in both classes of animals (in fact, in the embryos of all vertebrates) first forms out of *cartilage,* a connective tissue that is flexible and slightly elastic, yet moderately firm. In tetrapods (amphibians, reptiles, and mammals) *cartilage* is largely replaced by bone. Bone is a more rigid connective tissue than *cartilage,* owing its additional strength largely to the deposition of inorganic calcium salts.

In long bones (illustrated), such as those of the limbs, ossification (replacement of *cartilage* by bone) begins before hatching or birth. Cells invade from the membranes surrounding the solid cartilaginous skeletal precursor; *cartilage* is destroyed and specialized bone-forming cells begin laying down *spongy bone* in the center of the cylindrical shaft of the developing bone, the *diaphysis*. Layers of *compact bone* are laid down around the exterior of the *diaphysis*. *Compact*

and *spongy bone* differ only in the density and arrangement of their bone cells.

By the time of hatching or birth, ossification of the *diaphysis* is largely complete. *Spongy bone* in the interior has been destroyed to create a long *marrow cavity,* which is much larger in mammals than in reptiles.

Color the structures of the long bones. The epiphyses of mammals contain one or two secondary epiphyseal ossification centers (depending on the bone's size), whereas those of reptiles do not. Color the arrows showing the length of diaphysis and epiphysis.

In young mammals, the *cartilage* at the ends *(epiphyses)* of the long bones is gradually replaced by bone. Shown here is the bone of a young mammal, for ossification is not complete. Soon the *epiphyses,* except at the articular surfaces, will be composed entirely of bone tissue, and growth in the length of the bone will occur only at the *epiphyseal plate*. When a mammal reaches maturity, the *epiphyses* fuse with the *diaphysis* (closing the site of bone growth), and no further bone growth occurs. This complete ossification results in a system of bone growth involving strong structural support at the joints for meeting the more rigorous demands of active mammalian locomotion.

In contrast, the *epiphyses* of reptiles remain cartilaginous throughout their adult life. As bone replaces *cartilage* at the *epiphyseal plate,* new *cartilage* is continuously added to the *epiphyses*. Because the *epiphyses* of the long bones are of *cartilage* rather than bone, the articulations of the joints are relatively weak in comparison to mammals.

The mammalian pattern of bone growth considerably enhances the paleontologist's ability to estimate the age of fossilized individuals, for bone fossilizes, but *cartilage* does not. If the *epiphyses* of a mammalian long bone are missing from a fossil, we can conclude that the animal died before reaching adulthood. If a whole long bone is found, we can be certain that the fossil is of an adult.

BODY AND BONE GROWTH ★

GROWTH AND LIFE CYCLE ●

DEFINED STAGES A

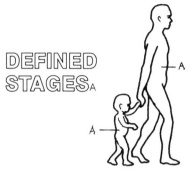

CONTINUOUS THROUGH LIFE B

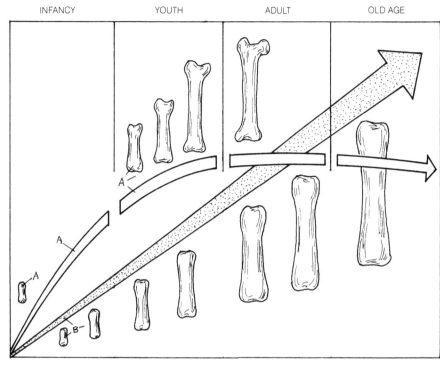

INFANCY   YOUTH   ADULT   OLD AGE

BONE GROWTH ●
CARTILAGE C
SPONGY BONE D
COMPACT BONE E
MARROW CAVITY F
EPIPHYSIS G
EPIPHYSEAL PLATE G1
SECONDARY EPI-
PHYSEAL OSSIFICATION
CENTER G2
DIAPHYSIS H

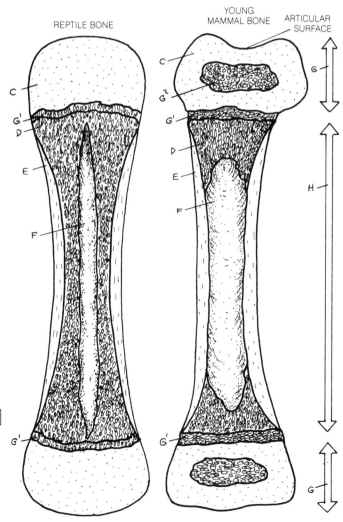

REPTILE BONE

YOUNG MAMMAL BONE   ARTICULAR SURFACE

DISTINGUISHING REPTILES FROM MAMMALS: SKULL AND DENTITION

Jaws and teeth of reptiles and mammals reflect their different feeding behaviors. Active and warm-blooded, mammals require a greater regular intake of "fuel" than do reptiles. Below, the skull of an opossum, a primitive mammal, is compared with that of *Dimetrodon,* a representative of the Permian mammallike reptiles.

Color the dentition, first in the reptile, then in the mammal. Color the tooth enlargements, which show the more complex shape of the mammal tooth.

The reptilian teeth are *homodont* (homo = same; dont = tooth) —conical in shape and vary only in size. Their teeth and jaws are adapted for seizing, puncturing, and holding food. As in fish and amphibians, reptiles replace their teeth continuously through life.

Mammalian teeth have become heterodont (hetero = other). Teeth of various sizes and shapes have different functions: *incisors* seize and hold, *canines* pierce; *premolars* and *molars* crush, tear, pulp, and grind. This intensive preparation of food before it reaches the gut speeds the absorption of nutrients for sustaining the mammal's high metabolism. Unlike reptilian teeth, mammalian upper and lower teeth fit together. To ensure this precise fit throughout the animal's life, mammals have just two sets of teeth: "milk" or baby teeth are replaced by permanent teeth after weaning. *Molars* are not replaced and are added from the front to the back of the jaw as an animal matures.

Color the bony elements of the lower jaw of both animals.

In reptiles, the lower jaw, or mandible, contains several bones: the *dentary,* which holds the teeth, the *angular, surangular, splenial,* and *articular.* The *dentary* contacts the skull, holds the teeth, and forms the entire lower jaw in mammals.

Color the ear bones, which are enlarged on the left. The arrow indicates their location within the skull, hidden from view.

The *stapes* is the only bone in the middle ear of reptiles; whereas the opossum, like all mammals, has three middle ear bones.

In evolution, new structures often appear by modifying existing ones to fill a new function. The *angular* bone of the reptilian mandible becomes the tympanic ring in mammals; the *articular* becomes the *malleus;* and the quadrate of the reptilian upper jaw becomes the *incus.* In mammals, the *stapes, malleus,* and *incus* form the ear ossicles that transmit sound vibrations.

Color the muscle attachments indicated on the side views by dotted lines.

In the reptile, there is one major muscle for closing the jaw, the *adductor mandibulae.* In mammals, this muscle is divided into two distinct muscles. One is called the *temporalis,* or temporal muscle, for it originates from the temporal region of the skull. This muscle is similar in origin to the *adductor mandibulae*; notice, however, that it has much larger areas of attachment. The second muscle, the *masseter,* contains fibers running at an angle to those of the *temporalis*; these pull the lower jaw forward and upward. This muscle inserts on the outside of the lower jaw rather than on the top, as in the reptile. This means this muscle can pull the jaws in the side-to-side motions necessary for grinding and crushing food.

Mammals have also evolved fleshy lips and cheeks for holding food in place while teeth and saliva do their jobs. The tongue of mammals is more mobile than that of reptiles; so it is more effective in manipulating food held within the mouth. One further adaptation is that in most reptiles (excluding crocodiles), the internal opening of the nasal passages is into the top of the mouth rather than into a special nasal cavity. This means they must transport food and breathe through the same passages. Mammals have evolved a bony partition—the secondary palate, which separates the nasal and food passages, allowing them to maintain a high breathing rate while chewing.

Color the braincase, which is inside the skull but is represented by the dotted circle on the top view.

One of the distinguishing characteristics of mammals is their enlarged brain. Plate 11 examines this evolutionary trend more closely.

SKULL AND DENTITION ★

DENTITION ●
HOMODONT A
HETERODONT ★
INCISOR B
CANINE C
PREMOLAR D
MOLAR E

LOWER JAW ●
DENTARY F
ANGULAR G
SURANGULAR H
SPLENIAL I
ARTICULAR J

EAR BONES ●
STAPES K
MALLEUS L
INCUS M

MUSCLE ATTACHMENTS ●
ADDUCTOR MANDIBULAE N
TEMPORALIS N'
MASSETER O
BRAINCASE P

REPTILE ●

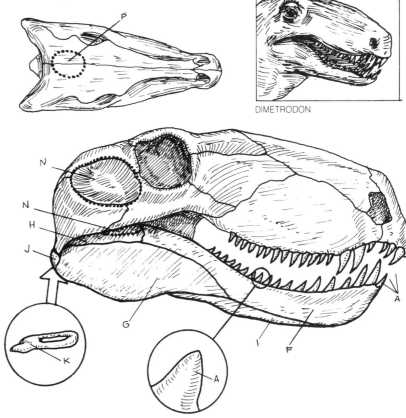

DIMETRODON

MAMMAL ●

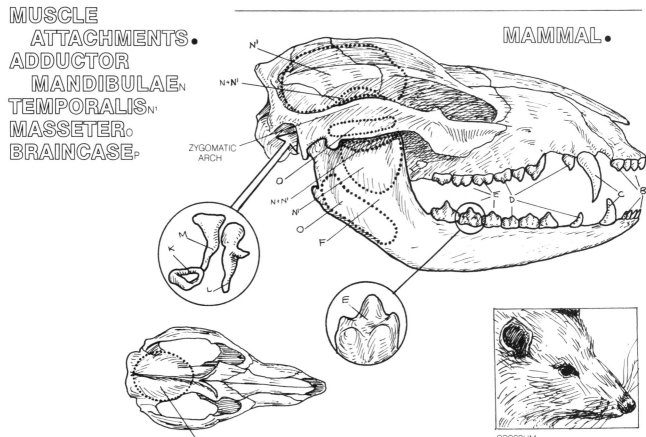

ZYGOMATIC ARCH

OPOSSUM

11

ANATOMICAL EVIDENCE FOR EVOLUTION: COMPARATIVE EMBRYOLOGY: ONTOGENY AND PHYLOGENY

Long before the theory of natural selection had been proposed, Ernst von Baer claimed that the more closely related any two species are, the more similar their development is. His treatise (1828) linked the study of *ontogeny* (the development of the individual through a single life cycle) to *phylogeny* (the development of the species through evolution). Von Baer noticed that animals are more similar during early stages of their embryological development than they are as *adults*. Arm *buds,* for example, from widely different species are virtually indistinguishable when they first form on the embryo; yet they may develop into a wing, an arm, or a flipper.

A comparison of developmental stages among vertebrates led Ernst Haeckel (1834–1919) to propose his famous principle "*ontogeny* recapitulates *phylogeny.*" An ardent supporter of Darwin, Haeckel claimed that the development of an individual *(ontogeny)* reflects the stages through which the individual's species has passed during its evolution *(phylogeny)*.

Today it is clear that "*ontogeny* recapitulates *phylogeny*" is an oversimplification and misleading. Although it is true that the early developmental sequences of all vertebrates are similar, there are important deviations from the general developmental plan.

The developmental stages of five species—an amphibian (the salamander), a bird (the chicken), and three mammals (the pig, monkey, and human) are examined on this plate.

Color the horizontal arrow representing ontogeny. Use six contrasting colors for each of the stages of development shown and color them as they are discussed.

The *fertilized eggs,* or zygotes, are very similar, though they differ slightly in the size of the cell nucleus. *Cleavage* refers to the orderly division of the single-celled zygote into a multicelled blastocyst. By the *late cleavage* stage, the embryos still look similar and differ only in their *cleavage* patterns, which vary due to the presence of differing amounts of yolk in the *egg*.

As the *body segments form,* all three mammals remain almost identical. Notice the ancestral gill slits, which in the mammals will later develop into parts of the ear. The mammals possess an umbilical cord that

leads to the placenta. The salamander and the chicken are nourished by yolk in the yolk sac.

As the forelimbs begin to develop, they appear first as *limb buds,* relatively indistinguishable between all the species shown. At the *late fetal* stage, the monkey and human are still surprisingly similar, with the main difference being the absence of a tail in the human fetus. (If this were an ape rather than monkey fetus, this difference would not exist.) Limbs have taken on their *adult* shapes. The chicken has developed its specialized shell breaker. The salamander has just hatched into its *larval* stage. It spends the first part of its life in the water, taking in life-giving oxygen through its feathery gill slits, using its limbs as flippers. Later the salamander undergoes metamorphosis, acquiring its *adult* form with terrestrial limbs and lungs for breathing air. Only then, as an *adult,* can it leave the water to live, but not reproduce, on dry land.

The offspring receive quite different treatment. The salamander mother abandons the *eggs* after she lays them, and the larvae receive no maternal care at all. The mother hen incubates her *eggs* with body heat, sitting on them in a nest. The *newly hatched* chicks receive some protection from the mother hen, but must begin immediately to find their own food. After gestation times of four (pig), six (monkey) to nine months (human), *newborn* mammals are nourished by their mother's milk and require extended care before they become independent *adults*.

Color the vertical arrow representing phylogeny.

More advanced species pass through developmental stages, in which they resemble the embryos, not the *adult* forms of their ancestors, as Haeckel proposed. This resemblance is much more apparent in the early stages of embryological development than in the later stages. Development, like evolution, is a conservative process, and descendants inherit a common developmental plan. Yet as development proceeds, it diverges from the common plan to accommodate new adaptive life-styles. We cannot look to *embryology* for a history of our *adult* ancestors—*ontogeny* does not recapitulate *phylogeny*. However, from *ontogeny,* we can find clues as to how evolution generates the diversity of life forms through time.

COMPARATIVE EMBRYOLOGY∗

ONTOGENY∗ AND∗ PHYLOGENY∗

FERTILIZED EGG∗
LATE CLEAVAGE∗
BODY SEGMENTS FORM∗
LIMB BUDS APPEAR∗
LARVAL FORM/LATE FETAL∗
NEWLY HATCHED∗ NEWBORN∗ ADULT∗

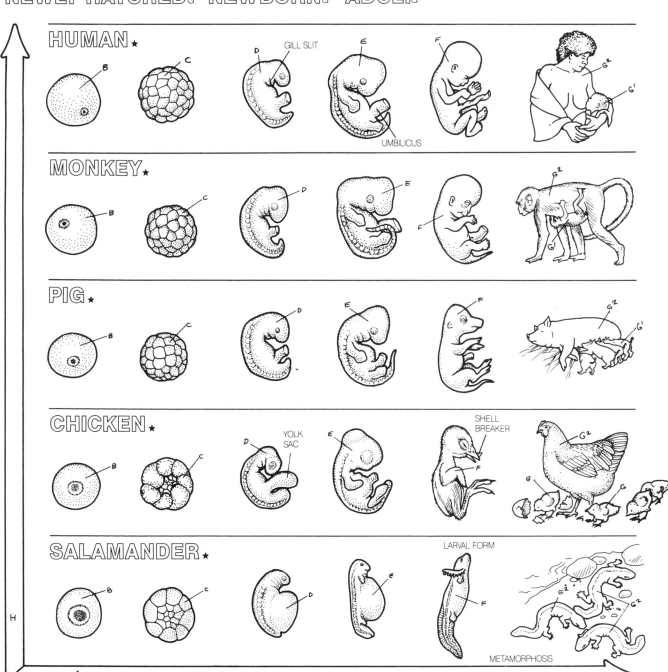

12

ANATOMICAL EVIDENCE FOR EVOLUTION: CONSERVATISM: MAMMALIAN EAR BONES

The structural similarity of all vertebrates is classically illustrated in the embryological development and evolution of the human jaw and hearing mechanism. Structures that are homologous, that is, arise from the same embryological precursors, can be traced back to the gill apparatus of the Devonian jawless fish. The slimy, eel-like hagfish pictured here descended from primitive filter feeders which obtained food by sifting microorganisms out of the seawater as it passed in through their mouths and out the *gill slits*. Evolution conserves the gill structures in the embryos of all vertebrates and puts them to use in a number of different ways in adult animals. This plate traces the phylogeny and embryology of the gill apparatus to illustrate how evolution modifies ''old'' structures to create ''new'' ones for new adaptations.

Begin at the top of the page with the jawless hagfish, and color all the structures in each animal as discussed. Color the five unlabeled arches and the four unlabeled gill slits of the jawless fish gray.

The hagfish has seven cartilaginous gill arches, separated by *gill slits*. Concentrate on the first two gill arches. Notice how the first gill arch—called the mandibular arch—consists of the *palatoquadrate* above and *Meckel's cartilage* below. The *first gill slit* is the opening between the mandibular arch and the second gill arch, called the hyoid arch. The hyoid arch, consisting of the *hyomandibular* above and the *hyoid* below, lies directly in front of the *second gill slit*.

In jawed fish, the first gill arch evolves into a pair of jaws for a more effective method of feeding. The remaining gills subsume respiratory functions in the exchange of waste carbon dioxide for needed oxygen as seawater passes over the gills and out through the *gill slits*. Notice how the *palatoquadrate* and *Meckel's cartilage* form the upper and lower jaws, respectively. Jawed fish have an inner ear, which functions primarily as an organ of balance. In some fish, pressure waves from the surrounding water are transmitted to the inner ear through the *hyomandibular*, which rests at one end against the skull and at the other against the bony chamber surrounding the inner ear. The *first gill slit* has become the *spiracle*, a tube that facilitates the pumping of water through the remaining gills, where gases are exchanged in respiration.

In mammallike reptiles, *Meckel's cartilage* contributes to several bones of the jaw and skull (see Plate 10 for a review of the reptilian jaw), including the *angular* of the lower jaw and the *articular*, at the jaw joint. The front view shows how the *first gill slit* forms the *middle ear cavity*, which houses the *stapes* (modified from the *hyomandibular*). The *Eustachian tube*, connecting the throat to the *middle ear cavity*, is also derived from the *first gill slit*. The *stapes* rests against the *eardrum*, which has now evolved from the *second gill slit*.

In mammals, the *eardrum* is sunk into the skull and an *outer ear*—homologous to the *second gill slit*—has evolved to aid in funneling air waves into the middle ear, now containing three bones. The *quadrate* and the *articular* have migrated inward to join the *stapes* as the *incus* and *malleus*, respectively. These three middle ear bones transmit sound waves that travel from the *eardrum* to the inner ear. The reptilian *angular* becomes the *tympanic ring*, encircling the *eardrum*.

Homologies are determined by following the fate of embryological structures. In humans, the gills are transitory structures appearing about four weeks after conception. The *gill slits* never open to the outside, as they do in fish, but are present as gill pouches, between the arches. The first arch becomes the *malleus*, the *incus*, and the *tympanic ring*.

Notice in the human fetal figure that *Meckel's cartilage* also forms the cartilaginous precursor of the lower jaw and a large part of the lower face. The second gill arch becomes the *stapes* and contributes to the *styloid process* and the *hyoid* bone, which functions in the support of the tongue and larynx. As in all mammals, the first gill pouch forms the *middle ear cavity* and the *Eustachian tube*, and the second gill pouch forms the ear canal and the *outer ear*. In an amazing example of conservatism, other parts of the ancestral filter-feeding apparatus become the thymus gland, which functions in the maturation of blood cells that produce antibodies and protect us against infection; the parathyroid gland, which regulates the metabolism of nutrient minerals; the tonsils; several lymph glands; and many muscles of the face and neck.

CONSERVATISM: MAMMALIAN EAR BONES.★

GILL ARCH 1
(MANDIBULAR).●
PALATOQUADRATE A
   QUADRATE A1
   INCUS A2
MECKEL'S CARTILAGE B
   ARTICULAR B1
   MALLEUS B2
   ANGULAR B3
   TYMPANIC RING B4
FIRST GILL SLIT C
   SPIRACLE C1
   MIDDLE EAR
     CAVITY C2
   EUSTACHIAN TUBE C3

GILL ARCH 2
(HYOID).●
HYOMANDIBULAR D
   STAPES D1
HYOID E
   STYLOID
     PROCESS E1
SECOND GILL SLIT F
   EARDRUM F1
   OUTER EAR F2

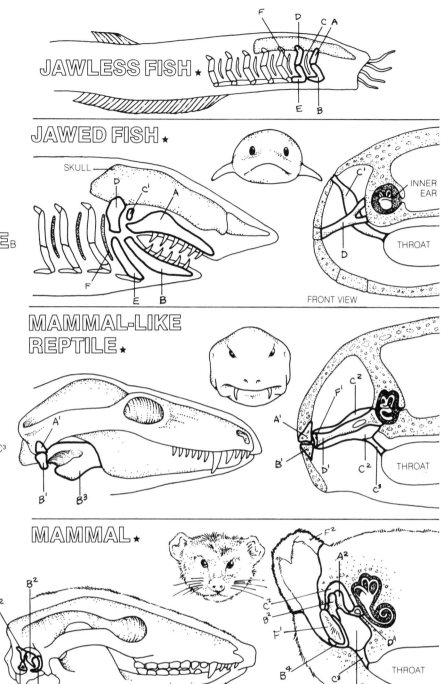

JAWLESS FISH ★

JAWED FISH ★

SKULL

INNER EAR

THROAT

FRONT VIEW

MAMMAL-LIKE REPTILE ★

THROAT

MAMMAL ★

THROAT

HUMAN FETUS ★

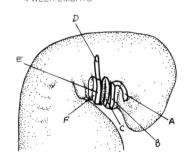

4 WEEK EMBRYO

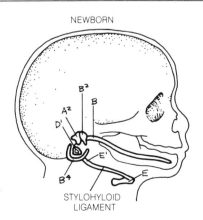

NEWBORN

STYLOHYOID LIGAMENT

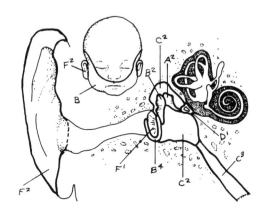

13
THE VERTEBRATE BRAIN: FROM FISH TO MAMMAL

The vertebrate brain provides another example of how anatomical structures can have a common origin, yet over time have evolved different functions. In all vertebrates, the brain first appears in development as three swellings at the anterior end of the neural tube, one of the first structures to appear in a developing embryo.

Color these swellings—called the forebrain, midbrain, and hindbrain—on the illustrations at the bottom left of the plate, first in the embryo, then in the adult reptile.

The reptile in the lower left corner illustrates the adult brain structures derived from these three embryological precursors. In the illustrations on the right, the way the common structural plan of the vertebrate brain is modified in different animals may be observed. These variations provide clues as to how different animals live.

Although brain evolution has been conservative, with old structures being retained and modified for new adaptive patterns, the adult brain of different species exhibits more variation than any other organ system of the body. Varieties of brain organization reflect differences in the kinds of sensory information and range of motor responses important to the survival of a species. Despite this great diversity, some general trends in brain evolution can be discerned.

Color each brain structure in the fish, reptile, bird, and mammal as they are discussed in the text.

The *olfactory bulbs,* structures of the *forebrain,* process smell information from the environment. In fish and reptiles, the *olfactory bulbs* are relatively large in comparison to total brain size—an indication of their heavy reliance on this sense for gathering information. The *olfactory bulbs* are smaller in most birds, except for certain scavenger birds that locate their food by smell. In mammals, the *olfactory bulbs* range from tiny, as in the advanced primates, to huge, as in the anteater, armadillo, and aardvark, which all use their noses to sniff out ants and termites.

Smell is the primary sense in fish. In addition to large *olfactory bulbs,* the *cerebrum* is almost exclusively devoted to the processing of smell information. The *optic nerve* transmits information from the eye to the brain. Thus, the *forebrain* of the earliest verte-brates was largely a "sense of smell brain." The *cerebrum* of reptiles is distinctively larger than in fish, marking the earliest appearance of the neocortex. This "new cortex" becomes progressively larger during subsequent evolution of the vertebrates; it takes over processing information involving vision, taste, and touch, which in fish is handled by *midbrain* and *hindbrain* structures.

Expansion of the neocortex correlates with the increased capacity for complex learning, for memory—for what we generally call "intelligence." This trend reaches its apex in the advanced mammals, where the *cerebrum* covers the entire surface of the *midbrain.* A new structure, called the corpus callosum appears. This bundle of several thousand nerve fibers connects the two lateral lobes of the neocortex and provides for rapid transfer of information from one side of the neocortex to the other.

The *optic lobes* of fish are small, but in reptiles and birds, they are almost always large and conspicuous, attesting to their excellent vision. The homologues of the *optic lobes* in mammals (the *superior colliculi*) are extremely reduced in size. Visual processing has been largely reallocated to the occipital lobes at the rear of the large *cerebrum.* Beneath the massive cerebral lobes, the *superior colliculi* of mammals appear as two tiny bumps on the exterior of the *midbrain.*

The *cerebellum* is the coordinator of voluntary muscle movements and plays a primary role in maintaining balance. The *cerebellum* is proportionately largest in birds, where it coordinates intricate movements necessary for flight.

The *medulla oblongata* is a direct extension of the spinal cord. It coordinates the autonomic, involuntary activities involved with breathing and blood circulation and integrates sensory impulses from all parts of the body for rapid reflex responses. In fish, the *medulla oblongata* is the only processing center for touch, temperature, taste, and balance. In higher vertebrates, much of the sensory processing functions have been taken over by the neocortex. Yet even in humans, with their very large neocortices, the *medulla oblongata* retains the capacity to initiate reflex responses. An accessory *pons* appears in birds and mammals. It acts as an integration center for increasingly complex neuronal transmissions between the *cerebrum, cerebellum,* and spinal cord.

FROM FISH TO MAMMAL★

FOREBRAINA
   OLFACTORY
      BULBB/**TRACT**B1
   CEREBRUMC
   OPTIC NERVED

MIDBRAINE
   OPTIC LOBEF
   SUPERIOR
      COLLICULUSF1

HINDBRAING
   CEREBELLUMH
   MEDULLA
      OBLONGATAI
   PONSJ

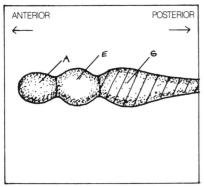

BRAIN OF A
VERTEBRATE EMBRYO

BRAIN OF A MATURE
VERTEBRATE REPTILE

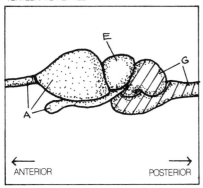

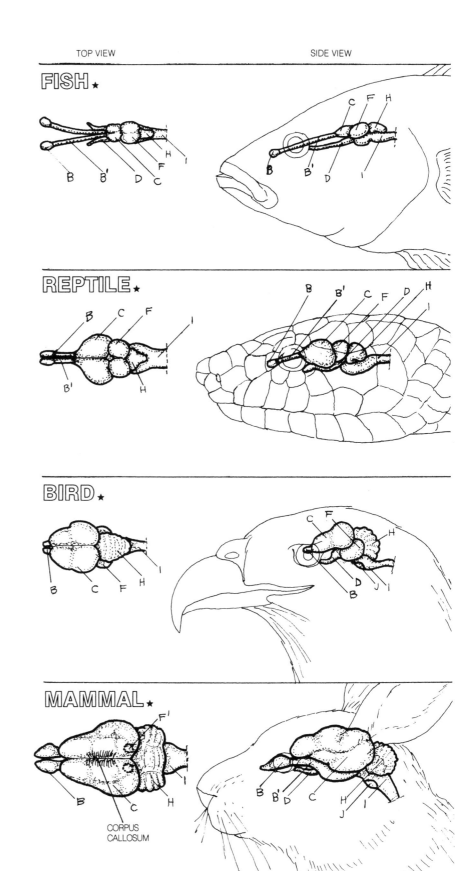

TOP VIEW   SIDE VIEW

FISH★

REPTILE★

BIRD★

MAMMAL★

CORPUS
CALLOSUM

14
ADAPTIVE RADIATION: MAMMALIAN FORELIMBS

The variety of mammalian forelimbs—the bat's wings, the seal's flippers, the elephant's huge supportive columns, the human's arms and hands—provides a lesson in the process of evolution. Comparative anatomy of homologous structures like the forelimbs, together with comparative embryology, provided early clues that evolutionary processes have created all these different mammals. Despite the obvious differences in shape, all mammalian forelimbs are composed of similar bones arranged in a comparable pattern. These homologies testify to a common evolutionary history. They also illustrate the phenomenon known as adaptive radiation. The external shape of the forelimb varies with its function, primarily with the animal's mode of locomotion. By adapting to environments as different as forest, plains, air, water, and underground, mammals have been able to radiate (like the sun's rays) into an incredible range of habitat milieus and ways of life.

Color the opossum scapula in the center of the page. Color the scapulae on all the surrounding species: the mole, bat, wolf, and around to the human. Continue to color in this manner first the opposum, then all other species.

The opposum represents the generalized ancestral mammalian forelimb pattern. The surrounding pictures show specializations of the forelimb for uses in different environments. The forelimb consists of the *scapula* (shoulder blade), which connects the rest of the forelimb to the trunk; *humerus* (upper arm bone); *radius* and *ulna* (lower arm); *carpal* (wrist) bones; *metacarpals* (hand bones of humans); and *phalanges* (finger bones of humans).

When you have colored all the individual bones, color the opossum's shadowed forelimb gray.

The opossum is much like the early (marsupial) mammals that originated more than 100 million years ago; so its forelimb is the kind from which all others were derived. The opossum is small bodied, moves easily on the ground or in trees, and has a flexible forelimb for these functions.

Turn now to the mole, which lives underground in moist soil and eats insects. Its forelimbs are short, close to the body, with large broad paws well suited for digging or "swimming" through the soil medium in which it lives. Its *scapula* is long and slender and anchors the forelimb against the trunk—a unique pattern, as you can see by comparison with the others. Its *ulna* is relatively long and robust, the *radius* somewhat reduced. The shovel-like paw constitutes almost half the length of the limb, and the elbows are rotated so that the palms face backward. The short, robust *metacarpals* and *phalanges,* with an extra bone (the falciform) make the short, broad paw an effective digging implement.

The bat's forelimb has been modified into a wing for flying. Its *humerus* is short relative to the long, slender *radius,* and the *ulna* is reduced—a pattern the reverse of that found in the mole. Remarkably long *metacarpals* and *phalanges* stretch out the skin like a sail. The broad wings are wider than the length of the bat's body.

The wolf is a swift runner, the better to pursue its prey. Its *humerus, radius,* and *ulna* are relatively long, providing a lengthy stride. The *metacarpals* are closely packed together for bearing weight—in marked contrast to those of the bat. The wolf walks somewhat up on its toes, on bent *phalanges.*

For its life in a predominantly watery environment, the sea lion has evolved a relatively short forelimb. It has a broad paw, modified into a paddle, with robust *metacarpals* and *phalanges* that stretch out the skin in a manner similar to the bat. The robusticity of the bones is necessary for supporting the sea lion's weight when it hauls out on land.

This evolution for support of a bulky mass is taken to the extreme in the 5-ton elephant. The bones are not only robust, but are arranged like an architectural column, from the *scapula* down to the *phalanges.*

Humans have long, mobile upper limbs that, unlike those of the other animals shown, are used primarily to manipulate objects rather than for weight bearing and locomotion. In common with the apes, the ball and socket joint of the *humerus* in the *scapula* allows the arm a 360° range of motion.

The similarity of a bat's wing and an elephant's leg may not be readily apparent until we examine homologous bone structures in all mammalian forelimbs. It then becomes clear that a common basic design has been modified by evolutionary processes into the particular pattern of each species.

ADAPTIVE RADIATION: MAMMALIAN FORELIMBS ⋆

SCAPULA A   RADIUS C   CARPALS (WRIST) E
HUMERUS B   ULNA D   METACARPALS (HAND) F
                                     PHALANGES (FINGERS) G

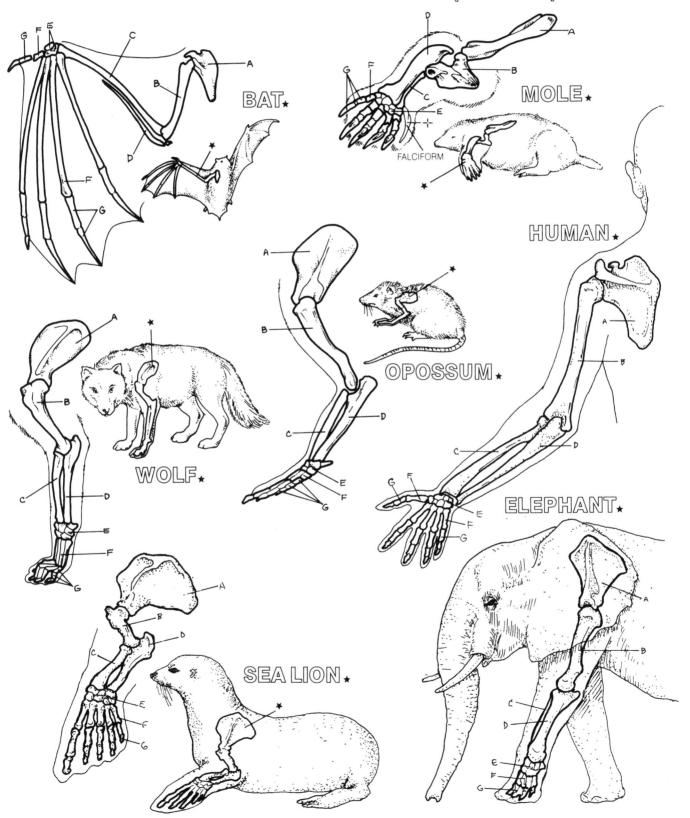

BAT ⋆

MOLE ⋆

FALCIFORM

HUMAN ⋆

WOLF ⋆

OPOSSUM ⋆

ELEPHANT ⋆

SEA LION ⋆

15
ADAPTIVE RADIATION: DENTITION

Mammals eat a wide range of foods and obtain them in diverse ways. The teeth, or dentition, serve to secure food from the environment and to hold and prepare it for digestion by slicing and grinding. The teeth also serve as offensive and defensive weapons, for cleaning fur, and for social purposes.

To accomplish these functions, mammalian teeth are specialized into four kinds: *incisors, canines, premolars,* and *molars.* In the same way the forelimb has changed in different species, basic tooth structures have been modified in form and number as adaptations to different diets and behaviors.

Teeth are only part of the dietary picture. Muscles are needed to put them into action by producing movement at the jaw joint. There are variations in ways of chewing, in the arrangement of muscles and their relative sizes, and in the shape of the jaw joint from species to species. The main chewing muscles are the *temporalis,* which attaches on the top and side of the skull, and the *masseter,* which attaches to the zygomatic arch and prominently on the lower jaw. The main action of the *temporalis* is fast jaw closure. The *masseter's* action is strength in jaw closure, but not speed. This plate examines the dentition and chewing muscles of four types of mammals: opossum, beaver, bobcat, and human.

Color each type of tooth: the incisors in all four species, then the canines, and so on. Then color the temporalis and masseter muscles.

The opossum has 50 teeth, 13 in the quadrant (one-fourth) of the upper jaw, as you see on the top picture, which shows half of the upper jaw. The lower jaw has 12 in each quadrant. The *incisors* and *canines* are suitable for seizing and holding insects and other small animals. *Premolars* and *molars* are used to pulp insects, soft-bodied animals, and soft plant foods.

The beaver, a member of the rodent order of mammals, has four grinding teeth on each side, and large continuously growing, chisel-like *incisors,* which have a self-sharpening mechanism. Rodents in general are very versatile in their feeding habits. The *incisors,* when used edge to edge, are effective for pinching, cropping, and puncturing, in addition to gnawing, the activity we tend to associate with beavers. The

canine teeth have been lost, and there is a large space (diastema) between the teeth. The posterior teeth, which consist of a *premolar* and three *molars* on each side, are square-shaped, powerful grinding and crushing teeth. They are involved in preparing the food for swallowing and digestion. Rodents have two kinds of jaw movements. In one, the lower jaw (mandible) is pulled forward; so the *incisors* come edge to edge for gnawing, pinching, and such. In the other, the jaw is pulled backward in order to use the posterior teeth for grinding. Note that their *temporalis* muscle is relatively small; the *masseter* is very large, for strength in grinding.

The order Carnivora is a diverse group of flesh-eating mammals including hyenas, dogs, otters, and badgers. It is represented here by the feline bobcat. The *incisors* are small and unimportant in function; the *canine teeth* are well developed: they are used to kill and hold the resisting or struggling prey. The *premolars* and *molars* are large, with pointed cusps. When they meet, they shear like a pair of scissors. Notice that both the *premolars* and *molars* are sharp; they are called the "carnassials," meaning flesh, for which the order is named. The *temporalis* muscle is large and powerful for closing the jaw quickly.

Humans have 32 teeth, 8 to each quadrant. The outstanding feature is that all four types of teeth are quite similar. *Incisors* are used to cut up food for ingestion, as in biting an apple. Our *canines* are used much like the *incisors,* though this is not the case for monkeys and apes. The *molars* and *premolars* are effective for grinding and crushing food to facilitate digestion and swallowing.

Humans are omnivorous, eating both plant and animal foods. Like all other primates, humans are hand feeders, bringing food to the mouth rather than taking their teeth to the food, as do beavers, bobcats, and opossums. *Temporalis* and *masseter* muscles are about equally developed for biting and for grinding.

As in the many kinds of upper limbs, so with the dentition—mammals have diversified during the past 60 million years. The relative sizes and shape of teeth, which ones dominate, and how each is used through the workings of the jaw joints and its muscles and processing food all reflect the variety of dietary patterns found among mammals.

ADAPTIVE RADIATION: DENTITION ★

TEETH AND CHEWING MUSCLES ★

TEETH ●
INCISOR A
CANINE B

PREMOLAR C
MOLAR D

MUSCLES ●
TEMPORALIS E
MASSETER F

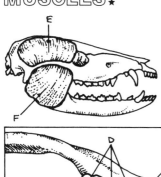

OPOSSUM ★

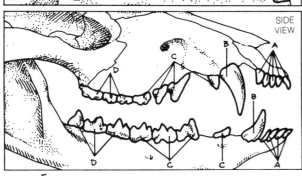

UPPER JAW, FROM BELOW

SIDE VIEW

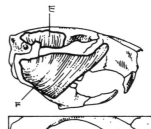

BEAVER ★

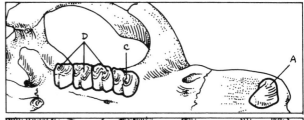

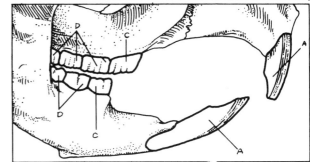

BOBCAT ★

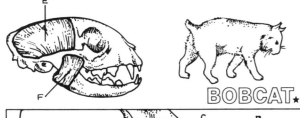

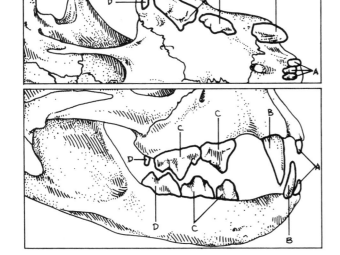

HUMAN ★

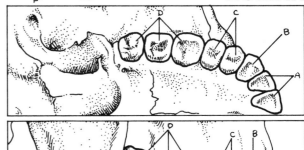

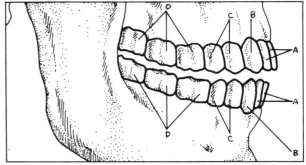
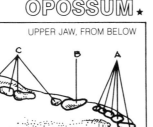

16
CONVERGENCE: THE SWIMMING NICHE

Just as there are differences among closely related animals, there are similarities among distantly related animals. Why does this happen? The phenomenon called convergence occurs because the environment is an important shaping force in evolution. The environment presents the opportunities and the challenges to organisms, which then respond by changing or by becoming extinct. Species of animals, even though distantly related, come to resemble each other when they live in a similar habitat or environment. Paleontologist George Gaylord Simpson pointed out that the expansion of life follows the opportunities presented rather than a preset plan.

Numerous examples of convergence exist, for example, between animals living in the tropical rainforests of West Africa and South America or among animals living in the grasslands of North America, South America, Africa, and even Australia. In this plate, animals of the ocean provide an excellent example of convergence in which the major groups of vertebrates are represented: fish, reptile, bird, and mammal.

Color the body shapes of the four animals.

The bodies of these animals are so streamlined that they appear not to have a neck; they all have fins or flippers and some form of tail. This *body shape* moves easily through the water with minimal resistance.

Color the forelimbs, first on the drawing of the individual animal, then on the enlargement on the right.

The shape on all four animals is very similar—elongated and flexible to act as a powerful paddle. The internal structures among the four animals are different. The similarity in external shape reflects the similarity in function; the differences in the internal structure reflect their different genetics and divergent evolutionary histories. Compare this to forelimb adaptive radiation on Plate 14.

Color the name of the animal and the internal structure of the forelimb.

The *shark*, which represents the fishes here, has no true bone; the supporting structure of its *forelimb* is cartilage, which is flattened and expanded into a fin.

Sharks come in all sizes and are major predators in the oceans around the world. They have been in existence since Devonian times, over 350 million years ago.

The swimming reptile is represented here by an *ichthyosaur,* a highly specialized marine reptile that appeared in the Triassic, flourished during the Jurassic, then became extinct. It was up to 3 meters long and had a streamlined body, legs modified into paddles, and a well-developed tail. The bones in its *forelimbs* were relatively inflexible. It looked like a big fish, but we know it is a reptile because of details in its morphology.

Although we think of birds as flying through the air, some birds, like the *penguin,* are adept at swimming. The wings of its ancestors were modified into flattened and fused bones, forming a compact and powerful fin *(forelimb). Penguin* feet are placed far back and webbed for swimming but also provide support on land, where their eggs are laid. *Penguins* live mainly in the oceans of the southern hemisphere and evolved from flying birds over 60 million years ago.

The mammal represented here is a *dolphin,* a member of the Cetaceans, which includes the whales and porpoises. This group successfully invaded the oceans more than 60 million years ago, evolving from a land-dwelling animal. The humerus, radius, and ulna are distinguishable and are homologous with those of other mammals on Plate 14. Most modification has taken place in the carpals, metacarpals, and phalanges to form a broad paddle. Although the *dolphin* looks like a fish, it is warm-blooded, as are other mammals. One young at a time grows within the mother's body. After birth, it is nourished by its mother's milk. A thick layer of blubber keeps *dolphins* warm.

Although the *shark, ichthyosaur, penguin,* and *dolphin* are only distantly related, they have converged by developing similar methods of locomotion and of obtaining food, both of which involve moving rapidly through water. Therefore, the modification of *body shape* and *forelimbs* in the four animal groups serve similar functions. Such marked similarities of ocean-living vertebrates suggests that there are a limited number of ways animals can effectively exploit the oceans and thus emphasizes the importance of the environment in shaping the course of evolution.

CONVERGENCE: THE SWIMMING NICHE★

BODY SHAPEₐ

FORELIMB_B

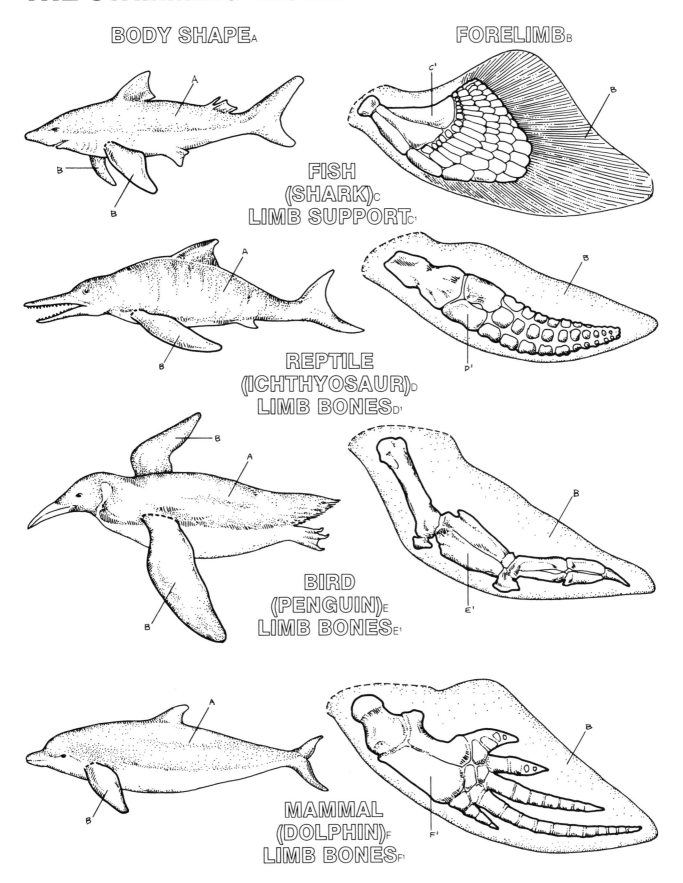

FISH
(SHARK)_C
LIMB SUPPORT_C¹

REPTILE
(ICHTHYOSAUR)_D
LIMB BONES_D¹

BIRD
(PENGUIN)_E
LIMB BONES_E¹

MAMMAL
(DOLPHIN)_F
LIMB BONES_F¹

17
CONVERGENCE: MARSUPIALS AND PLACENTALS

One of the most dramatic examples of convergence in evolution is that between marsupial and placental mammals. They diverged from a common ancestor about 120 million years ago during the Cretaceous, and each group continued to evolve independently. But despite their temporal and geographical separation, marsupials in Australia and placentals in North America have remarkable similarities in body shape, diet, and locomotion.

Color in gray the placental embryo within the map of North America.

Placental mammals are so called because of the placenta, through which the developing embryo is nourished by its mother. If you need to refresh your memory, refer back to Plate 7.

Color in gray the marsupial in its pouch shown within the map of Australia.

Marsupial young begin their development in the uterus but are born very immature. They complete their development within the mother's pouch and receive nourishment from the teat. Australia, in the southern hemisphere, is a continent the size of North America: 200 million years ago it was part of Gondwana, the large southern continent that included Africa, Antarctica, and South America. As Gondwana split up, Australia became isolated and has remained so for over 100 million years. The marsupials entered Australia before isolation; so their evolution has been independent for over 100 million years.

In Australia, there are well over 200 species of marsupials, varied in habitat, diet, and locomotion. There are also placentals, but many fewer species. The marsupials have undergone an adaptive radiation to occupy the diversity of habitats in Australia, just as the placentals have radiated in North America.

Begin coloring the animals at the top of the plate. Color each pair of animals (marsupial and placental) as they are discussed before moving on to the next pair.

Marsupial mice, like *mice* in North America, are small, agile climbers inhabiting low shrubs. They are similar in size and body shape to the common *mouse* in North America. There are numerous species of both kinds of *mouse.* They often live in dense ground cover and are active at night.

The *flying phalanger* resembles the *flying squirrel;* both are gliders with skin stretched between forelimbs and hindlimbs to provide planes for gliding from one tree to the next. They are similar in body size and shape, and both eat insects and some plants.

The *marsupial mole,* like the common *mole* in North America, is streamlined in body shape, burrows through soft ground, and eats insects. Its forelimbs are modified for digging with shovellike claws on the toes; the velvety fur is white to orange in the *marsupial mole* and gray in the North American *mole.*

The *wombat,* very much like the North American *ground hog* (also called a wood chuck), has rodentlike teeth; it eats roots and other plants. It also excavates burrows, as does the *ground hog.*

The *rabbit-eared bandicoots* are a group of marsupials that are varied in their diet—some are insect eaters; others eat plants. In comparison, the *rabbit* is primarily a vegetarian. The long feet in both these animals and their well-developed lower limb reflect their convergence in a hopping form of locomotion. The long ears emphasize the importance of the sense of hearing in these species.

The *Tasmanian wolf* is a carnivorous marsupial resembling the true *wolf.* The skull and the teeth (used for tearing meat) are similar, and the limb bones are long, adapted for running. The *Tasmanian wolf,* native to Tasmania in recent times, also inhabited Australia until it was settled by people thousands of years ago. Tasmania was once part of the Australian continent, but in the last ten thousand years, with the rising of the sea level, it became an island. The striking similarities of animals that have evolved thousands of miles apart, and more than a hundred million years distant, genetically demonstrates again the power of similar lifestyles in similar environments to shape, through natural selection, species form and function.

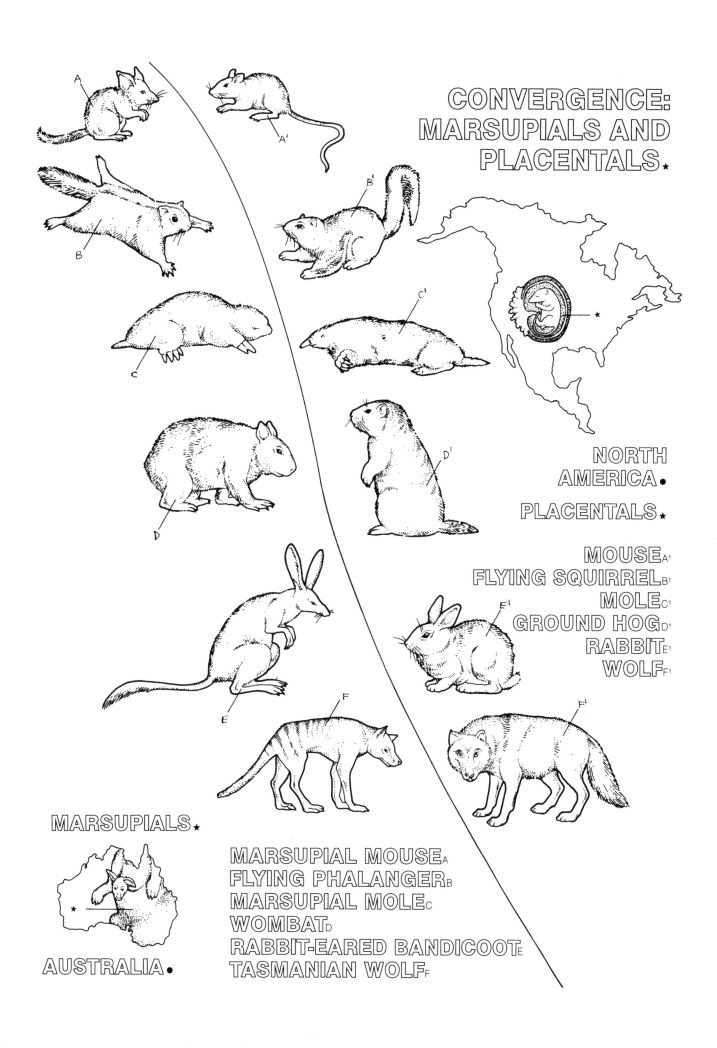

CONVERGENCE: MARSUPIALS AND PLACENTALS ★

NORTH
AMERICA ●

PLACENTALS ★

MOUSE A¹
FLYING SQUIRREL B¹
MOLE C¹
GROUND HOG D¹
RABBIT E¹
WOLF F¹

MARSUPIALS ★

MARSUPIAL MOUSE A
FLYING PHALANGER B
MARSUPIAL MOLE C
WOMBAT D
RABBIT-EARED BANDICOOT E
TASMANIAN WOLF F

AUSTRALIA ●

18

DARWIN'S FINCHES: NATURAL SELECTION AND ADAPTIVE RADIATION

The thirteen *Galapagos Islands* lying 600 miles off the western coast of *Peru* and *Ecuador* played a major role in shaping Darwin's thinking about the processes of evolution. During the voyage of the Beagle, Darwin noticed that many species found on the South American mainland were not represented on any of the islands. The island plant and animal species had arrived there by chance, probably rafting there on floating pieces of vegetation. Birds probably found their way to the *Galapagos* after being caught by westward winds.

Color the maps of the Galapagos Islands. Notice their distance from the mainland. Then color the three boxes relating to each finch as it is discussed in the text before moving on to the next bird. Note that the titles may not be listed in the order that you are coloring. Color the picture first; then find the corresponding title and color that. Color the boxes for habitat with light colors.

Fourteen species of finches are found on the *Galapagos Islands,* with up to ten species on any one island. Though similar in their plumage, nest-building techniques, and calls, the different species of finches could easily be distinguished by the size and shape of their *beaks.*

Here five of the ten species from the largest island, Albemarle, are examined. *Camarhynchus crassirostris,* a "vegetarian" finch, prefers tall *trees* in the dense, humid forest and eats *buds,* leaves, and *fruit* in the *trees* and occasionally descends to the *ground* for *seeds.* Notice the thick and short *beak.*

Geospiza fortis, a medium-sized ground finch, feeds on *seeds.* This species coexists with two other ground finches, one slightly larger and one somewhat smaller. Apparently the differences in *beak* and body size have adapted each of the *ground* feeders for eating slightly different sized *seeds.*

Geospiza scandens, the "cactus" finch, is also a *ground* dweller, but it is restricted to the more arid lowlands, where the prickly pear *cactus,* its main food source, grows. This finch has evolved a long and straight *beak* and a long, forked tongue for probing the *flowers* of the *cactus* to extract nectar and soft pulp.

Certhidea olivacea, the "warbler" finch, so resembles the unrelated warbler that even Darwin mistook them at first. Its *beak* is thin for eating both flying and *ground*-dwelling *insects.*

Perhaps the most remarkable of the finch adaptations is found in *Camarhynchus pallidus,* the "woodpecker" finch. Found chiefly in *trees* of more humid zones, this finch has converged with the true woodpecker: its *beak* is somewhat long and straight, chisel-shaped for excavating *insects* from beneath the bark. To feed, the "woodpecker" finch uses a *tool,* poking a small twig or *cactus* spine into the cracks and crevices, and then waiting for the *insects* to emerge.

British ornithologist David Lack visited the *Galapagos* in 1938–1939 and published *Darwin's Finches* in 1947. He proposed that an ancestral mainland flock of finches populated the islands very soon after they formed. With no other species of birds there to compete for available food sources, the finches were free to occupy even those niches that on the mainland were occupied by other kinds of birds. Ecology varied from one island to the next; so on each, the finch population evolved slightly different adaptive patterns.

In order for two or more species to exist on the same island, they must have a slightly different niche. If two species were competing for exactly the same food resources, the better adapted species would eventually replace the other. Darwin's finches alleviated the problem of interspecific competition by evolving different feeding habits. These are reflected in the structure of their *beaks.*

Color the adaptive radiation of finches diagrammed in the bottom left corner.

Darwin's finches are an elegant example of two important aspects of the evolutionary process: geographic isolation and adaptive radiation. Over an extended period of time, geographic isolation allows natural selection to produce new species. The evolution of distinctive structural adaptations and behavior patterns in a group of closely related descendant species, or adaptive radiation, is a means by which organisms respond to competition for resources. Natural selection favors those individuals whose characteristics are most suited for survival and reproduction for a particular way of life.

DARWIN'S FINCHES: NATURAL SELECTION AND ADAPTIVE RADIATION ★

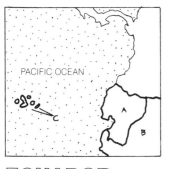

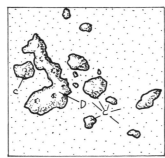

GROUND FINCH
GEOSPIZA FORTIS

"CACTUS" FINCH
GEOSPIZA SCANDENS

ECUADOR A
PERU B
GALAPAGOS ISLANDS C
  ALBEMARLE D

FINCHES ●
HABITAT ★
  TREES E
    GROUND F
DIET ★
  BUDS, FRUIT, FLOWERS G
    CACTUS H
    SEEDS I
    INSECTS J
TOOL K
BEAK L

"VEGETARIAN" FINCH
CAMARHYNCHUS CRASSIROSTRIS

"WARBLER" FINCH
CERTHIDEA OLIVACEA

"WOODPECKER" FINCH
CAMARHYNCHUS PALLIDUS

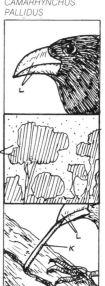

ADAPTIVE RADIATION ★

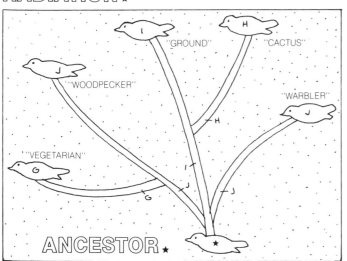

"GROUND" "CACTUS"

"WOODPECKER"

"WARBLER"

"VEGETARIAN"

ANCESTOR ★

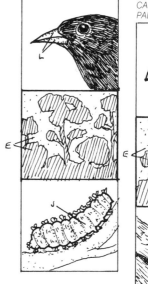

19
THE MOLECULAR BASIS OF LIFE: DNA: THE DOUBLE HELIX

Cells are the fundamental units of living things; they are like miniature machines or tiny chemical factories that arrange molecules into living matter. Most plants and animals are made up of millions of cells, each specialized for its own particular role: red blood cells carry oxygen; nerve cells conduct electrical impulses; and the cells lining the stomach secrete digestive enzymes. All cells have a full set of the genetic information that directs all the processes of life: growth, development, and metabolism. This genetic information, passed from parent to child, is carried in code in the molecule called *deoxyribonucleic acid* or DNA. The DNA, which directs the construction of molecules necessary for cellular function, codes for genetic information and is identical in all living organisms, indicating a common origin for all life on earth.

Begin in the upper right-hand corner and color each picture as it is mentioned in the text. First color the cell membrane, the structure that encloses and protects the cell's components and controls what leaves and enters the cell. Color all the components of the cell.

Here we picture a typical eukaryotic or nucleated cell. In eukaryotic cells, the DNA is found in packages called *chromosomes,* located within the nucleus, which is surrounded by the *nuclear envelope.* (In prokaryotic cells, which lack a nucleus, the DNA is located in the *cytoplasm.*)

Notice also the *nucleolus,* a nuclear organelle that produces ribosomes. (We will come across ribosomes again in Plate 22.) The porous nuclear membrane allows the selective exchange of molecules between the nucleus and *cytoplasm;* it remains intact as long as the cell is not dividing.

Color the enlarged chromosome to the left of the cell.

Most of the time, the DNA in the nucleus resembles a loosely tangled mass and distinct *chromosomes* cannot be detected. When the cell prepares to divide (in prophase), the DNA duplicates and forms two duplicate *chromosomes* joined by a *centromere.* Each duplicate *chromosome* is called a *sister chromatid.* Distinct *chromosomes* can be seen under the microscope.

If you move to a greater level of magnification, you can see the helical shape of the DNA molecule. It is a *double helix,* looking somewhat like a twisted ladder. The ladder is composed of *strand 1* and *strand 2,* twisted in opposite directions around one another.

Color the magnified double helix. Color each strand of the large helix on the left.

The backbones of these two strands are held together by the *base pair* "rungs."

Begin with the top rung and color all the base pairs that occur between the strands or backbone. Color the hydrogen bonds as well.

A *base pair* is formed by two nucleotide bases attached to one another by the partial charges of *hydrogen bonds.* (The structure of DNA will be more fully explained in Plate 21.) A single *hydrogen bond* is weak, but the multiple bonds down the rungs of the DNA molecule make it very strong. The multiple bonds also give the DNA molecule the capacity to open and close like a zipper, which is important for replication.

As the master controller of the cell, DNA controls its own replication and governs the synthesis of structural proteins, out of which our bodies are built. It governs the production of enzymes needed for the control of the thousands of chemical reactions that occur in the cell and contains the complete instructions for making a new organism.

The DNA *double helix* structure was discovered in 1953 by Francis Crick and James Watson. Along with Maurice Wilkins, they won the Nobel Prize for physiology and medicine in 1962. The story of this exciting discovery is told in Watson's *The Double Helix.* Once the structure of DNA was known, it became possible to unravel further secrets of the hereditary material: how it replicates itself, how it finds its way from one generation to the next, and how the information recorded on the *double helix* is "read" within the cell.

DNA: THE DOUBLE HELIX ⋆

LOCATION IN CELL ●
CELL MEMBRANE A
CYTOPLASM B
NUCLEAR ENVELOPE C
NUCLEOLUS D
CHROMOSOMES E
   CENTROMERE E¹
   SISTER CHROMATIDS E²

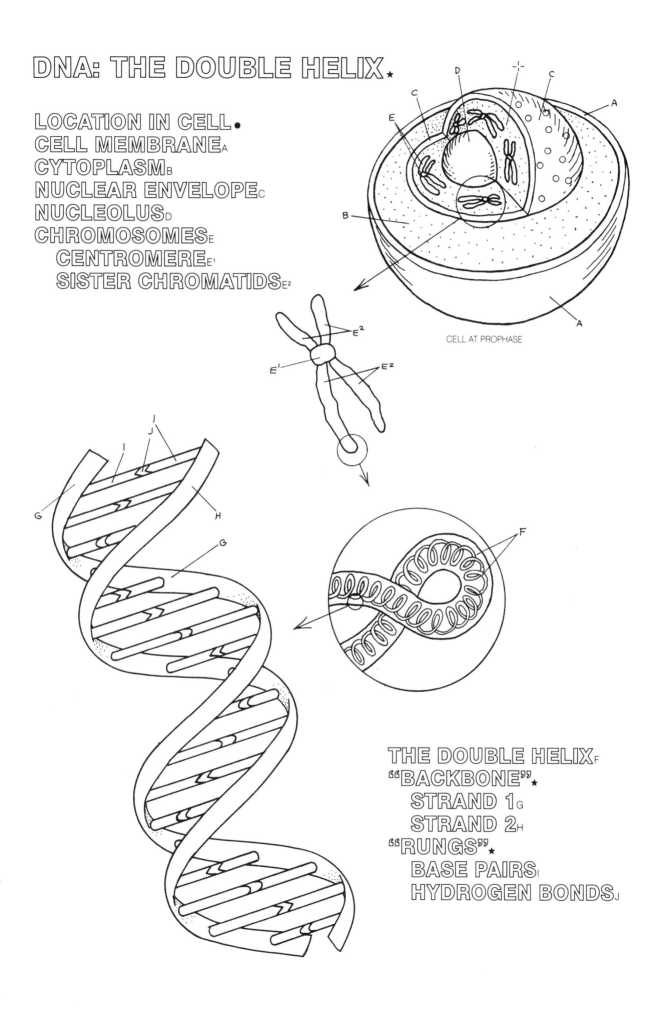

CELL AT PROPHASE

THE DOUBLE HELIX F
"BACKBONE" ⋆
   STRAND 1 G
   STRAND 2 H
"RUNGS" ⋆
   BASE PAIRS I
   HYDROGEN BONDS J

20
THE MOLECULAR BASIS OF LIFE: DNA REPLICATION

The unique structure of the DNA double helix allows for the accurate duplication of genetic information coded into the molecule. Each time a cell divides through mitosis (Plate 29), as the organism grows, or in the formation of sex cells through meiosis (Plate 30), each new daughter cell gets an identical copy of the genetic material. This plate shows how this exact duplication of information occurs.

Continue to use the same colors as on the previous plate for G through J. Begin at the top of the plate and color each structure in the first section: parent strand 1, parent strand 2, the base pairs, and the hydrogen bonds holding them together. You have now colored the DNA molecule as it appears prior to replication, in its double helix formation.

Proceed to the middle of the plate and color parent strands 1 and 2 and the bases, where they have pulled apart.

The replication process begins when the two *parent strands* separate, or "unzip," as the *hydrogen bonds* between two *base pairs* are broken in the presence of specific enzymes. The arrows represent each of the *parent strands* pulling apart. The bases are now exposed to the many molecules floating within the nucleus. Some of these molecules are *free nucleotides*.

Color all the free nucleotides that are floating around the DNA strands. All parts of the free nucleotides are colored the same.

A *free nucleotide* is a three-part molecule composed of a *phosphate (K^1)*, a *sugar (K^2)*, and a *base (K^3)*. Each *base* is complementary to the newly separated *bases* in the *parent DNA strands*. The *free nucleotides* and the free ends of the *bases* on the unzipped DNA *parent strands* attract complementary *bases*

(see Plate 21), as indicated by the small arrows. *Hydrogen bonds* form. These can be seen at the bottom of the middle section.

At the bottom of the plate you can see that as a *free nucleotide* forms a *hydrogen bond* with a base on the *parent strand,* a new ladder begins to form: the *base pairs* form the "rungs," and the *phosphates* and *sugars* bond to form a new "backbone" for the *daughter strand.*

Color the daughter backbone on the diagram to the left. The stippling indicates the actual structure of the molecules. Coloring the entire area gives an indication of the three-dimensional space forming the backbone. Color the rest of the plate.

Replication is finished when a complete *daughter strand* is joined with each *parent strand,* thus forming two new double helices. One intact strand from the parent helix, and one newly synthesized complementary *daughter strand* form each new double helix.

From one DNA molecule, two identical molecules have been formed, each carrying the same genetic information. After the DNA has replicated, the cell is ready to divide into two daughter cells. The accurate duplication and transmission of genetic information from generation to generation is essential. Mistakes in this replication process, called mutations, regularly occur. Many mutations are disadvantageous, although some are neutral or advantageous. The mistakes provide the variation at the genetic level necessary for change and evolution. We will see (in Plates 41 and 42) an example of how a mutation in the human DNA changes a blood protein so that it is advantageous in certain environments.

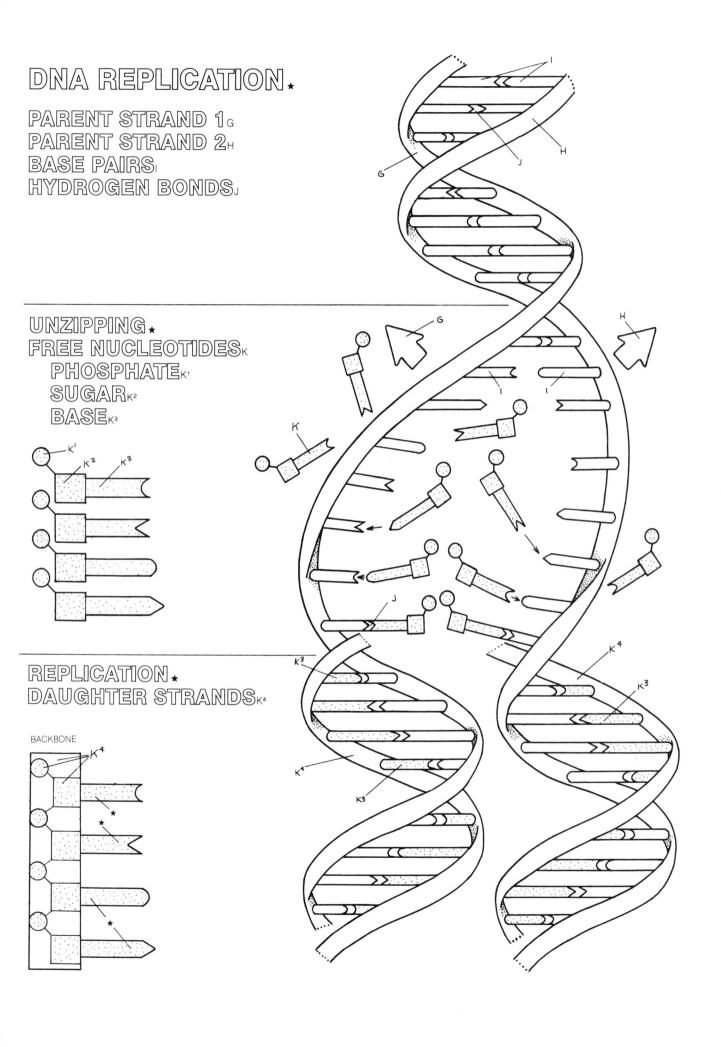

DNA REPLICATION ★

PARENT STRAND 1 G
PARENT STRAND 2 H
BASE PAIRS I
HYDROGEN BONDS J

UNZIPPING ★
FREE NUCLEOTIDES K
   PHOSPHATE K¹
   SUGAR K²
   BASE K³

REPLICATION ★
DAUGHTER STRANDS K⁴

BACKBONE

21

THE MOLECULAR BASIS OF LIFE: COMPONENTS OF DNA & RNA MOLECULES

We will now look even closer at the molecules that carry all our genetic information. We learned in the previous plate that the *nucleotide* is the fundamental unit of the DNA molecule.

Begin by coloring the dotted line around the single nucleotide on both the molecules located on the DNA and the messenger RNA molecules.

Note again that a *nucleotide* consists of a *phosphate, sugar,* and a *base.*

Color all the components of the DNA molecule on the left, including those within the dotted line.

Notice that the *phosphate* and *sugar* of each *nucleotide* bind to form what we have been calling the *backbone* of strands 1 and 2. Note that the *sugar* of the DNA *backbone* is *deoxyribose,* hence the name *de*oxyribose *n*ucleic *a*cid.

Color the stippled area (G) gray so that the impression of the three-dimensional phosphate-sugar backbone stands out more clearly.

There are two kinds of bases: *purines* and *pyrimidines,* which are complementary and joined by the *hydrogen bonds.* The *purines* are the "longer" molecules whose ends point outward; the *pyrimidines* are "shorter" and are diagrammed here with notches. (On this part of the plate, they are colored the same.) The notches and points indicate a "lock and key" complementarity.

After you have colored all the components of the DNA molecule, proceed to RNA on the right.

In contrast to DNA, messenger RNA consists of a single strand. Its *nucleotides* consist, as in DNA, of a *phosphate,* a *sugar,* and a *purine* or *pyrimidine base.* The RNA *sugar* is *ribose* rather than *deoxyribose sugar,* which has one less oxygen than *ribose.* Thus, *RNA* stands for *r*ibose *n*ucleic *a*cid.

Using two shades of one color for the purine bases and three shades of another, contrasting color for the pyrimidine bases, color the complementary base pairs at the bottom of the plate. Begin with *DNA.*

As noted on the two previous plates, the rungs of the DNA ladder are formed by complementary base pairs. Note that there are two *purine bases, adenine* and *guanine,* and two *pyrimidine bases, cytosine* and *thymine.* A single rung of the DNA double helix must consist of one *purine* and one *pyrimidine,* never two *purines* or two *pyrimidines.* Again, notice how the ends on each base are shaped so they fit together.

In *DNA, adenine* and *thymine* always go together, as do *guanine* and *cytosine.* During replication (as we saw on the previous plate), the free *nucleotides* conjoin with their complementary *bases* of the parent strand. *Adenine* on the parent strand would attract only a nucleotide containing *thymine,* and vice versa. *Guanine* is always found on the same rung as *cytosine.*

Proceed to color the *RNA.*

RNA also has four bases, three identical with those of DNA—*adenine, guanine,* and *cytosine.* But the fourth base in RNA is *uracil,* which is chemically very similar to *thymine* (which it replaces). Notice that *uracil* is the same shape as *thymine.* RNA is able to carry the genetic "message" out of the nucleus to the ribosomes in the cytoplasm. The message carried on a single strand of DNA is encoded on the mRNA (messenger RNA) in a complementary fashion. If the base pair on DNA is *guanine,* the mRNA will carry a *cytosine* at that point. For each *adenine* base on the DNA, the mRNA will carry a *uracil.* Remember that in RNA, *thymine* is replaced by the closely related *uracil.*

The next plate shows how the messenger RNA relays the information from the nucleus to the cytoplasm—in particular to the ribosomes—where it provides the manufacturing plans for protein production.

COMPONENTS OF DNA & RNA MOLECULES ⋆

NUCLEOTIDE_A
DNA ●
PHOSPHATE_B
SUGAR (DEOXYRIBOSE)_C
BASES_D ( )
   PURINE_D1
   PYRIMIDINE_D2
HYDROGEN BONDS_E

MESSENGER RNA ●
PHOSPHATE_B
SUGAR (RIBOSE)_F
BASES_D ( )
   PURINE_D1
   PYRIMIDINE_D2
BACKBONE_G ⋆

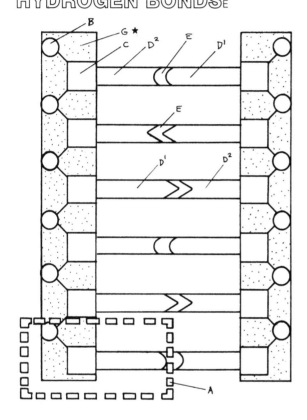

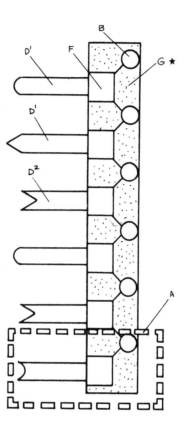

COMPLEMENTARY BASE PAIRS ⋆
PURINE
BASES_D1
   ADENINE_H
   GUANINE_I
PYRIMIDINE
BASES_D2
   CYTOSINE_J
   THYMINE_K
   URACIL_L

22

THE MOLECULAR BASIS OF LIFE: PROTEIN SYNTHESIS

In the three previous plates we learned that the genetic information necessary to direct all the processes of life is contained in coded form in the DNA. How does this information direct cellular activity? One important way is through the DNA-directed synthesis or manufacture of proteins. This plate provides an overview of the process of protein synthesis. In Plate 23, you will learn how the exact "language" of the DNA code specifies each particular protein.

Proteins, from the Greek meaning "primary," are molecules that perform many functions in the body. Proteins are the major structural components of skin, muscle, and organs. Enzymes, which control the rates of chemical reactions, and antibodies, which provide immune defense against foreign substances, are also proteins. There are probably over 30,000 kinds of proteins in the human body.

All proteins are composed of one or more chains of *amino acids*. Each chain is specifically coded for by a particular length of the DNA molecule called a gene.

To see how DNA directs protein production, begin by coloring the cutaway of the cell in the upper right. Color all the structures in the cutaway before moving on. Choose light colors for the ribosomes, messenger RNA and the transfer RNA.

Within the *nuclear envelope*, the DNA and the *free RNA nucleotides* combine to form the *messenger RNA*. The *mRNA* then leaves the nucleus and combines with *ribosomes, transfer RNA,* and *amino acids* in the cytoplasm to make the *amino acid* chains that form proteins. The plate shows this process in great detail.

Proceed to the diagram on the left, which is an enlargement of the nucleus. Color the structures —the nuclear envelope, the free mRNA nucleotides, the DNA, and the mRNA.

Protein synthesis begins when the DNA strands separate and the *free mRNA nucleotides* attach to the complementary bases of DNA to form the single strand of the *messenger RNA*. This part of protein synthesis is called transcription because the order of the bases on the DNA is transcribed directly onto the *mRNA* molecule.

The *mRNA* molecule passes through the *nuclear envelope* into the cytoplasm of the cell, where it attaches to a *ribosome*. A *ribosome* is composed of protein and RNA and is produced in the nucleolus. The *ribosome* can be thought of as a factory where the *amino acid* chains are manufactured.

Now color the enlarged ribosome and the mRNA, including the codons that run through the ribosome.

As the *mRNA* attaches to the *ribosome,* a "docking station" is formed where the bases are exposed three at a time, as the stippling pattern indicates. Each set of three *mRNA* bases, called a *codon,* is like a code word that corresponds to a particular *amino acid* in the protein chain to be formed (see Plate 23).

The translation of that code word into the appropriate *amino acid* in the chain is accomplished by the use of *transfer RNA* molecules (tRNA).

Color the transfer RNA and its anticodon.

At one end, each *tRNA* molecule has three base pairs complementary to an *mRNA codon*. These *tRNA* codes are called *anticodons*. On the other end, a specific *amino acid* is attached.

Color the amino acids and note that the line pattern on the tRNA corresponds to the attached amino acid. Each tRNA carries a specific amino acid. Also, color the peptide bonds formed between linked amino acids.

The job of the *tRNA* molecule is to transport its particular *amino acid* to the *ribosome*, where it attaches by hydrogen bonds to its complementary *codon* on the *mRNA* molecule.

When two *tRNA*s are both attached to the *mRNA*, a *peptide bond* is formed between adjacent *amino acids*. The *tRNA* then detaches itself from both the *amino acid* and the *mRNA* and leaves the "factory site" to find another *amino acid*. The *ribosome* moves along the *mRNA*, reading the *codons* sequentially, forming the growing *amino acid* chain in the bottom of the picture. When the message is completely read, the chain of *amino acids* is released and proceeds to carry out its specified job in the cell. When its job is completed, the short-lived *mRNA* disintegrates through enzyme action and its nucleotides are recycled in later RNA synthesis.

The number and kind of *amino acids* and their precise order distinguishes one protein molecule from another. How the precise sequences of bases in the *codons* specifies a particular *amino acid* in the chain is described in the next plate.

PROTEIN SYNTHESIS ★

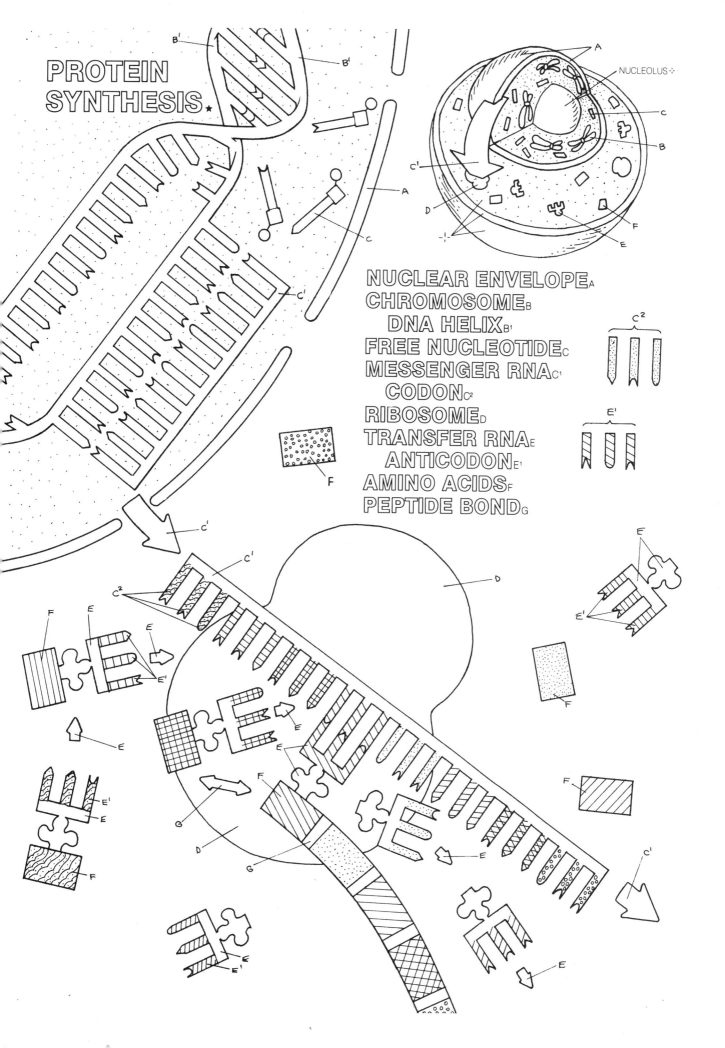

NUCLEOLUS ⋮

NUCLEAR ENVELOPE A
CHROMOSOME B
DNA HELIX B¹
FREE NUCLEOTIDE C
MESSENGER RNA C¹
CODON C²
RIBOSOME D
TRANSFER RNA E
ANTICODON E¹
AMINO ACIDS F
PEPTIDE BOND G

23
THE MOLECULAR BASIS OF LIFE: THE GENETIC CODE

Using the analogy of letters, words, and sentences, we can understand how vital information lies encoded in the linear sequence of nucleotide bases on the DNA molecule. In doing so, we are learning the language of the genes.

The genetic code is the "dictionary of life." The "words" of the genetic code consist of "letters," each representing the nucleotide bases in messenger RNA —*U, C, G,* and *A*. Each word is three letters long; that is, each word is a length of mRNA that is three bases long. We call these three-letter words codons. In their various combinations of three letters, these words are each a code for a specific amino acid.

The "sentence" in our dictionary of life consists of a long string of codons. The sentence spells out the structure for an amino acid chain of a protein molecule.

In the upper left-hand corner, the bases that comprise the codons, or base triplets—uracil, cytosine, adenine, and guanine—are to be colored light gray. Each base is designated by a particular line pattern. In the upper right-hand corner, color the first codon, UCC, light gray. One particular amino acid is coded for by the sequence of bases UCC—uracil, cytosine, cytosine. Color the amino acid. In the list to the lower left, find the matching letter (P) and you discover the amino acid is serine. Follow the same process for the next two amino acids at the top right. Thus, UGG codes for tryptophan and UUG, for leucine.

There are 64 possible three-letter combinations of the four bases: $4 \times 4 \times 4 = 64$.

Color the titles black: first RNA nucleotide base, second RNA nucleotide base, and third RNA nucleotide base. Color the bases gray.

There are only 20 amino acids rather than 64. This means that there is a redundancy in the code (in other words, several of the twenty amino acids may be coded for by more than one codon). In addition, certain codons represent instructions for "start" and for "stop" in the manufacturing instructions.

Now color the title for the first amino acid, alanine. Find the matching letter (A) in the boxes to the right.

Note that the manufacturing instructions for *alanine*

may be given in four different ways: the bases guanine and cytosine must always be in the first and second positions, but the third position of the codon can be filled by *uracil, cytosine, adenine,* or *guanine*. The words for *alanine,* then, are *GCU, GCC, GCG,* and *GCA.*

Proceed to color each amino acid title. Then find the matching letter in the box or boxes on the right and color them the same.

Continue to note the code word for each amino acid. Some amino acids, such as *methionine* and *tryptophan,* have but a single codon. Others have several, such as *leucine,* with six possible codons.

Notice that the codon *AUG* (adenine, uracil, guanine) serves as a "start" codon to signal with its amino acid *methionine* the beginning of the RNA message; all proteins are originally constructed starting with methionine. Note too that *UAA, UAG,* and *UGA* serve as "stop" codons, signaling the end of the message and causing the completed protein to be released from the ribosome.

You can now appreciate more fully how a mistake in the order of base pairs can result in a mutation. If the mutation occurs in the formation of sex cells, during meiosis, then the mistake may be transmitted to offspring. For example, if the first base in the codon *UCC* (for *serine*) is changed to *A,* the resulting codon *(ACC)* will code for threonine instead. This change may affect the protein's ability to function normally in the body.

Because of the redundancy of the code, some point mutations do not necessarily change the amino acid sequence. For example, *UAU,* which codes for *tyrosine,* may become *UAC*—where, at the third RNA nucleotide base, *cytosine* is substituted for *uracil*. In spite of this mutation, the amino acid coded for remains *tyrosine.* In this case there would be no change in function and the mutation is said to be "silent." In contrast, Plate 41 shows how a mutation in the arrangement of base pairs can have a profound effect on a human trait: a single reversal of base pairs on the sequence of DNA that carries the gene for the hemoglobin molecule can result in the disease of sickle cell anemia.

The 20 amino acids coded for by the four nucleotide bases in various combinations of three make up the thousands of proteins that carry out our bodily functions. Subsequent plates discuss such proteins as hemoglobin, cytochrome *c,* and the fibrinopeptides.

THE GENETIC CODE .

BASES FOR
CODON (BASE TRIPLET)•
URACIL
CYTOSINE
ADENINE
GUANINE

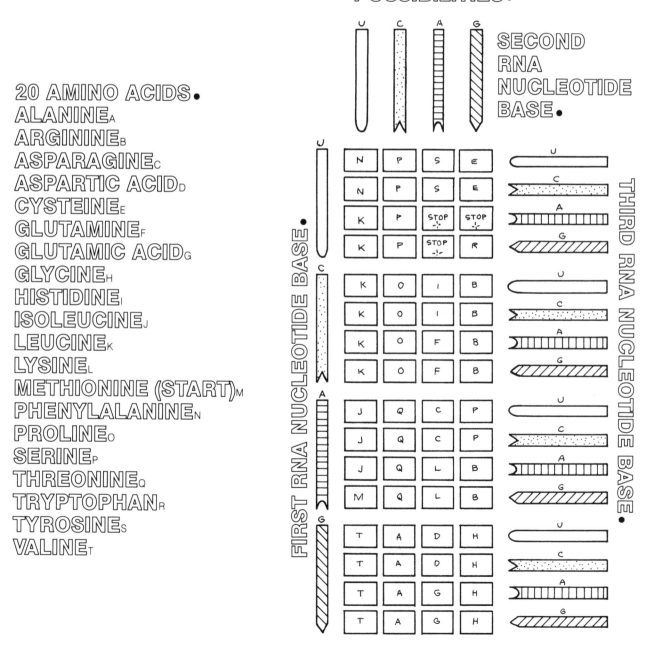

U★
C★
A★
G★

AMINO
ACIDS.

P R K

64 CODON
POSSIBILITIES•

SECOND
RNA
NUCLEOTIDE
BASE•

20 AMINO ACIDS•
ALANINE A
ARGININE B
ASPARAGINE C
ASPARTIC ACID D
CYSTEINE E
GLUTAMINE F
GLUTAMIC ACID G
GLYCINE H
HISTIDINE I
ISOLEUCINE J
LEUCINE K
LYSINE L
METHIONINE (START) M
PHENYLALANINE N
PROLINE O
SERINE P
THREONINE Q
TRYPTOPHAN R
TYROSINE S
VALINE T

FIRST RNA NUCLEOTIDE BASE•

THIRD RNA NUCLEOTIDE BASE•

24
BIOCHEMICAL EVIDENCE FOR EVOLUTION: DNA HYBRIDIZATION

In the first section of the book, we examined how anatomical comparisons can be used to deduce the evolutionary relationships between both living and extinct species. In the past twenty years, accompanying rapid advances in *DNA* research, techniques have been developed that enable us to compare the *DNA* of a wide variety of living species and, in so doing, provide quantitative information on their evolutionary background. Species differ in their genes—that is, in the sequence of nucleotides in their *DNA*. The degree to which the species differ genetically will be reflected in the sequence of amino acids in their proteins. The next four plates show how this kind of information contributes to our understanding of the genetic relatedness of humans to other primates.

One method of comparing the sequences of *DNA* for a number of species, *DNA* hybridization, was developed in the 1960s. Hybridization allows us to estimate whether two species have segments of *DNA* that are similar and, if so, to what degree. To illustrate the technique, we will compare *humans* and *chimpanzees*.

Begin at the top of the page by coloring human DNA strands and their complementary base pairs. Now color the chimpanzee DNA and its complementary base pairs. Proceed to color the human and chimpanzee DNA in the middle of the page.

The *DNA* of *chimpanzee* and *human* can each be separated or "melted" into single strands by heating. The heat breaks the hydrogen bonds between the *complementary base* pairs. Note that the strands from each of these species "melts" at *86°C*.

Color the melting temperatures (E) in a light color.

Next, the biochemist uses enzymes to "snip" a single homologous *DNA* strand for each species into fragments of about 500 nucleotides in length.

Color these fragments in the petri dish.

The hybridization technique relies on the tendency

for single strand segments of *DNA* to find complementary segments and to rewind into a double helix when the solution cools and hydrogen bonds reform. Hybrid *DNA* is formed when *human* and *chimpanzee* single strands wind together. The degree to which any two single strands rewind into hybrid *DNA* complexes tells us the degree to which their sequences are complementary.

On the bottom of the plate is a diagrammatic enlargement showing how well the two single strands mix. Color each strand and the complementary bases.

The correct *complementary bases* between the two species line up and attach themselves to each other by a hydrogen bond. The two strands then form a hybrid DNA helix—one strand of *human DNA* matched up with one strand of *chimpanzee DNA*.

Note that not all the pairs match up. Color these noncomplementary bases.

It is possible to estimate the proportion of *noncomplementary* bases by observing the temperature at which the two strands of the hybrid molecule separate.

Color the melting temperature of the hybrid DNA at 83.6 °C, which is 2.4 °C lower than for each of the "pure" DNA strands.

Since not all the bases on the chimpanzee and human single strands are *complementary* and therefore do not form hydrogen bonds, the hybrid double helix "melts" more easily, at a lower temperature. The data from a number of hybridization experiments suggest that each 1 °C reduction in *melting temperature* indicates approximately 1 percent mismatched or *noncomplementary bases*. The 2.4 °C temperature difference indicates about 2.4 percent of the bases between *chimpanzee* and *human DNA* are *noncomplementary*.

DNA HYBRIDIZATION.★

HUMAN_A DNA_{A¹}
CHIMPANZEE_B DNA_{B¹}
COMPLEMENTARY
    BASES_C
NONCOMPLEMENTARY
    BASES_D
MELTING
    TEMPERATURE_E

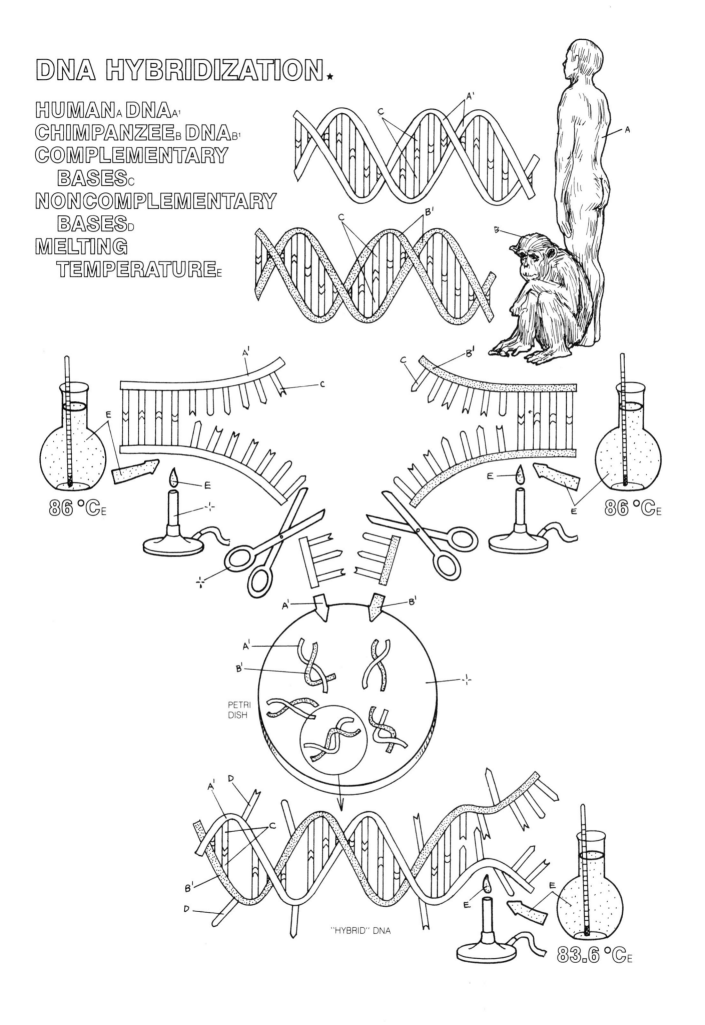

86 °C_E

86 °C_E

PETRI
DISH

"HYBRID" DNA

83.6 °C_E

25

BIOCHEMICAL EVIDENCE FOR EVOLUTION: DNA: SIMILARITY AMONG PRIMATES

Human DNA has been hybridized with DNA from a number of other primate species. DNA is the base level at which a genetic change takes place through time as evolution occurs. Therefore, the percentage of DNA similarity between species is a measure of their genetic relatedness. Comparison of species' DNA is a way of showing evolutionary relationships. This method is most useful for closely related species.

The graph shows degrees of similarity between 50 and 100 percent. Begin by coloring the human figure and corresponding bar. Color the percentage bars and the comparison species as they are discussed.

Human DNA represents 100 percent similarity; that is, if two strands of *human* DNA were hybridized, all the bases would be complementary and the melting temperature of the "hybrid" would be 86 °C.

When *chimpanzees* are compared to *humans,* they show *97.6 percent* similarity (or 2.4 percent dissimi-larity, as noted on the previous plate). *Gibbon* is *94.7 percent* similar to human. Next is the comparison of *human* DNA to that of two Old World monkeys—a *rhesus monkey,* a kind of macaque from Asia, and a *vervet monkey,* or guenon, from Africa. Note that both are about equally similar to humans: *rhesus monkey* at *91.1 percent* and *vervet monkey* at *90.5 percent.* The *capuchin monkey* from South America, a New World monkey, at *84.2 percent,* is less similar to humans than to either the apes or Old World monkeys. Finally, the *galago,* an African prosimian, and humans show much less similarity in the fit of their DNAs—only *58 percent.*

The results of DNA hybridization show that *humans* are most similar to the *chimpanzee,* an African ape, somewhat less to the *gibbon,* and least to the *galago.* This type of molecular data can be used to construct family trees for living species. Examples of these can be seen in Plates 71, 77, 79, and 82.

DNA: SIMILARITY AMONG PRIMATES ★

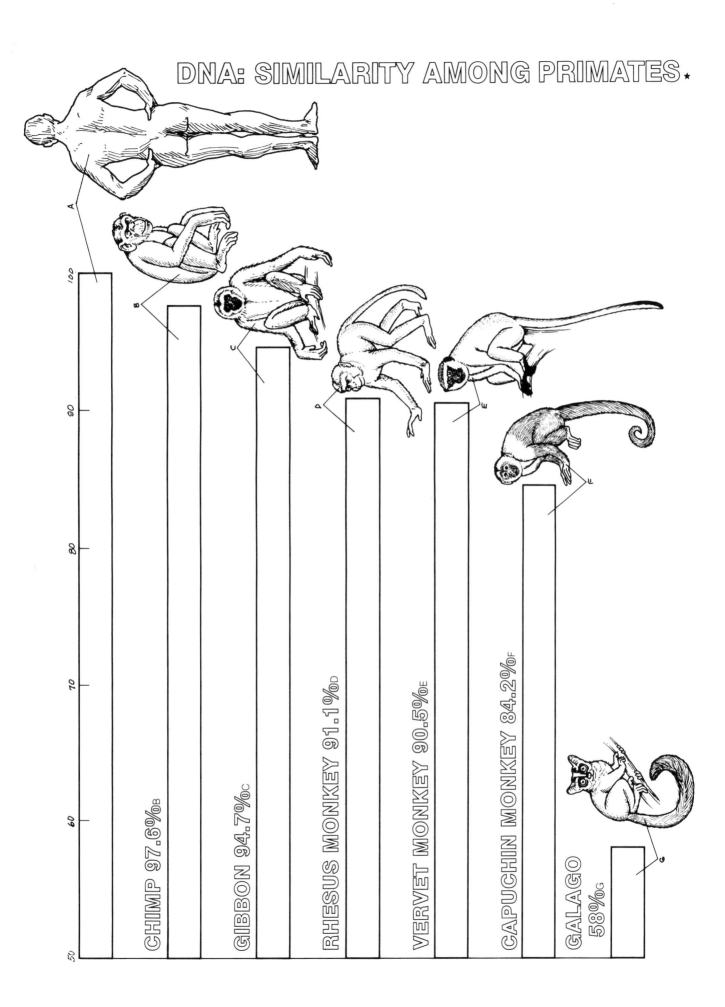

CHIMP 97.6% B

GIBBON 94.7% C

RHESUS MONKEY 91.1% D

VERVET MONKEY 90.5% E

CAPUCHIN MONKEY 84.2% F

GALAGO 58% G

26

BIOCHEMICAL EVIDENCE FOR EVOLUTION: AMINO ACID SEQUENCES: HEMOGLOBIN BETA CHAIN

Another way to study the genetic similarity between a number of species, in even finer detail, is by comparing the sequence of amino acids for a particular protein. As noted in Plate 21, each of our proteins has a specific number of amino acids arranged in a particular order, determined directly by the sequence of codons on the messenger RNA. The mRNA, in turn, is a complementary copy of the sequences of bases along the gene, or DNA.

Hemoglobin is the protein in red blood cells that is responsible for picking up oxygen in the lungs and transporting it to all bodily tissues. Adult hemoglobin is composed of four chains: two alpha chains and two beta chains. Each type of chain is coded for by a different gene. The beta chain of hemoglobin is one of several proteins whose sequence of amino acids has been determined for each of several primate species. There are 146 amino acids in the entire hemoglobin chain.

Begin by coloring the long horizontal chain of the amino acids in the center of the plate; only the amino acid positions that vary from one of these species to the next are listed here. The numbers below each amino acid indicate its position in the protein chain. Color the single amino acid below the gorilla.

Note that human and chimpanzee (color these and the other figures gray) have an identical sequence of amino acids in the beta chain of their hemoglobins. The *gorilla*, however, has a beta chain that differs from *human* and *chimpanzee* by one amino acid. At position 104, the *gorilla* hemoglobin contains *lysine* rather than *arginine*. The genetic "code words" that specify *lysine* are *AAA* or *AAG*; that is, the sequence of bases on the messenger RNA responsible for specifying position 104 on the hemoglobin is adenine-adenine-adenine (AAA) or adenine-adenine-guanine

(AAG). (Refer back to Plate 23.) Note also that the codon for *arginine* is *AGA* or *AGG*. Thus, we can infer that a single mutation—a change in the second nucleotide base of the codon determining position 104 of the protein chain—can account for the difference between human and gorilla hemoglobin. This indicates a very close genetic relationship between human, chimpanzee, and gorilla.

Now color the three amino acids that differ in gibbon from human, chimpanzee, and gorilla: aspartic acid at position 80 rather than asparagine, lysine at position 87 rather than threonine, and glutamine rather than proline at position 125.

Next, color the eight amino acids that differ in the rhesus monkey and the nine in the squirrel monkey. Note that the sites differ except for positions 87 and 125, all of which have glutamine. Note, too, that the amino acid sequences of the apes are much more similar to humans than are those of the monkeys.

The results of amino acid sequencing for hemoglobin and for many other proteins are consistent with those from DNA hybridization and comparative anatomy and offer ways of discovering just how closely we are related to the apes. Although the closeness of humans to the African apes—the chimpanzee and gorilla—has been recognized since Darwin and Huxley, only with the methods of molecular biology can we quantify that relationship.

It is still not well understood how protein sequences lead to anatomical features. Ape-human anatomical differences seem much greater than their small protein difference. In contrast, Old and New World monkeys look very much alike, but their proteins differ by about seven times as much as those of humans and African apes.

AMINO ACID SEQUENCES: HEMOGLOBIN BETA CHAIN ★

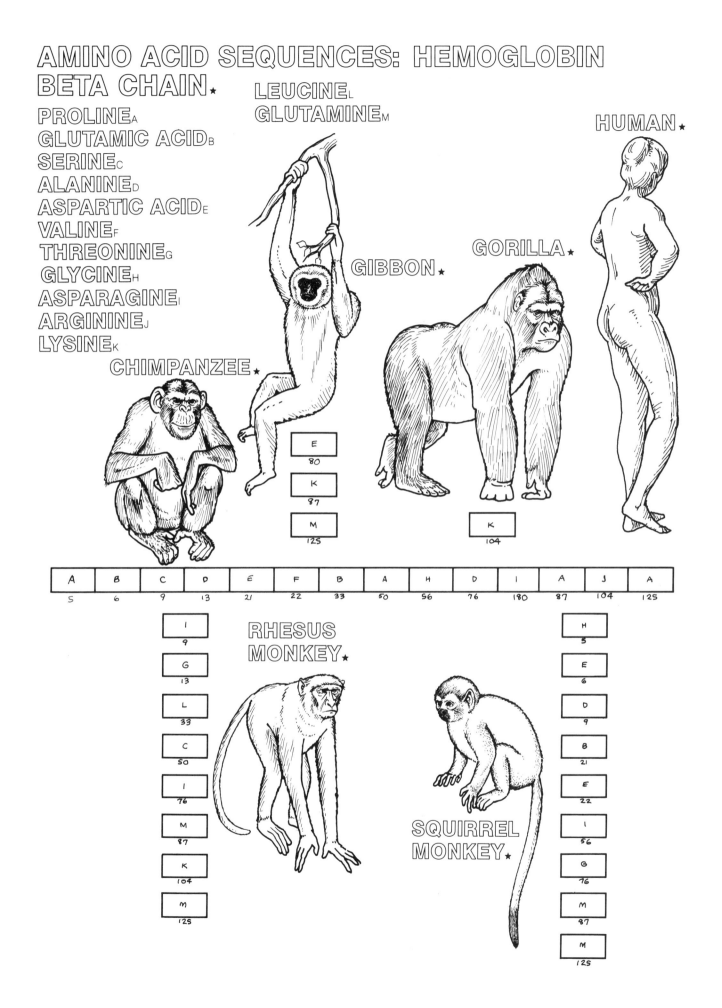

PROLINE A
GLUTAMIC ACID B
SERINE C
ALANINE D
ASPARTIC ACID E
VALINE F
THREONINE G
GLYCINE H
ASPARAGINE I
ARGININE J
LYSINE K
LEUCINE L
GLUTAMINE M

HUMAN ★

GIBBON ★

GORILLA ★

CHIMPANZEE ★

Gibbon: E 80 — K 87 — M 125

Gorilla: K 104

A	B	C	D	E	F	B	A	H	D	I	A	J	A
5	6	9	13	21	22	33	50	56	76	180	87	104	125

Chimpanzee: I 9 — G 13 — L 33 — C 50 — I 76 — M 87 — K 104 — M 125

RHESUS MONKEY ★

SQUIRREL MONKEY ★

Squirrel Monkey: H 5 — E 6 — D 9 — B 21 — E 22 — I 56 — G 76 — M 87 — M 125

27
BIOCHEMICAL EVIDENCE FOR EVOLUTION: IMMUNOLOGICAL STUDIES

Another method for estimating the degree of similarity of proteins in different species uses immunological techniques. Immunological studies are an indirect method for detecting differences in proteins and so reflect differences in the genetic material in different species. The method can be used on specific isolated proteins, such as albumin, transferrin, and collagen, or on complex mixtures of proteins. This plate illustrates the technique using *human serum*, the fluid part of the blood, and comparing it with serum from other mammals to see how closely related they appear.

The procedure requires several steps. Starting at the top of the plate, use a light color to color the serum from human blood that is injected into a rabbit.

The many proteins in the *human serum* act as a foreign material (antigen) in the *rabbit*. The *rabbit's* immune system makes antibodies that recognize the antigens, attach to them, and render them harmless to its body. Antibodies are protein molecules made by an organism in response to an antigen, usually a foreign protein. Antibodies stick to their specific antigens and thus can be used to identify molecular structure. Once the *rabbit* makes antibodies specific to *human serum* proteins (or *antihuman antibodies*), these can be extracted from the rabbit's blood.

Color the antihuman antibody extract in the upper right.

The next step in the procedure is to combine the *antihuman antibodies* with serum from different species. If we start by combining it with *human serum*, the picture on the middle left shows the reaction of the antibodies to the human antigens.

Color the human serum and the rabbit antihuman antibodies as they are poured in the test tube. Then using both colors, one on top of the other, color the darker stippled area occupying most of the mixed solution. This area represents the precipitate that forms when an antibody recognizes the antigen against which it is made. The inset enlargement to the right shows the antigens (from the human serum) and the antibodies fitting together in a lock-and-key forma-

tion, **which then sinks out of solution and forms a precipitate.**

The more similar the antigen is to the one against which the antibodies were made, the stronger the reaction and so the larger the amount of precipitate. Because the antibodies were made against *human serum* proteins, mixing them with *human serum* forms the largest amount of precipitate possible.

When one mixes serum from other species with the *rabbit antihuman antibodies*, smaller amounts of precipitate form. This is because the proteins are somewhat structurally different from the human and so are not identical antigens.

The next picture to color shows the rabbit antihuman antibodies reacting with gorilla serum. A large amount of precipitate forms, but not as much as with human serum.

Next color the baboon serum and the antihuman antibody and note that there is less reaction than with the gorilla but more than between the lemur serum and antihuman antibodies. Color the rest of the plate.

The least reaction is between the rat serum and the antihuman antibodies. The rat is a rodent, not a primate, and less closely related to humans than are the other primates. The insert to the left of the rat shows that there is little fit between the *rabbit antihuman antibodies* and the *rat serum*; so they both remain in solution and form very little precipitate or "recognition" reaction.

Immunological studies have provided a great deal of information on the evolutionary relationships of humans and many other species. The greater the reaction of antigens and antibodies, the greater the similarity of protein structure and the more recently the two species shared a common ancestor. These studies, which have flowered in the last two decades, support conclusions about evolutionary relationships derived from other kinds of evidence, such as comparative anatomy and other molecular approaches, including DNA hybridization and amino acid sequencing. Immunological studies show that humans are most closely related to the apes, less closely related to the monkeys, and even less closely related to prosimian primates, represented here by the *lemur*.

IMMUNOLOGICAL STUDIES.★

HUMAN SERUM A
HUMAN SERUM PROTEIN A'
RABBIT B
RABBIT ANTIHUMAN ANTIBODY C
GORILLA SERUM D
BABOON SERUM E
LEMUR SERUM F
RAT SERUM G
RAT SERUM PROTEIN G'

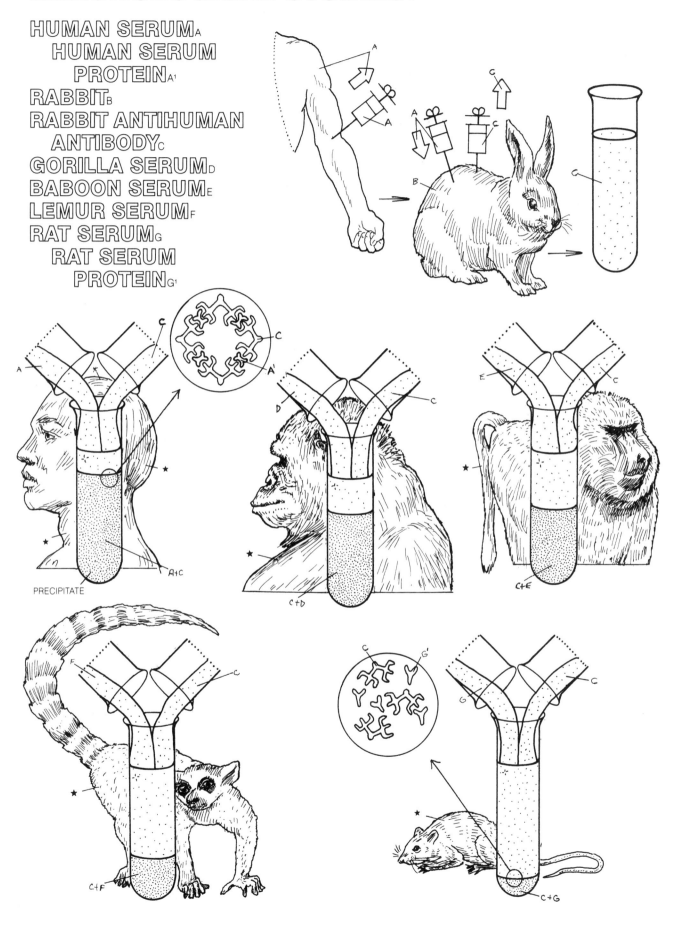

PRECIPITATE

28

THE MOLECULAR CLOCK: PROTEINS AND THEIR RATES OF CHANGE

The longer species have been separated evolutionarily, the more amino acid differences there are in their protein structure, but the degree of similarity between two species is not the same for all proteins. One protein may be nearly identical, whereas another protein may differ significantly. Each protein has a specific rate of change through time.

Changes in protein structure reflect mutations in the gene at the DNA level. The mutation rate is probably similar for all genes, but the amount of change seen in the amino acid sequence depends upon the selection pressures for maintaining the exact structure and function of any particular protein molecule. To illustrate this point, we consider four different proteins with different rates of change. Natural selection has operated such that *histones* have changed very little and *fibrinopeptides* have changed considerably.

Begin by coloring the hourglasses in the upper right-hand corner that contain histone, cytochrome c, hemoglobin, and fibrinopeptides.

The hourglasses represent time, and the sandgrains represent amino acids of a protein. These "clocks" have been set to "zero" at ninety million years ago, when the major groups of placental mammals diverged, based on the fossil record—a horse, a human, a bat, and a squirrel. (The human figure represents the primate lineage.) The kangaroo, a marsupial animal, diverged one hundred twenty million years ago, before the radiation of placental mammals.

In the enlarged hourglasses, we compare the horse and human. Begin at the left with histone. As you color it, you can see that none of the sandgrains have dropped, showing there has been no change in the amino acid sequence in ninety million years. Color the zero percent below and note that no part of the horse or human is colored.

Histones are basic proteins that are wrapped in long fibers around the DNA in the chromosome, providing structural support and regulating DNA activities, such as replication and RNA synthesis. Its intimate relationship with the DNA makes its particular structure critical, allowing almost no mutations to be maintained. The 103 amino acids in this protein are identical for almost all plants and animals.

Now color cytochrome c and the hourglass sand

that represent its 104 amino acids.

The few fallen sandgrains represent the twelve amino acid differences between horse and human *cytochrome c*.

Color the 12 percent difference in horse and human in the figures at the bottom.

On average, *cytochrome c* changes faster than *histones*. *Cytochrome c* is a molecule involved in cellular respiration, participating in the utilization of oxygen for the production of energy. *Cytochrome c* is found in all aerobic cells (those that use oxygen to produce energy) and has changed slowly because of its vital role in energy production.

Now color hemoglobin, the molecule at the top, all the sandgrains in the hourglass, the percentages, and the figures at the bottom. Note the larger amount of change in the amino acid sequence compared to cytochrome c.

The *hemoglobin* beta chain has 146 amino acids; 26 of them differ in horse and human, representing about an 18 percent difference. *Hemoglobin* transports oxygen from the lungs to other tissues throughout the body; the exact sequence of amino acids is not so important as long as it can bind and release oxygen molecules. Natural selection has allowed more of the substitutions to stay because they do not interfere with function.

Color the fibrinopeptides.

Horse and human amino acids differ by 86 percent: they are almost entirely different. The *fibrinopeptides* change about 1 percent of their amino acid sequence every 1.1 million years. *Fibrinopeptides* are segments of the fibrinogen molecule, which is important in blood clotting. The segments simply have a spacing function, to keep the molecule inactive when clotting is not needed. When clotting action is required, the *fibrinopeptides* are cut out of the molecule and discarded, leaving the "sticky" surface to aid in the clotting. The actual sequence of amino acids is unimportant for their job of spacing; so selection has allowed many mutations to remain in the sequence.

The study of molecular difference not only gives us information about the relative relatedness of species, but also provides a clock to discover when the divergence between two species occurred.

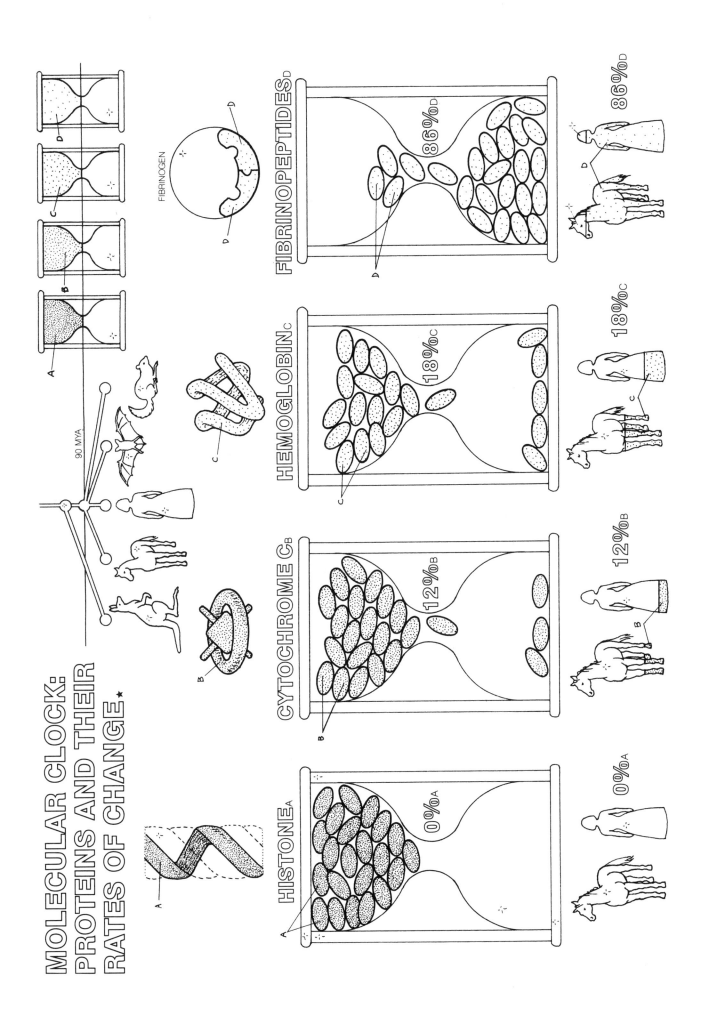

MOLECULAR CLOCK: PROTEINS AND THEIR RATES OF CHANGE★

90 MYA

FIBRINOGEN

HISTONE_A   0%_A   0%_A

CYTOCHROME C_B   12%_B   12%_B

HEMOGLOBIN_C   18%_C   18%_C

FIBRINOPEPTIDES_D   86%_D   86%_D

29

SOMATIC CELL DIVISION: MITOSIS

The molecular clock implies that genetic information is passed on from one generation to the next and that the changes at the DNA level occur at a fairly constant rate. Each cell in the body contains all the genetic information, beginning with the first single cell formed at fertilization. The initial cell grows and divides throughout the body's development. All cells, except for nerve cells and female germ (sex) cells, constantly grow and divide to replace dying cells. During these divisions, the genetic material replicates and is passed on to every new cell of our bodies through a process called mitosis.

Eukaryotes, or organisms with nucleated cells, evolved about a billion years ago (see Plate 1). The mitotic divisions occurring over evolutionary time enabled genetic material to become more complex so that several long chains of DNA or chromosomes could be replicated and sorted efficiently into daughter cells. With an increase in chromosome number and amount of DNA through time, evolution produced the diversity of complex plant and animal life observed today.

This plate illustrates mitosis—somatic cell division (from the Greek word, "soma," meaning body). Mitosis refers to the replication of the cells in our bodies. Germ or sex cell formation (the ova in the female and spermatozoa in the male) are a special case and are shown in the next plate.

Begin with interphase. Color all the structures in each phase as discussed, before moving on to the next. Use a light color for the cytoplasm and a dark color for the centrioles and the mitotic spindles. The mitotic spindles are shown as a series of dots. To color them, draw a thin line connecting the dots so the spindles appear as thin fibers emanating from the centrioles.

Mitosis is a continuous process, but it is usually described in four stages, as pictured here: prophase, metaphase, anaphase, and telophase. During interphase, before mitosis begins, *chromatin*, from the Greek word for color, referring to its affinity to a certain dye, lies in a jumbled mass in the cell nucleus. The *nucleolus*, where the ribosomes are manufactured, is intact. The *nuclear envelope*, a porous membrane, encloses the *nucleoplasm* and *chromatin*. Outside the *nuclear envelope* in the *cytoplasm* are the *centrioles*. These important structures produce *mitotic*

spindles, small tubular structures that play a critical role in cell division.

In humans there are forty-six *chromosomes*, twenty-three homologous or "matching" pairs. Each *chromosome* consists of a long chain of DNA with histone proteins wound along its length (Plate 19).

For the sake of illustration, only four *chromosomes*, or two homologous pairs, are shown here. As the long, thin strands of *chromosomes* coil up and condense, they become detectable under an electron microscope during prophase. In prophase the *chromosomes* appear as *sister chromatids*, which are duplicates of the same *chromosome* connected at some point along their length by a *centromere*. *Sister chromatids* are the product of DNA replication during interphase. During prophase, the cell has twice the normal amount of chromosomal material. Notice that the *nucleolus* has disappeared and that the *nuclear envelope* has begun to disintegrate in preparation for cell division. Also at this stage the *centrioles* begin to migrate toward the opposite ends of the cell. As they migrate, the *mitotic spindle fibers* begin to form from within the cell.

During metaphase, the *centromeres* line up along the *cell equatorial plane*, and the identical *sister chromatids* lie on opposite sides of the plane. The *centrioles* have found their positions at opposite ends of the cell, and the *mitotic spindles* have attached themselves to the *centromeres*. Notice that the *nuclear envelope* has completely disappeared and *mitotic spindles* are well developed.

During anaphase, the cell begins to elongate as the *mitotic spindles* seemingly pull the *sister chromatids* apart. The *centromeres* separate and each chromosomal pair migrates toward opposite ends of the cell, pulled along by the contracting *mitotic spindles*.

During telophase, mitosis is completed: *chromosomes* cluster around the *centrioles* at opposite ends of the dividing cell. The *cytoplasm* pinches off in the center and a new *cell membrane* forms along the *equatorial plane*. By late telophase, a *nuclear envelope* has formed in each new daughter cell; a *nucleolus* reforms, and the *mitotic spindles* disintegrate. Each new daughter cell contains the full genetic complement of the parent cell.

At the completion of mitosis, the *chromosomes* uncoil and return to their random state as *chromatin* until each of these new daughter cells divides again into more somatic cells.

SOMATIC CELL DIVISION: MITOSIS ★

CYTOPLASM A	SISTER CHROMATIDS F1
NUCLEOPLASM B-¦-	CHROMOSOME F2
CELL MEMBRANE C	CENTROMERE G
NUCLEAR ENVELOPE D	CENTRIOLES H
NUCLEOLUS E	MITOTIC SPINDLE FIBERS I
CHROMATIN F	CELL EQUATORIAL PLANE J

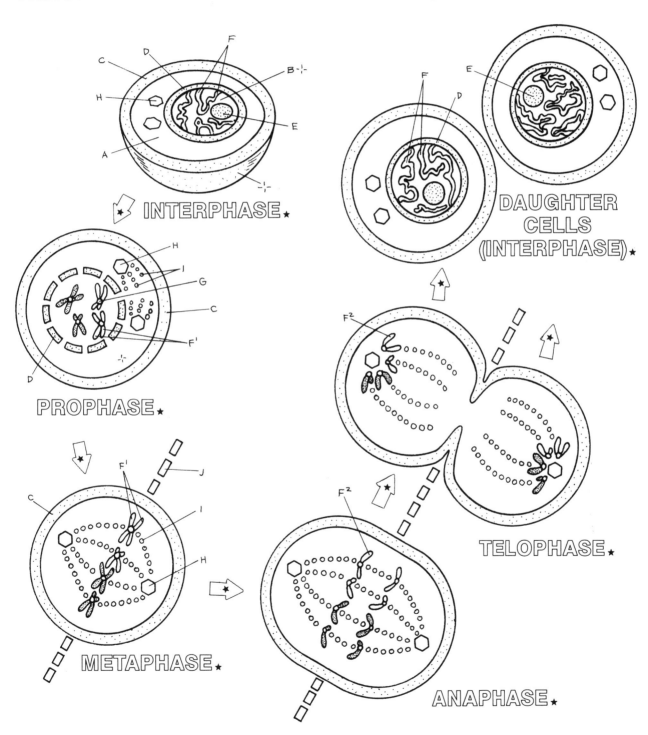

★ INTERPHASE ★

PROPHASE ★

DAUGHTER CELLS (INTERPHASE) ★

METAPHASE ★

ANAPHASE ★

TELOPHASE ★

30

SEXUAL REPRODUCTION: MEIOSIS

Sexual reproduction, which appeared about a billion years ago (see Plate 1), produces offspring through the union of male and female gametes, or sex cells. Gametes, a term that includes both female ova and male spermatozoa, contain only half the number of *chromosomes* of a somatic cell; thus, they are said to be *haploid*. When the ovum and *spermatozoan* unite, the *chromosome* number is returned to its normal duplicate or *diploid* state. In humans, then, the *haploid* state is twenty-three *chromosomes*.

Meiosis, from the Greek word meaning to reduce, is the process through which these *haploid* sex cells are produced.

In this plate we trace the genetic material during the process of gamete formation, or gametogenesis. We will pay particular attention to *chromosome* number, since the goal of meiosis, unlike mitosis, is the production of *haploid* cells with a reduced number of *chromosomes*.

Choose four contrasting colors to follow the changes in chromosome number through meiosis and color the band surrounding the cell at each step to represent its chromosome number at that point.

For the sake of simplicity, we will follow the fate of two matching pairs of *chromosomes*. For humans, meiosis involves twenty-three pairs of *chromosomes*, or forty-six total.

Begin with oogenesis, the production of ova. Color the oogonium and then color the arrow and the parts in the primary oocyte.

Oogonia are ovary cells that may become ova. Two million are present in the human female at birth, but only around four hundred actually mature during a woman's lifetime.

The oogonium contains the full *diploid* number of *chromosomes*. The DNA replicates, forming identical *sister chromatids* characteristic of primary oocytes.

The chromosome number at this stage is *pseudotetraploid*, indicating 2x the *diploid* amount of chromosome material.

Now the first meiotic division occurs. Color the arrow to represent that process. Now color the secondary oocyte.

Matched *chromosomes* line up along the equatorial plane, and one of each pair goes to a daughter cell. Notice that this division differs from mitosis in that the centromere remains intact so that the *sister chrom-*

atids remain attached to one another. Thus, the secondary oocyte is *pseudodiploid*. The actual amount of genetic material is the same as in a somatic cell, but only half the *chromosomes* are present, although in duplicate form. Notice also the unequal distribution of cytoplasm at this cell division. The secondary oocyte receives most of the cytoplasm; the other daughter cell, the *first polar body*, receives virtually none.

At this point the egg, with its *first polar body*, is released from the ovary. If fertilization takes place, the oocyte undergoes a *second meiotic division*.

Color the arrow to represent this process and color the ovum.

This division is similar to mitosis and results in the formation of a *second polar body* and the ovum, which now has the desired *haploid chromosome* number. The ovum receives most of the cytoplasm, ensuring that the zygote (fertilized egg) will have a rich supply of nutrients

Inside the ovum, notice the *male pronucleus* contributed by the *spermatozoan*.

Color each stage of spermatogenesis, the production of sperm.

Now we turn to spermatogenesis and the production of male sex cells in the male testes. The spermatogonia have a *diploid chromosome* number. *DNA replication* (represented by the arrow) produces primary spermatocytes with a *pseudotetraploid chromosome* number. The first meiotic division produces two *pseudodiploid* secondary spermatocytes with equal amounts of cytoplasm and one chromosome from each homologous pair. The *second meiotic division* results in four spermatids of identical size and *haploid chromosome number*. These spermatids grow a long tail and develop a specialized head that enables them to penetrate and fertilize the egg.

In summary, several important differences exist between the production of egg and sperm. The primary oocyte produces only one ovum with unequal distribution of the cytoplasm; the primary spermatocyte gives rise to four viable spermatozoa. In humans, the female is born with all the ova she will ever produce and she usually produces one egg every twenty-eight days if not pregnant or lactating. Males produce millions of spermatozoa continuously from the onset of puberty. Over a quarter of a billion spermatozoa are released at one time. In the next plate we shall examine the union of the ovum and spermatozoa in fertilization.

SEXUAL REPRODUCTION: MEIOSIS.

CHROMOSOMES_A
  SISTER CHROMATIDS_{A¹}
CHROMOSOME NUMBER ★
  DIPLOID_B
  PSEUDOTETRAPLOID_C
  PSEUDODIPLOID_D
  HAPLOID_E
DNA REPLICATION_F
CENTRIOLE_G

MEIOTIC SPINDLE_H
FIRST MEIOTIC DIVISION_I
SECOND MEIOTIC DIVISION_J
FIRST POLAR BODY_K
SECOND POLAR BODY_L
SPERMATOZOAN_M
  MALE PRONUCLEUS_{M¹}

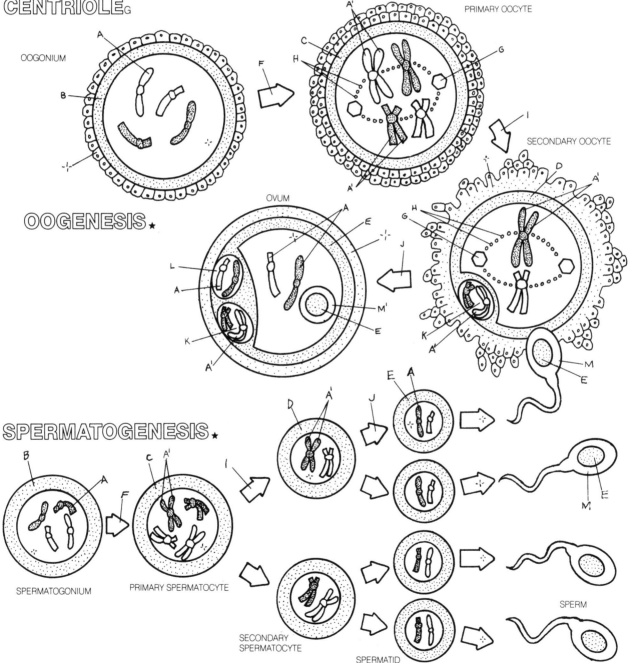

OOGENESIS ★

SPERMATOGENESIS ★

31

SEXUAL REPRODUCTION: FERTILIZATION

The importance of meiosis to evolution lies in the possibility for recombination of genes from two different parents through sexual reproduction. Fertilization is the process through which sexual reproduction is carried out. Asexually reproducing organisms produce offspring that are exactly like themselves. But animals (and plants) that produce offspring through sexual reproduction, or the union of *haploid* sex cells, enhance the reproductive potential of the species by introducing new genetic combinations, and variation is an essential component of the evolutionary process.

This plate examines the events preceding and following fertilization, beginning with ovulation followed by the union of *sperm* and egg as it normally occurs in the reproductive tract of the human female. As illustrated in the smaller diagram in the lower right-hand corner of the plate, *sperm* enter the female through the vagina, using their long, whiplike tails to swim through the uterus and up to the fallopian tubes. From there, tube wall contractions move the *sperm* along the hornlike ends of the fallopian tubes, where fertilization can take place.

Begin by coloring the ovary in the center of the plate. As you read below, color your way to the left and up around to the right, through events leading to the implantation of the five-day-old blastocyst in the uterine wall. As you color the oocytes, the zygote, the morula, and blastocyst in the larger picture, also color them in the smaller drawing in the lower right-hand corner of the plate.

Each month one *oogonium*, or female sex cell, begins developing within a microscopic mass of cells called a *follicle*. The *oogonium* enlarges until it is ready to undergo meiotic division. (Refer back to the previous plate to review how the *oogonium* first undergoes DNA replication.) The first meiotic division produces the *secondary oocyte* and *first polar body.*

Color the secondary oocyte in a light color and its first polar body a darker, contrasting color. Also, color all the parts of the secondary oocyte on the far left, depicting fertilization.

When this *secondary oocyte* bursts from the *ovary*, it emerges surrounded by *follicle* cells and is *pseudo-diploid* in *chromosome* number. This event is called ovulation.

Fertilization usually takes place in the horn-shaped openings of the fallopian tubes. Several *sperm* reach the ovum simultaneously, but only one is able to digest its way into the *secondary oocyte*. The entrance of the *sperm* into the ovum stimulates its second meiotic division.

Color all parts of the fertilized ovum.

The result of the second division is a *second polar body* devoid of cytoplasm. The ovum now contains the *male pronucleus* and *female pronucleus*. The pronuclei carry the *haploid* set of *chromosomes*, which unite and will form the genetic material of the new offspring.

Color all parts of the zygote.

Soon after the *sperm* enters the ovum, the two *pronuclei* fuse to form a single *diploid* nucleus. Now we have a single-celled *zygote* with a new combination of *chromosomes*, half from one parent and half from the other.

Within twenty hours, the *zygote* begins to divide by mitosis into two cells, each half the size of the ovum, called blastomeres. (Refer back, if needed, to Plate 29 to review mitosis.) Here, metaphase of mitosis is illustrated.

Color the rest of the plate.

Notice that the polar bodies soon disintegrate and disappear. After four mitotic divisions, the *zygote* will contain sixteen cells. The *zygote* is now called a *morula*, from the Latin word "morus," meaning mulberry, which it resembles.

Now the *zygote* begins to hollow out from within, as shown in cross section. The *blastocyst* begins to dig into the inner wall of the uterus, the endometrium. It completely buries itself in the wall of the uterus, where it begins to develop into another human being.

SEXUAL REPRODUCTION: FERTILIZATION ★

DAY-3 MORULA

DAY-5 IMPLANTED BLASTOCYST

2-CELL STAGE O

ENDOMETRIUM

UTERINE WALL

VAGINA J

FALLOPIAN TUBE

FERTILIZATION

OVULATION

OVARY$_A$
FOLLICLES$_B$
OOGONIUM$_C$
PRIMARY OOCYTE$_D$
SECONDARY OOCYTE$_E$
FIRST POLAR BODY$_F$
CHROMOSOMES$_{G(\ )}$
PSEUDODIPLOID$_{G^1}$
HAPLOID$_{G^2}$
DIPLOID$_{G^3}$
CENTRIOLES$_H$
MEIOTIC SPINDLE$_I$

SPERM$_J$
MALE PRONUCLEUS$_{J^1}$
FERTILIZED OVUM$_K$
FEMALE PRONUCLEUS$_L$
SECOND POLAR BODY$_M$
FUSED PRONUCLEI$_N$
ZYGOTE$_O$
MORULA$_{O^1}$
BLASTOCYST$_{O^2}$
MITOTIC SPINDLE$_P$

32
MENDEL'S PEAS AND PARTICULATE INHERITANCE

Modern views of inheritance date back to Gregor Johann Mendel (1822–1884), a retired Augustinian monk who practiced amateur botany in the quiet garden of his monastery. In 1866 he published a report of his breeding experiments on the garden pea plant. Only seven years after Darwin had set forth his theory of natural selection Mendel stated his famous laws of inheritance. Perhaps because the attention of the scientific community was directed toward Darwin's exciting new theory of natural selection, Mendel's publication remained obscure until 1900. At that time, three scientists working independently rediscovered Mendel's laws. They were Hugo deVries of Holland, Karl Erich Correns of Germany, and Erich von Tschermak of Austria.

Since then, the science of genetics has virtually exploded in complexity. Although Mendel knew nothing of genes, chromosomes, or DNA, his laws of heredity provide the backbone of the modern synthetic approach to the study of evolution.

Out of Mendel's studies came his idea of particulate inheritance, which challenged the popular nineteenth-century notion of blending inheritance. Blending inheritance maintained that traits passed on from the parents through sex cells are blended in the offspring. Darwin and his colleagues adhered to this idea, assuming that offspring manifest characteristics that are intermediate between those possessed by the mother and father. This is illustrated by the analogy of mixing two colored liquids in a test tube.

Choose two primary light colors to represent the parental characteristics and color the two test tube drawings. Color the liquid in the receiving test tube with both parent colors.

Mendel proposed fundamental units of inheritance called "factors," which are distinct entities that do not blend during fertilization. Thus, the offspring contained the distinct factors or particles from each parent.

Mendel's experiments with the pea plants are an excellent example of sound scientific method, even by today's rigorous standards. He carefully chose the plants whose characteristics varied in discrete and measurable ways. After testing hundreds of species of garden plants, he chose the common sweet pea, genus *Pisum*. He found that the pea plant had seven easily observable characteristics that came in two, and only two, varieties.

Now look at the traits of the pea plants Mendel studied. Begin by coloring the terms for Mendel's pea plant traits gray: seed shape, seed interior, seed coat, and so forth. Now using two contrasting colors, color the dominant characters one color and the recessive characters another color. Suggested colors are yellow for dominant and green for recessive characters.

Seed shape was either *smooth* or *wrinkled*; seed interior was either *yellow* or *green*; seed coat, either *gray* or *white*; ripe pods were either *inflated* or *constricted*; unripe pods, either *green* or *yellow*; flowers were positioned either *axially* or *terminally* on the plant stem; and stems were either *long* or *short*.

Other botanists in Mendel's day merely described the characteristics of parent and offspring plants they had bred. Mendel meticulously recorded the number of plants possessing a given characteristic in each generation. He believed that the ratios of plant varieties in a generation of offspring would yield clues to the mechanisms of inheritance. Appropriately, he based his conclusions on large samples from his breeding experiments. Mendel was diligent in his experimental technique: he examined the characteristics of each plant individually and took the greatest care to prevent plants from uncontrolled cross-breeding and self-pollination.

Thus, Mendel's discovery of the principle of particulate inheritance followed from his carefully controlled experiments. When he crossed plants producing *smooth seeds* with those producing *wrinkled seeds*, the result was always the same—*smooth seeds*. No plants contained both *smooth* and *wrinkled* seeds, and none of the seeds was just a little bit *wrinkled*. No blending had occured. The same thing was noted for all the characters. As we shall observe on the next plates, the *wrinkled seeds* did reappear in later crosses. Mendel discovered that some traits were *dominant* and some, *recessive*. For his seven pea plant characters, one variety was found to be *dominant*. In order to explain how *recessive* traits reappeared in later generations, Mendel proposed that factors contributed by both parents were maintained in the offspring in discreet units that could be inherited and passed on, regardless of whether they were expressed in the outward appearance of the offspring. This is particulate inheritance.

In the next two plates we will further examine how Mendel developed his ideas of inheritance.

MENDEL'S PEAS AND PARTICULATE INHERITANCE.★

FIRST PARENT_A
SECOND PARENT_B

OLD IDEA:●
BLENDING INHERITANCE.★

MENDEL'S IDEA:●
PARTICULATE INHERITANCE.★

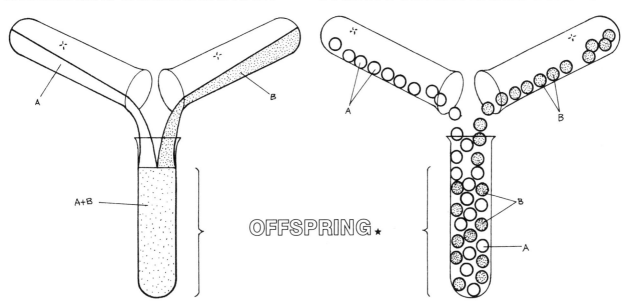

A+B

OFFSPRING.★

MENDEL'S PEA PLANT CHARACTERS.●

DOMINANT_C
SEED SHAPE.★
SMOOTH_{C¹}
SEED INTERIOR.★
YELLOW_{C²}
SEED COAT.★
GRAY_{C³}
RIPE PODS.★
INFLATED_{C⁴}
UNRIPE PODS.★
GREEN_{C⁵}
FLOWER.★
AXIAL_{C⁶}
STEM.★
LONG_{C⁷}

RECESSIVE_D

WRINKLED_{D¹}

GREEN_{D²}

WHITE_{D³}

CONSTRICTED_{D⁴}

YELLOW_{D⁵}

TERMINAL_{D⁶}

SHORT_{D⁷}

33

MENDEL'S PEAS: LAW OF SEGREGATION

What had happened to the recessive traits as they were passed on from one generation to the next? Mendel's Law of Segregation explains how the sorting of "factors" during meiosis, and their subsequent recombination during fertilization, can account for the sudden reappearance of recessive traits in the second generation of hybrid offspring.

The discrete inheritable units Mendel called "factors" are now called genes. We now know that genes are lengths of the DNA molecules that carry the manufacturing instructions for one protein (see Plate 22) and are arranged on a chromosome like beads on a string. The varieties of each plant character Mendel observed are due to the presence of alleles, or alternative forms of the same gene. Alleles differ from one another in just a few nucleotide bases; so they correspondingly code for slightly different sequences of amino acids. They are found on the exact same position along *homologous chromosomes*: they are said to occupy the same gene locus. Humans and sweet peas, like all diploid organisms, possess two genes for every trait. One was contributed by the female parent, the other, by the male parent.

Mendel invented the system of referring to alleles with letters: capital letters stand for the dominant allele and small letters stand for the recessive allele. On this plate we use the example of seed shape: *R* stands for the allele for *smooth* seeds, and *r* represents the allele for *wrinkled* seeds.

It will be helpful at this point to master a few more new terms. "Genotype" refers to the genetic makeup of an individual for any given trait or traits. If you have inherited the same allele for a trait from both your parents, your genotype is *homozygous* for that trait. If you have inherited a different allele from each parent, your genotype for that trait is *heterozygous*. "Phenotype" refers to the outward appearance of the individual and should be thought of as the observable expression of the genotype. For example, pea plants with a *smooth* seed phenotype may have either of two genotypes: *RR* or *Rr*. The genotype *RR* is *homozygous,* for such an individual has inherited identical alleles from each parent. The genotype *Rr* is *heterozygous:* one parent contributed the allele *R* and the other, the allele *r*. Since the *R* allele for *smooth* is dominant to the *r* allele for *wrinkled* seeds, the phenotype of both these genotypes is *smooth* seeds.

These principles are illustrated on this plate. Choose six complementary colors and follow the fate of the alleles for seed shape through two generations of cross-breeding, just as Mendel did. Start at the top of the plate, coloring the parent pea plants gray. Color the chromosomes and the two forms of the seed shape gene. Color the *phenotypes* of the parent plants two different colors.

Below the parent plants, the pairs of *homologous chromosomes* that carry the alleles for seed shape in each individual are shown. One chromosome in each pair is dotted and one is left plain to remind you that one is contributed by each parent. The alleles for the seed shape gene are located in the same position on four chromosomes. Each of the parents has a *homozygous* genotype. The female plant is *homozygous* for the *smooth* trait, and the male plant has a *homozygous* recessive genotype possessing two alleles for *wrinkled* seeds.

Refer back to Plate 30 to review how *homologous chromosomes* become packaged into separate sex cells after the first meiotic division during gametogenesis. As *homologous chromosomes* segregate, they carry with them all the genes along their length.

Color the genotypes and the arrows to represent the segregation of the seed shape alleles during meiosis. Continue coloring down the plate. Color the gametes, each of which contains one allele for seed shape. Follow the recombination of genes in sexual reproduction.

Now we can see why Mendel observed that 100 percent of the first generation of hybrid plants had *smooth* seeds. Each pea plant contains a recessive allele for *wrinkled* seeds, but the allele for *smooth* seeds is completely dominant and prevents the *wrinkled* allele from being expressed in the phenotype.

Observe what happens if this first generation of *heterozygous* genotypes is allow to interbreed. Gametes are produced and recombined, as the arrows indicate. Random mating produces plants with three different genotypes, one of which is a recessive *homozygote, rr*. This one genotype causes the *wrinkled* phenotype to reappear in this generation of offspring. Mendel harvested over seven thousand peas and observed the ratio of three smooth peas to every one wrinkled pea, a phenotypic ratio of 3:1. Through this observation he deduced that maternally and paternally derived "factors" segregate into separate gametes. The Augustinian monk knew nothing of centrioles or meiotic spindles, but with our more sophisticated understanding of cellular division, we can trace the mechanisms for his observed patterns of inheritance to the precise manner by which chromosomes are sorted into daughter cells during meiosis.

LAW OF SEGREGATION.

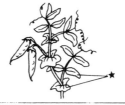

PARENT PLANTS.

HOMOLOGOUS CHROMOSOMES$_A$
ALLELES: R$_B$ r$_C$

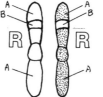
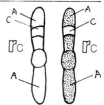

R   R   r$_C$ r$_C$

PHENOTYPE.
SMOOTH$_D$
WRINKLED$_E$

X
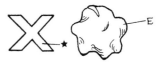

GENOTYPES.
HOMOZYGOUS RR$_B$
HOMOZYGOUS rr$_C$
HETEROZYGOUS$_C$ R$_B$r$_C$

R$_B$R$_B$ r$_C$r$_C$

SEGREGATION DURING MEIOSIS.

     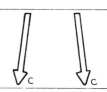

B   B   C   C

GAMETES (HAPLOID SEX CELLS).

     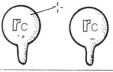

R$_B$ R$_B$ r$_C$ r$_C$

RECOMBINATION OF GENES (DURING FERTILIZATION).

B C   B   C   B   C   B   C

FIRST GENERATION.
PHENOTYPE.

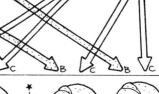

D   X   D

GENOTYPE. $_B$Rr$_C$ $_B$Rr$_C$ $_B$Rr$_C$ $_B$Rr$_C$

RECOMBINATION OF GENES (DURING FERTILIZATION).

B B B   C   C   B   C   C

SECOND GENERATION.
RATIOS: 3$_D$:1$_E$ PHENOTYPE.

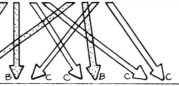
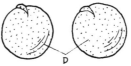

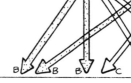

D   D   E

1$_B$:2$_C$:1$_C$ GENOTYPE. $_B$RR$_B$ $_B$Rr$_C$ $_B$Rr$_C$ $_C$rr$_C$

34

MENDEL'S PEAS: LAW OF INDEPENDENT ASSORTMENT

Now let us take a broader look at how two of Mendel's characteristics are inherited simultaneously, seed shape and seed coat color. Though Mendel did not know it, the genes for these two traits are carried on two different pairs of homologous *chromosomes*.

Color the four chromosomes. One chromosome of each pair is dotted, indicating that one chromosome is contributed by each parent. Choose four contrasting colors (other than yellow and green) to color the alleles: R = smooth, r = wrinkled, T = yellow, t = green.

Now use yellow (Y) and green (G) to color all the possible phenotypes you can derive by various combinations of these four alleles.

Now follow the fate of these four alleles in a first generation hybrid mating. As always, this includes the sorting of alleles into haploid gametes, followed by their recombination in the diploid zygote. Each parent is *heterozygous* for both seed color and seed shape. In other words, their genotypes are RrTt; their phenotype is *smooth yellow* seeds.

Color these alleles on the chromosomes. Color the phenotypic appearance of each of the parents.

The arrows represent the sorting of alleles during gametogenesis. Notice that homologous alleles always become segregated from one another in the gametes. *R*, for example, is never found with *r* in a gamete. This is Mendel's Law of Segregation. (Review Plates 30 and 32.)

Notice that the *chromosome* carrying *R*, the allele for *smooth* seed coat, may end up in a gamete containing either *T* or *t*. Likewise, the allele *t*, for *green* seed coat, may find its way into a gamete with a *chromosome* carrying either allele *R* or *r*.

Color your way through the segregation of each pair of homologous alleles to produce four possible combinations of alleles in the male and female gametes.

Mendel found that for each of his seven pea characteristics, the inheritance pattern of one trait did not depend in any way on the inheritance pattern of the other traits. He concluded that "factors" for different characteristics segregate independently of one another in gamete production. Actually, it is the *chromosomes*, and not the genes, that segregate independently during meiosis. So we would predict from our knowledge of meiosis that each of Mendel's characters is located on a different pair of chromosomes.

Scientists examining *Pisum* have ascertained that pea plants have just seven pairs of *chromosomes*—or fourteen in all. Mendel was extremely lucky to choose seven characteristics whose gene loci all lie on different homologous pairs.

Now fill in the grid. First use your four contrasting allele colors for the genotypes of all the possible offspring. Each square on the grid represents the union of a male and female gamete. The genotypes of the offspring are calculated by combining alleles from the male gamete at the left of the grid with alleles contributed by the ovum at the right.

Once you have colored all the genotypes, you can determine what the offspring pea plants will look like—their phenotype.

Seed shape will be determined by the alleles *R* and *r*. Offspring that are homozygous *RR* or heterozygous *Rr* will all be *smooth* seeded, as illustrated. Only those offspring containing two alleles for *wrinkled* seed coat, *rr*, are phenotypically *wrinkled*.

Color the homozygous TT and the heterozygous Tt peas yellow. Color the homozygous tt peas green.

In the lower left-hand corner, use your allele colors to color the genotypes for seed shape and seed coat color, respectively. You can count the occurrence of offspring genotypes RR, Rr, rr, and TT, Tt, and tt on the grid to ascertain that they occur in the 1:2:1 genotype ratio, as indicated. Color the phenotype ratios.

The phenotype frequency for both seed color and seed shape is 3:1, just as we found when examining the inheritance of a single trait. Our cross-breeding yielded nine *smooth, yellow-seeded* plants, three *smooth, green-seeded* plants, three *wrinkled, yellow-seeded* plants, and one *wrinkled, green-seeded* plant. Mendel observed the *9:3:3:1 ratio* whenever he examined the inheritance of any two of the pea plant traits simultaneously. This observation led him to formulate the Law of Independent Assortment in 1886. Only during the twentieth century has Mendel's description been linked to the cellular process of gametogenesis.

LAW OF INDEPENDENT ASSORTMENT.★

CHROMOSOMES:A
ALLELES. Rc rD Te tF

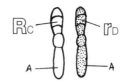 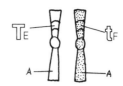

POSSIBLE PHENOTYPES:●

SMOOTH YELLOWY
SMOOTH GREENG

WRINKLED YELLOWY
WRINKLED GREENG

SEGREGATION OF ALLELES DURING MEIOSIS.★

MALE.★   GAMETES.●   FEMALE.★

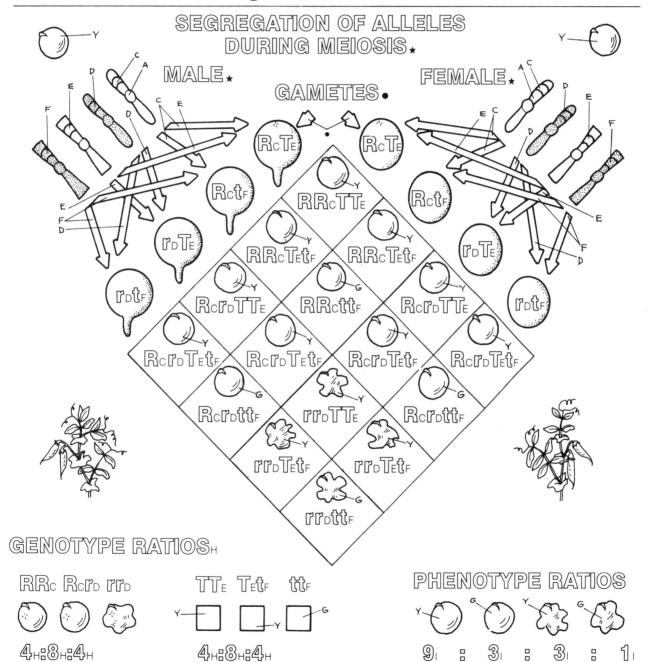

GENOTYPE RATIOSH

RRc RcrD rrD

4H:8H:4H
(1)H:(2)H:(1)H

TTE TetF ttF

4H:8H:4H
(1)H:(2)H:(1)H

PHENOTYPE RATIOS

9I : 3I : 3I : 1I

35
SEX DETERMINATION AND X-LINKED CHARACTERISTICS

The 46 chromosomes in each human cell are found in 23 pairs: 22 homologous pairs and the sex chromosome pair. The two kinds of sex chromosomes, X and Y, determine an individual's sex. Females have two X chromosomes, and males have one X and one Y chromosome.

Start at the top of the plate and use contrasting colors for the male and female figures and for the X and Y sex chromosomes.

The X and Y chromosomes are nonhomologous; that is, they do not carry the same genetic material. The Y chromosome is very small and apparently carries a small number of genes. One or more of these genes determines that the reproductive organs of the developing embryo will become testes. After this, phenotypic *male* development is controlled by hormones produced by the testes.

Color the gametes each parent can produce and the two types of offspring.

Like the homologous pairs of chromosomes, the sex chromosomes segregate into different gametes during the first meiotic division. Thus, during gametogenesis, females always produce ova containing an X chromosome and males can produce sperm containing an X or a Y chromosome.

The father's gamete determines the sex of the offspring: a sperm containing an X combined with the mother's X ovum will develop into a baby girl, and a sperm containing a Y will develop into a baby boy. This segregation of X and Y in approximately equal proportions in the sperm results in approximately equal numbers of male and female babies.

The X chromosome carries many genes that affect normal development at a variety of levels and does not simply have a sex-determining function. There is no counterpart for these genes on the Y chromosome. Therefore, females, with two X chromosomes, are diploid for genes on the X chromosome. Males, with only one X, have just one copy of the genes on the X chromosome. Therefore, females can carry an abnormal recessive allele and not express it, but a male with an abnormal allele on the X will phenotypically express it. This has led to the identification of over one hundred X-linked characteristics, the abnormal genes for which are carried in the female and are found expressed in their male children. Common X-linked diseases are hemophilia, a disease in which the blood does not clot properly, and Duchenne's muscular distrophy, a disease in which the skeletal muscles deteriorate.

To illustrate how X-linked characteristics are carried by females and are expressed in males, we will use red-green color blindness. A female with one allele for *color blindness* and one *normal* allele will not be color blind because the dominant *normal* allele on the other X chromosome will be expressed. Such a woman is said to be a "carrier" of *color-blindness,* for she can pass the trait on to her offspring.

Color the alleles for normal color vision, A, and for color blindness, a, on the X chromosomes. Notice the female has a heterozygous genotype. The male has one normal allele; so he has normal color vision. Next color the gametes produced by these individuals. Then color the offspring.

The ova all contain an X chromosome: half will carry the *normal* allele and half, the allele for *color-blindness*. Half the sperm will contain an X with the *normal* allele; the other half will contain the Y, which carries no homologous genes.

These four gamete types can come together to form four possible offspring genotypes. Therefore, there is a one in four chance of having a *normal* female and a one in four chance of having a female that carries the gene for *color blindness*. None of the female offspring in this kind of mating will express the X-linked characteristics. There is also a one in four chance of having a *normal* male and a one in four chance of having a male with *color blindness*.

SEX DETERMINATION AND X-LINKED CHARACTERISTICS.

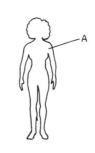
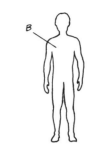

PHENOTYPES.
FEMALE$_A$ MALE$_B$

SEX CHROMOSOMES.

X_C Y_D

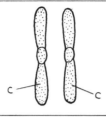

GAMETES.

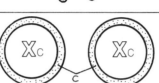
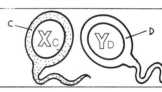

OFFSPRING.

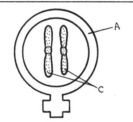
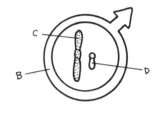

TRANSMISSION OF COLOR BLINDNESS.
ALLELES.
NORMAL$_E$ A$_E$
COLOR BLIND$_F$ a$_F$

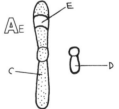

GAMETES.

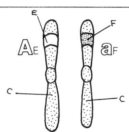

OFFSPRING.

NORMAL$_E$ FEMALE$_A$ A$_E$A$_E$
CARRIER FEMALE$_A$ A$_E$a$_F$
NORMAL$_E$ MALE$_B$ A$_E$
COLOR BLIND MALE$_B$ a$_F$

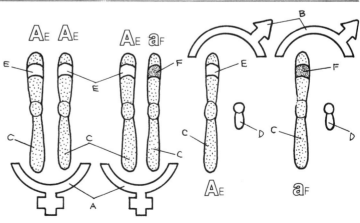

36

LINKED GENES AND CROSSING OVER

Mendel's Law of Independent Assortment applies only to genes located on different pairs of homologous chromosomes. In contrast, linked genes, which are located near one another on the same chromosomes, sort together. And the closer two genes lie on a chromosome, the more likely they are to "stay together." Genes that lie in close proximity to one another are called "linkage groups." Most of the research on linkage groups has focused on X-linked traits (those carried by genes on the sex chromosomes). The pattern of inheritance of a certain genetic defect involving the nails and knee caps has been found to follow the same pattern of inheritance as ABO blood groups (Plate 37 and 38).

Color the top half of the plate. Color the two alleles for trait 1 (S and s) and the two alleles for trait 2 (T and t) with four contrasting colors.

The first cell shown is an oogonium (sex cell in the female) or a spermatogonium in the male (see Plate 30). This individual is heterozygous for both traits 1 and 2.

In preparation for the *first meiotic division*, the *DNA replicates* (represented by the arrow) and produces two identical sister chromatids. Notice how the *first meiotic division* segregates homologous chromosomes into separate primary gametocytes. In the *second meiotic division,* four haploid gametes are produced. Notice that alleles for traits 1 and 2 have not sorted independently during meiosis. Alleles S and T, for example, are "packaged" on a single chromosome in the parent and remain exclusively together in the ova or spermatids.

Do genes located on the same chromosome always stay together? Not necessarily. Meiosis is an extremely complex process, involving two cellular divisions and forty-six strands of DNA, each several million nucleotides in length. All this replicating and sorting takes place within the microscopic confines of a tiny cell. It should come as no surprise then that meiosis is not the perfectly executed process we have described. During the *first meiotic divison,* chromosomes frequently "exchange" parts. This process is called the crossing over of linked genes.

Use the same four allele colors to follow the fate of linked genes through meiosis, when crossing over occurs.

Notice that our hypothetical traits are determined by genes whose loci are farther apart than in the above example, but they are still linked on the same chromosome. During anaphase of meiosis I, nonsister homologous chromatids may overlap, as you see here. Through rearrangements in molecular bonding, segments of the DNA are exchanged. The place where crossing over occurs is called the chiasma. There may be several chiasmata along a single homologous pair. This means that the farther apart two genes are located on a chromosome, the more likely it is that crossing over will take place at some point between them and the greater the possibility that they will become "unlinked" during the *first meiotic division.* Genes that are very close together, however, have a much greater chance of staying together and of segregating nonindependently, as in the first example at the top of the plate. With crossing over, alleles S and s exchange places; so after the *second meiotic division,* two gametes contain chromosomes that carry a novel combination of alleles for traits 1 and 2.

Notice that these two new combinations of alleles were not present in the genome of the individual who produced them. Crossing over contributes new varieties of individual genotypes in the next generation.

We have seen that mutations (Plate 23), the independent segregation of chromosomes during meiosis (Plate 30), the recombination of genes during sexual reproduction (Plate 31), and the crossing over of linked genes all introduce variability into the gene pool. This vast pool of variation between individuals becomes ordered into patterns of evolutionary change by natural selection. Those individuals, or those genetic variants, best suited to the environment into which they are born will survive and will contribute a larger portion of their genes to the next generation. Darwin made this proposal in the nineteenth century, knowing nothing of the genetic mechanisms underlying heredity. He had no explanation for the origin of new variation. The twentieth-century biological sciences have provided many answers. Now we understand how the genetic information is shuffled, reshuffled, and maintained over generations. The biochemical behavior of DNA during replication and sex cell division is apparently introducing genetic variation to the gene pool at a fairly constant rate, hence the molecular clock. The missing pieces of Darwin's theory have fallen into place.

LINKED GENES AND CROSSING OVER.

GAMETES.

LINKED GENES: NONINDEPENDENT ASSORTMENT.

ALTERNATE ALLELES.
   TRAIT 1: $S_A S_B$
   TRAIT 2: $T_C t_D$
DNA REPLICATION$_E$
CENTRIOLES$_F$
MEIOTIC SPINDLE$_G$
FIRST MEIOTIC DIVISION$_H$
SECOND MEIOTIC DIVISION$_I$
CROSSING OVER.
NEW VARIABILITY$_J$

CHIASMA

CROSSING OVER.

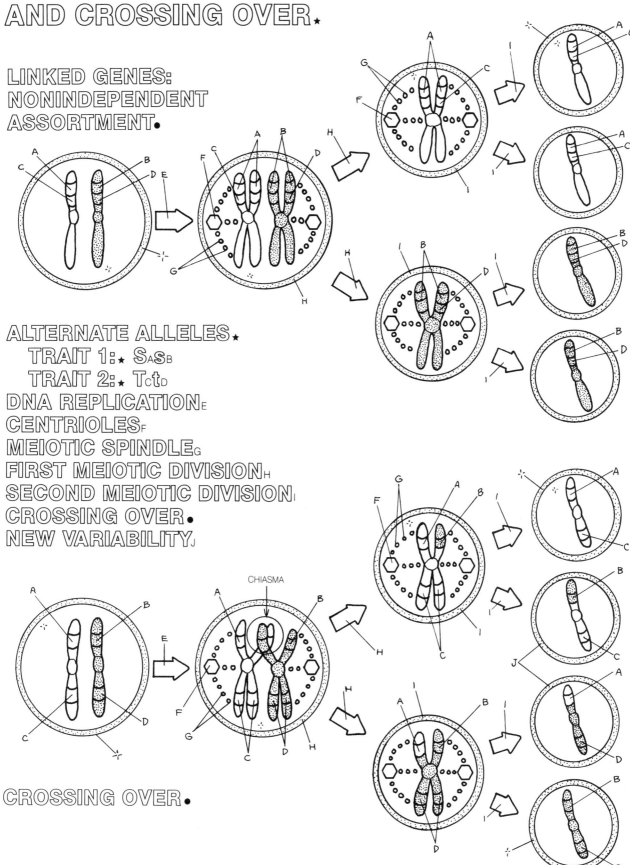

37

ABO BLOOD GROUPS: BIOCHEMISTRY AND INHERITANCE

Having looked at how genetic traits are transmitted and how molecular and cellular processes introduce variation into the gene pool, let us examine more closely variation in human populations. The blood groups are perhaps the most thoroughly studied genetic systems in humans. Differences in blood groups first became apparent in the nineteenth century, when blood transfusions were attempted. Transfusions between some individuals were successful but between others they were lethal because the transfused blood caused clotting that blocked blood vessels. Karl Landsteiner's recognition of the ABO blood groups early in this century made transfusions relatively safe. Of the more than twenty blood groups, ABO is the major system and is the common term used for human blood types—*A, B, O,* or *AB.*

We will first look at the genetic basis for the ABO blood group. Three major alleles can be present at the ABO locus: *A, B,* and *O.*

In the upper left of the chart, use three contrasting colors to color the six possible genotypes, AA, AO, BB, BO, OO, and AB.

The *A* and *B* alleles are both dominant (or co-dominant); the *O* allele is recessive. Co-dominance means that both the alleles on homologous chromosomes are expressed phenotypically. There are thus four possible phenotypes: *A, B, O,* and *AB.*

Color the phenotypes, noting that A type blood and B type blood antigens can be produced by two different underlying genotypes: a person with the A phenotype could genetically be AA or AO, since O is recessive to A.

The same is true for a person who is phenotypically *B.* A person who is phenotypically *O* must have two *O* alleles because *O* is recessive. A person who is phenotypically *AB* must be genetically *AB* because *A* and *B* are co-dominant.

In the body, the difference between these three alleles lies in the structure of the sugar molecules on the surface of the red blood cells. In an individual with the *A* phenotype, the sugar molecules on the red blood cell surface end with one sugar, acetylgalactosamine, whereas an individual with the *B* phenotype has the sugar molecule end with galactose.

Color the terminal sugars for the A and B molecules. Note that a person with the O phenotype has neither terminal sugar on the molecule; the AB person has both.

These molecular differences have a medical effect during blood transfusions because these terminal sugars of *A* and *B* blood types act as antigens, against which *antibodies* are produced. (Review Plate 27 for antigen-antibody reactions.)

Now color the final section on the chart.

You can see that persons with the *A* antigen have anti-*B antibodies* in their bloodstream. Persons with the *B* antigen have anti-*A antibodies* in their bloodstream. A person who is *O* phenotypically—and so has neither *A* nor *B* antigens—has both anti-*A* and anti-*B antibodies.* A person who is *AB* has both *A* and *B* antigens on his cell surfaces and no *antibodies* in his blood. This antigen and *antibody* status is the basis for adverse transfusion reactions and clotting. If an *A* person, who has anti-*B antibodies,* receives blood from a *B* person, her *antibodies* will "recognize" the *B* antigens and destroy or damage transfused cells. This antigen-antibody reaction can be lethal if a large amount of blood is transfused.

The bottom half of the plate shows the inheritance patterns for ABO blood types. Start by coloring the phenotypically A father and the phenotypically B mother. Notice that both parents have heterozygous genotypes of AO and BO, respectively. Follow the segregation and recombination of blood type alleles to see how offspring of four different genotypes can be produced by such a mating. Color these genotypes and then color the four different phenotype possibilities that would occur in equal proportions in the offspring of this mating.

This plate explains the molecular and genetic basis of the ABO blood group system. The next plate shows how the alleles are distributed in populations around the world.

ABO BLOOD GROUPS.

BIOCHEMISTRY AND INHERITANCE.

RED
BLOOD CELL.

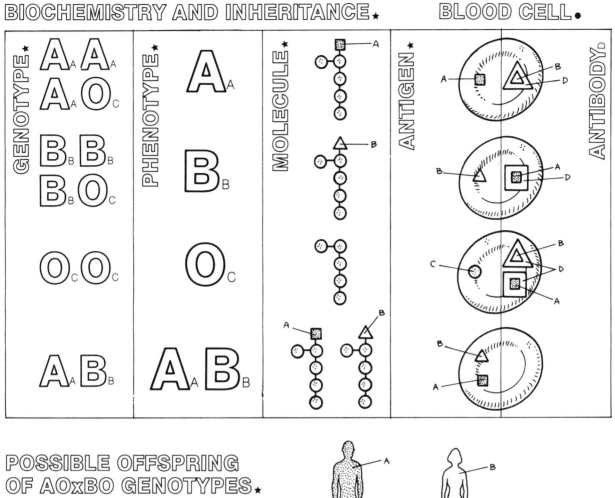

POSSIBLE OFFSPRING
OF AOxBO GENOTYPES.

PARENTS'
PHENOTYPE. A$_A$ B$_B$

PARENTS' GENOTYPE. A$_A$O$_C$ B$_B$O$_C$

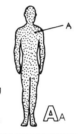 

OFFSPRINGS'
GENOTYPE.   A$_A$B$_B$ A$_A$O$_C$ B$_B$O$_C$ OO$_C$

OFFSPRINGS'
PHENOTYPE.   A$_A$B$_B$ A$_A$ B$_B$ O$_C$

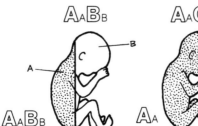 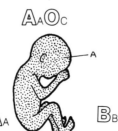 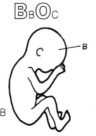 

38
ABO BLOOD GROUPS: WORLD DISTRIBUTION

We begin with a map of the world in order to examine a small sample of populations and the variation in the frequency of the three blood group alleles. Hundreds of populations have been sampled in order to determine the blood group distribution. An important finding is that ABO blood types are not distributed randomly.

Start at the upper left, using the same colors for A, B, and O that you used in the previous plate. The pie slices show graphically the allele frequencies for each allele in the particular population mentioned. As you color each population group, color the percentages and then color the circle in the graph on the upper right that shows the world distribution of A and B allele frequencies.

O is the allele with the highest frequency in nearly all populations. Some populations, such as the *Chinese,* have large amounts of both *A* and *B* alleles. Several of the populations depicted have no *B* at all: *Navahos, Bedouins,* and the *Australian Aborigines.* The *Xavante Indians* of South America have only the *O* allele, no *A* or *B*.

Why do different populations have different *ABO* allele frequencies? Why are there several alleles at this locus? In other words, why is it polymorphic? When more than one allele in a population occurs at a locus (that is, is polymorphic) and cannot be explained by the mutation rate, it implies some selective pressure operating to keep the alleles in the population at a relatively high frequency. There are two evolutionary mechanisms that help explain the polymorphism in the *A B O* locus: genetic drift and natural selection.

Genetic drift is the chance random loss or appearance of an allele in a small population. Genetic drift may occur when a small group migrates away from the parent group and remains isolated. This probably explains the loss of the *B* allele in the *Navaho* populations. The ancestors of the American Indians migrated through the Bering Straits from eastern Asia, where the frequency of the *B* allele may have been relatively high, as it is among the *Chinese.* By chance, the small migrant population had a lower frequency of the *B* allele than the parent group. With time, the *B* allele was completely lost in many of the American Indian populations. The *Xavante Indians* in South America probably lost the *A* allele in a similar fashion, as small groups of peoples migrated down from North America.

If genetic drift were the only mechanism operating on gene frequencies, you would expect all populations to have retained only one allele of the *ABO* locus. The pie graphs show that this certainly is not the case; so there must be another evolutionary force keeping the alleles at relatively high frequencies in a population. This force is natural selection: under different environmental conditions, different alleles will be selectively advantageous. For example, susceptibility to certain diseases has been statistically correlated with different alleles. Individuals with the *A* allele appear to be more susceptible to stomach cancer, pernicious anemia, and smallpox than individuals with the *B* or *O* alleles. Individuals who are *O* are more susceptible to gastric and duodenal ulcers and bubonic plague. The different frequencies of the alleles in populations may result from selection by infectious diseases to which the population has been exposed for long periods of time. People with the resistant alleles would be more likely to survive; so that gene would increase in frequency over time. This type of balanced selective advantage in the *ABO* system is difficult to substantiate, but the nonrandom distribution of allele frequencies, as depicted in the graph in the upper right, suggests that forces other than genetic drift are involved in determining the world distribution of *ABO* allele frequencies. In Plates 41 and 42 we will see how an infectious disease, malaria, has been the selective force in the evolution of the sickle cell gene.

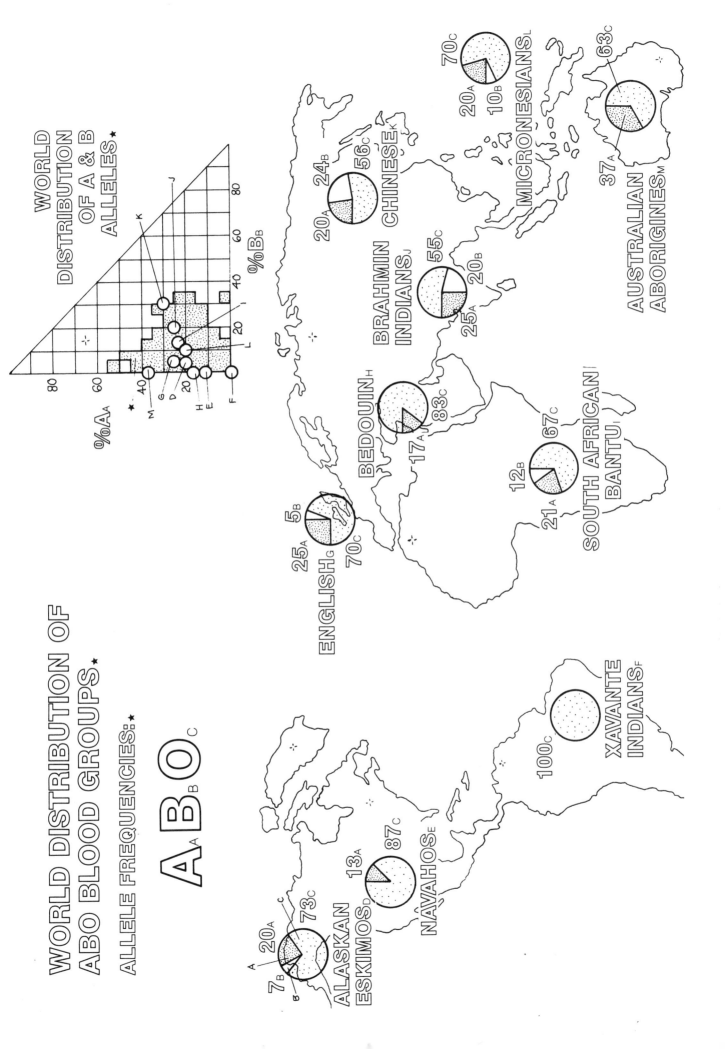

WORLD DISTRIBUTION OF
ABO BLOOD GROUPS ★

ALLELE FREQUENCIES: ★★

A_A B_B O_C

WORLD
DISTRIBUTION
OF A & B
ALLELES ★

%A_A

%B_B

★ M
G
D
K
L
I
H
E
F
J

ALASKAN
ESKIMOS_D
20_A
7_B
73_C

NAVAHOS_E
13_A
87_C

XAVANTE
INDIANS_F
100_C

ENGLISH_G
25_A
5_B
70_C

BEDOUIN_H
17_A
83_C

BRAHMIN
INDIANS_J
25_A
20_B
55_C

CHINESE_K
20_A
24_B
56_C

MICRONESIANS_L
20_A
10_B
70_C

AUSTRALIAN
ABORIGINES_M
37_A
63_C

SOUTH AFRICAN
BANTU_I
21_A
12_B
67_C

39

Rh FACTOR: MATERNAL-FETAL INCOMPATIBILITY.

Another set of red cell antigens that vary in the human population is the "Rh factor," so called because it was first found in rhesus monkeys. Whereas ABO blood group incompatibilities are responsible for most transfusion reactions, the Rh factor may result in maternal-fetal incompatibility; and babies may be stillborn or born with anemia, jaundice, or brain damage—a condition known as erythroblastosis fetalis.

The Rh factor, too, is genetically determined. In order to have an *Rh positive* phenotype, the allele for the *Rh antigen* must be present. The allele that determines the presence of the *Rh antigen* is *dominant;* just one allele can produce the Rh phenotype. If the antigen is absent (that is, neither allele carries the antigen code), the individual is phenotypically *Rh negative (Rh –).*

Color the top half of the plate.

The female parent here has two *(recessive) Rh –* alleles; so she is *Rh –.* The male parent has an *Rh +* allele and an *Rh –* allele; so he is *Rh +.* The offspring has an even chance of being *Rh –* (rr) or *Rh +* (Rr).

Now proceed to the bottom of the plate to Rh incompatibility and color the pregnancy where mother and baby are compatible and the baby is born unafflicted.

If the baby is *Rh –,* there is no problem.

Now color all the structures in pregnancies II and III.

If the baby is *Rh +,* maternal-fetal incompatibility exists and the *mother* may make *antibodies* against the fetal red cells. However, this usually does not happen with a first incompatible pregnancy, for only a few fetal red cells get into the maternal circulation prior to delivery. The placenta (the spongy tissue that connects the fetus to the maternal uterus) acts as an effective filter to keep most red cells out until the time of delivery. Then the placenta breaks away from the uterus, there is a great deal of bleeding, and many fetal cells may enter the maternal circulation. At this time, the mother's lymphocytes (the cells that make *antibodies*) may become *sensitized* to the fetal *Rh antigen.*

The sensitization persists, but not the actual *antibodies:* the triangles in frame II_2 represent the persisting *sensitized lymphocytes.* When a later pregnancy occurs with an *Rh +* fetus (frame III), even the few fetal red cells that enter the maternal circulation may be sufficient to evoke a strong antibody reaction from the existing *sensitized lymphocytes.* These *antibodies* will cross the placenta into the fetal circulation and damage or destroy the fetal red cells, causing the anemia, jaundice, and other complications of this condition.

Sensitization may be prevented and subsequent offspring spared this serious disease by injecting the mother with anti-Rh antibodies within a day or two after delivery of her first *Rh +* baby. This antibody "covers" the *Rh antigen* on the fetal cells and prevents the maternal *lymphocytes* from "seeing" it and becoming *sensitized.* In a case where an infant is born with erythroblastosis, it is possible to minimize the damage (or at least prevent further damage) by an "exchange transfusion," in which the damaged fetal blood as well as the *maternal antibodies* are completely replaced by normal blood free of *antibodies.*

Rh FACTOR: MATERNAL-FETAL INCOMPATIBILITY ★

PHENOTYPE ●
Rh POSITIVE
  (Rh ANTIGEN PRESENT) A
Rh NEGATIVE
  (Rh ANTIGEN ABSENT) B
ALLELES ★
  DOMINANT C R_c
  RECESSIVE D r_D

INHERITANCE ●

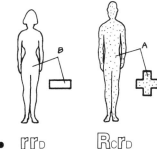

GENOTYPE ● rr_D $R_c r_D$

GAMETES ★

POSSIBLE OFFSPRING ★

GENOTYPE ● $R_c r_D$ rr_D $R_c r_D$ rr_D

PLACENTA E
Rh ANTIGEN
  ON FETAL RED BLOOD CELLS A ⊙ ⊙ ⊙ A
MOTHER'S ANTIBODIES G △ △ △ G
  SENSITIZED LYMPHOCYTES G¹
BABIES ★
  UNAFFECTED H / AFFECTED I   Rh INCOMPATIBILITY ●

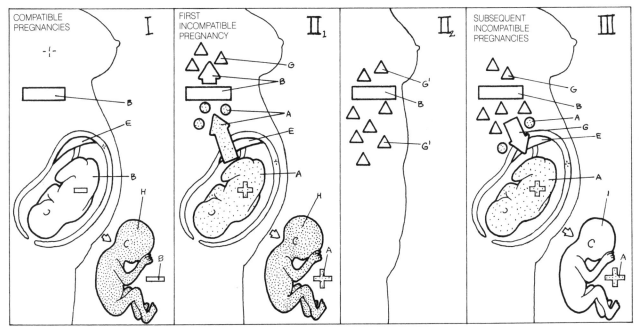

I — COMPATIBLE PREGNANCIES
II₁ — FIRST INCOMPATIBLE PREGNANCY
II₂
III — SUBSEQUENT INCOMPATIBLE PREGNANCIES

40

THE BIOLOGY OF MILK DRINKING: LACTOSE DIGESTION

Lactose is the principal sugar in milk; *lactase* is the *enzyme* that digests it so it can be absorbed into the body and used for energy. Populations throughout the world vary tremendously in the degree to which adults have or lack the *lactase enzyme,* and therefore the degree to which they can consume milk. This variability has enormous economic and cultural implications.

Begin by coloring the top picture on the digestion of lactose.

The lactose molecule consists of two other sugars, *glucose* and *galactose,* into which lactose is split by the *enzyme lactase. Glucose* and *galactose* are readily absorbed through the *small intestinal wall* into the *blood vessels* and so used by the body. In the large intestine, the normal *bacteria* live in harmony with their host.

Color the bottom picture.

In the "indigestion" of lactose, lactose intolerant adults do not possess sufficient *lactase* to break the lactose into *glucose* and *galactose.* A small amount may be absorbed into the *small intestinal wall,* but most of it will pass into the *large intestine* undigested. There, lactose drains water from surrounding tissues and *bacteria* ferment the milk sugar to produce *acid* and *carbon dioxide gas.* Shortly after drinking milk, a lactose intolerant adult will usually experience symptoms of gastric distress: intestinal cramping, bloating, flatulence, and diarrhea. Nondigestion of lactose means "indigestion."

In almost all mammals, the cells of the intestinal wall produce plenty of *lactase* during infancy, when a major component of the diet is maternal milk. Mammalian *lactase* production decreases rapidly at the age of weaning and is minimal in adults. However, in some human populations, particularly those with a tradition of dependence on dairy products, adults continue to have a high production of *lactase* throughout life. A cultural practice—in this case dairying as a means of acquiring food—has created the selection pressures for a human adaptation that is unique among mammals.

A high incidence (80–100 percent) of adult lactose tolerance is found among North Europeans and among the Fulani and Tussi, African herders who have relied on dairying for several thousand years.

Several other African herding groups have poor lactose tolerance. Although they use milk in their diet, their cultural traditions prescribe that milk is first fermented and consumed mainly as cheese and as yogurt. In these forms, microorganisms provide the enzymes for breaking down lactose before it enters the human digestive tract.

Other African groups, such as the Ibo or Yoruba, do not keep dairy animals at all. Nor have Oriental human populations traditionally relied on milk. These groups do not have *lactase* as adults. When they try to consume milk, they experience the unpleasant gastrointestinal distress of lactose intolerance.

The example of adult lactose intolerance illustrates how culture can influence what is and what is not adaptive for human beings—how it in effect has created the selection pressures operating on the human gene pool. Ignoring the genetic adaptation of our species can have unfortunate consequences. For example, the staple of economic aid to underdeveloped countries has traditionally been powdered skim milk, a food containing lactose, to populations traditionally lacking in *lactase.* In order to devise successful strategies for meeting nutritional needs around the world, we must understand the physiological potentials and limitations conferred on us by our evolutionary past.

THE BIOLOGY OF MILK DRINKING: LACTOSE DIGESTION. ★

SMALL INTESTINAL WALL_A
LACTOSE SUGAR ●
   GLUCOSE_B
   GALACTOSE_C
LACTASE ENZYME_D
BLOOD VESSEL_E

LARGE INTESTINAL WALL_F
BACTERIA_G
CARBON DIOXIDE (GAS)_H
ACID_I

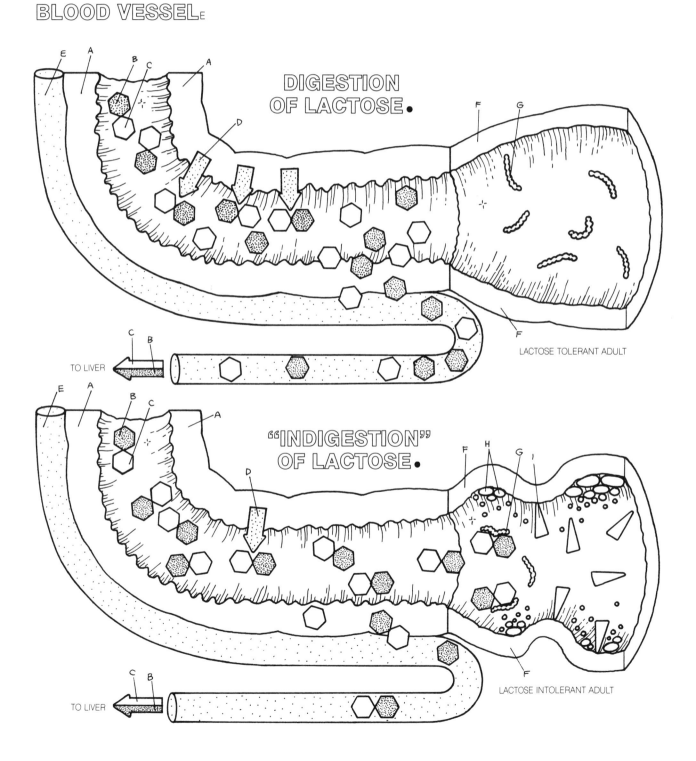

DIGESTION OF LACTOSE ●

TO LIVER

LACTOSE TOLERANT ADULT

"INDIGESTION" OF LACTOSE ●

TO LIVER

LACTOSE INTOLERANT ADULT

41

SICKLE CELL TRAIT: DEFENSE AGAINST MALARIA

Sickle cell anemia is a serious genetic disease that derives its name from the crescent or sickle-shaped appearance of the red blood cells of afflicted individuals. On the molecular level, the "sickling" can be traced to the presence of a variant hemoglobin—called hemoglobin S (HbS). *Sickle cell disease* is so pervasive throughout tropical Africa that investigators wondered how such a serious defect could survive. Studies by Anthony Allison in the 1950s suggested that the sickle cell allele remained in human populations generation after generation because it provided protection against what was, until recently, the greatest killer of the tropical world: malaria.

Hemoglobin is the main protein in red blood cells; each cell contains over 270 million hemoglobin molecules. The protein consists of four amino acid chains —two alpha chains (with 141 amino acids each) and two beta chains (with 146 amino acids). Each hemoglobin molecule transports oxygen from the lungs to all other parts of the body. (Recall that in Plate 26 we compared amino acid sequences in hemoglobin beta chains to obtain clues on evolutionary relationships among living primates.) There also exists "mutant" varieties of hemoglobin within human populations. The most common variant is sickle hemoglobin (HbS), in which the amino acid *glutamic acid* of normal hemoglobin (HbA) at position 6 on the two beta chains is replaced by *valine*. This single amino acid substitution has profound effects.

Begin at the top of the plate by coloring the red blood cells from a normal individual and those from a person with sickle cell disease. Use contrasting colors. Next color the amino acids at positions 5, 6, and 7 of the hemoglobin beta chains.

Note that *valine* rather than *glutamic acid* exists in HbS (sickle hemoglobin). Linus Pauling discovered this amino acid substitution in the 1950s.

Now color the nucleotide bases. Only one nucleotide base (uracil instead of adenine) on the messenger RNA codon produces valine rather than glutamic acid. This then is the genetic basis of the sickle cell hemoglobin.

How does a single substitution in the DNA cause such a major change in the shape of red blood cells? When HbS molecules release their oxygen in the capillaries, they stack up, one upon the other, in long rigid chains called polymers. Recent research links this abnormal

polymerization to the presence of *valine* in position 6 of the beta chain. Normal hemoglobin—with *glutamic acid* in position 6 of the beta chains—does not form polymers.

Now color the bottom illustration, showing the inheritance of the sickle cell gene.

Individuals with two HbA alleles *(AA)* will have all normal blood cells. Heterozygous *(AS)* individuals have the sickle cell trait: some of their red blood cells contain HbA, others, HbS; so only some of their cells are sickled. Except in extreme conditions, such as physical exertion at high altitudes, persons with *sickle cell trait* do not suffer the debilitating effects of *sickle cell disease.*

Homozygous *(SS)* individuals suffer from problems caused by blocked capillaries and reduced circulation to various parts of the body. Individuals with *sickle cell disease* experience intermittent bouts of severe pain, serious anemia, and injury to tissues resulting in joint and brain damage, kidney and heart failure, and abnormal growth. Although modern medicine still cannot cure sickle cell anemia, new approaches have been used to lessen the severity of many of the complications, and researchers are actively investigating methods to prevent red blood cells from sickling.

Now color the designated area of the African map gray. This is the region where malaria plagues human populations.

Malaria is a parasitic disease transmitted from the blood of one individual to the next by mosquito bites. The parasite, *Plasmodium falciparum,* infects and reproduces in human red blood cells, causing a high fever and severe tissue damage, especially in young children, where the malarial infection may also be fatal.

In malarial areas a person with *AA* genotype has *no protection* and is susceptible to severe malarial infections. A person with *SS* genotype has *sickle cell disease,* which, until the advent of modern medicine, was usually fatal before the age of twenty. A person with an *AS* genotype has the *sickle cell trait,* in which some red blood cells function normally but others sickle and provide protection against severe malarial infections. In an environment with endemic malaria, heterozygous individuals *(AS)* are most likely to survive and reproduce, so "evolutionarily speaking," they have the greatest fitness.

SICKLE CELL TRAIT: DEFENSE AGAINST MALARIA.

RED BLOOD CELLS.
NORMAL_A
SICKLE CELL_B

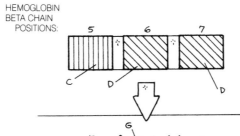

AMINO ACIDS.
PROLINE_C
GLUTAMIC ACID_D
VALINE_E

HEMOGLOBIN BETA CHAIN POSITIONS:

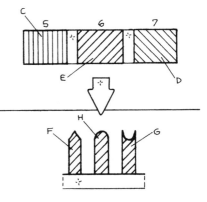

NUCLEOTIDE BASES.
GUANINE_F
ADENINE_G
URACIL_H
THYMINE_{H1}
CYTOSINE_I

mRNA CODON

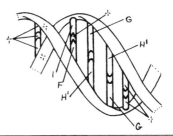

DNA

IN MALARIAL ENVIRONMENT.

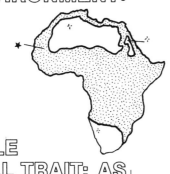

MATING

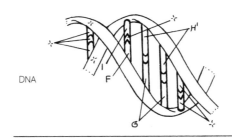

POSSIBLE OFFSPRING

SICKLE CELL TRAIT: AS_J (GREATEST FITNESS)_J
NORMAL GENO.: AA_{A1} (NO PROTECTION)_{A1}
SICKLE CELL: SS_{B1} (DISEASE)_{B1}

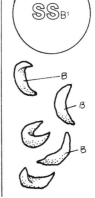

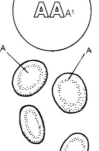

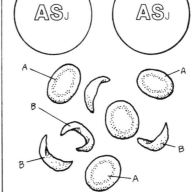

42
SICKLE CELL: THE INTERPLAY OF BIOLOGY, CULTURE AND ENVIRONMENT

Malaria is endemic in large parts of the world, especially in tropical areas near the equator. There is little doubt that the disease that has killed millions of people has been a selective agent on human populations.

Using shades of the same color, begin by coloring the top map of Africa, showing where malaria is endemic, seasonal, and sporadic. The endemic areas that suffer the highest incidence of malaria are tropical rainforests where anopheles mosquitos can find the necessary moisture for reproduction year-round.

Two thousand years ago the tropical rainforests of central Africa changed significantly after the introduction of *slash-and-burn agriculture* from Southeast Asia. With this technique of growing crops, small patches of forest are cleared and planted. After several years of use, they are abandoned and allowed to lie fallow to replenish soil nutrients. The abandoned fields provide open sunlet pools, which are the ideal breeding grounds for the anopheles mosquito. A change in environment due to human activities thus promoted an increase in the mosquito population.

Human populations were also affected. With the introduction of food cultivation, human groups became more sedentary and the domesticated food supply allowed an increase in population density. Prior to plant cultivation, peoples of Africa had been primarily gatherers and hunters, living in small, nomadic social groups. Now, with larger stable groups of people in a small area, transmission of the malarial parasite from one person to the next was unavoidable.

As you color the map on the right, notice how the endemic malarial area surrounds and includes the entire range of slash-and-burn agricultural practice. With the introduction of techniques for crop cultivation, malaria became a significant health hazard in central Africa.

Now color the left map, using shades of a color that contrasts with the other two.

In areas of *slash-and-burn,* individuals with sickle cell trait are less susceptible to malarial infection and are at a significant selective advantage over individuals whose blood cells contain only normal hemoglobin. The *Plasmodium* malarial parasite will infect the red blood cells containing HbS and those containing HbA equally. However, the infected cells containing HbS will sickle much faster than they would if they contained no parasite. Rapid sickling kills the parasite

before it completes its forty-eight-hour reproductive cycle and thus prevents it from proliferating and producing a more serious infection. Mutant hemoglobin is not completely protective, but it does lessen the severity of a malarial infection until the individual with sickle cell trait can build up a natural immunity to the parasite. Due to the malarial resistance it confers, the HbS allele evolved to very high frequencies—10–16 percent—in many African populations as a protection against malaria. Notice that the areas of highest HbS allele frequency are almost identical to the areas where *slash-and-burn* has been practiced.

Other hemoglobin variants have been identified in populations in the Mediterranean, the Middle East, and India, all regions plagued by malaria. These variants may also confer some protection against malaria. The severe parasitic disease has been a strong enough selective force, even outside of Africa, that populations around the world where malaria is continually a threat may have independently evolved genetic mechanisms to lessen the severity of the infection.

The ancestors of American blacks are primarily from the western coast of Africa, where the frequency of HbS is the highest. The frequency of the HbS allele among American blacks is approximately 4 percent—significantly lower than the allele frequencies of 10–16 percent in African groups. This decrease in the frequency is partially due to mixing of American blacks with other groups having a much lower frequency or absence of the allele. Furthermore, in temperate North America, the selective force of malaria is not as great. Where malaria is less of a threat, heterozygous individuals are not at a marked selective advantage over homozygous HbA individuals. The frequency of the HbS allele is gradually decreasing because of the higher mortality rate among homozygous individuals with sickle cell disease.

This example illustrates how selective pressures that influence the gene pool of evolving human populations emerge from an interplay of environmental conditions: in this case, the breeding habits of a particular mosquito, the reproductive cycle of the malarial parasite, and a cultural practice for obtaining food. A single point mutation in the DNA can cause a severe genetic disease, but that same point mutation confers resistance to an equally severe parasitic infection. The sickle cell allele and its relationship to health and disease provides a classic model for investigation of the human genetic response to specific environmental and cultural conditions.

SICKLE CELL: THE INTERPLAY OF BIOLOGY, CULTURE AND ENVIRONMENT.

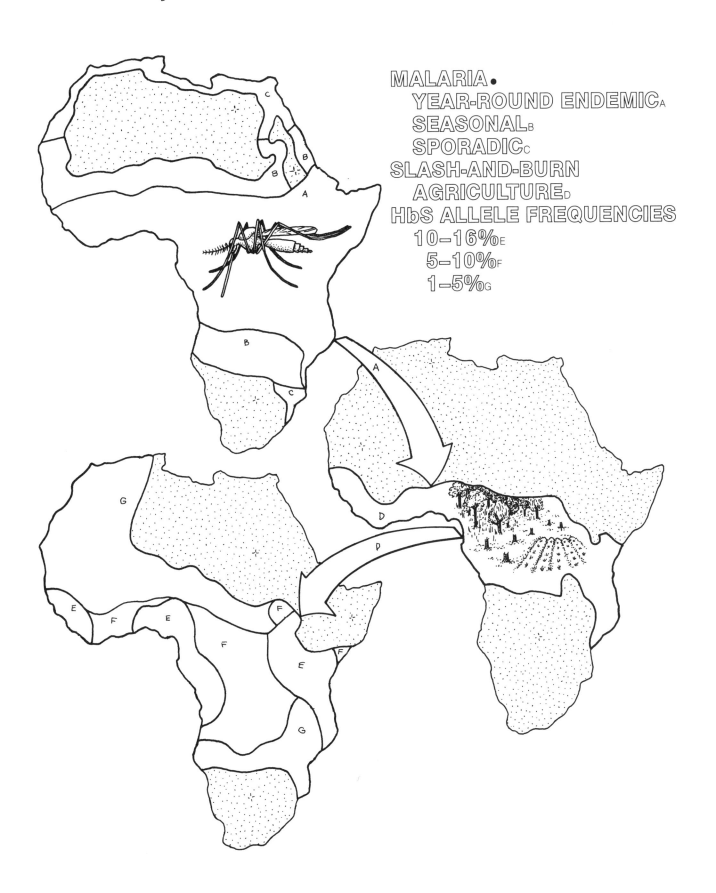

MALARIA.
YEAR-ROUND ENDEMIC$_A$
SEASONAL$_B$
SPORADIC$_C$
SLASH-AND-BURN
AGRICULTURE$_D$
HbS ALLELE FREQUENCIES
10–16%$_E$
5–10%$_F$
1–5%$_G$

43

EYES AND SEEING:
THE WORLD OF DAY AND COLOR

Human beings rely extensively on their sense of sight to inform them about their environment. This capability evolved long before early humans appeared on the African savanna and is a feature we share with other primates, especially the monkeys and apes. In this, the first of two plates on vision, we discuss light sensitivity, acuity and color vision, and behavioral correlates.

Vision is the ability to sense energy in part of the electromagnetic spectrum, the entire range of wavelengths (measured in meters) of electromagnetic radiation, from the shortest gamma rays to the longest radio waves. Visible light, so called because we "see" it, is only a small part of the entire spectrum. Other animals may not see the entire range from violet to red that we see, but they may be able to perceive wavelengths in the ultraviolet or infrared regions.

Begin by coloring the basic structures of the eye and vision at the top of the plate.

The eyeball is mobile and is moved around by a set of *eye muscles*. Light reflected from an object in the environment (here seen schematically as a tree) first encounters the cornea, bulging slightly outward from the surface of the protective white of the eye (the sclera). The cornea bends light rays toward the back of the *retina,* and they travel through the opening of the iris, called the pupil. The iris is a muscular structure that controls the size of the pupil, making it smaller in *bright light* and larger in *dim light*. The iris can also protect the *retina* by absorbing light in its pigment, which gives each of us our characteristic eye color.

Light rays cross each other on their path through the eye; so a tiny inverted image is projected onto the light-sensitive cells of the *retina*. Light induces a chemical reaction in the pigment of the photoreceptors, which in turn send electrical signals to the brain along the *optic nerve.*

Color the retina across the middle of the plate and the light-sensitive receptors (photoreceptors): the cylindrical rod cells and the more bulbous cone cells. Use a light blue or green for the rods and a yellow or orange for the cones.

The primate *retina,* like that of other mammals, contains both *rods* and *cones.* In nocturnal primates, like the bug-eyed galago pictured here, *rods* predominate. The *cones* are more numerous in day-living (diurnal) primates and are essential for visual acuity and color vision. In sheer numbers, *rods* far exceed *cones;* for example, the human *retina* contains 125 million *rods,* but less than 7 million *cones. Rods* are distributed around the outer edges of the *retina* and are important for peripheral vision.

Cones dominate the specialized central part of the *retina* surrounding the fovea. The fovea is a small depression in the *retina* and contains 30,000 densely packed *cones* (no *rods*). The fovea magnifies images of objects and "sees" images in greatest detail. It is the anatomical basis for recognition of patterns such as faces or words on a printed page.

Complete the plate by coloring the galago and the range of visible light it can see with its rodlike eye (the left hump on graph), then the mangabey, a day-living primate with color vision. Also color the color spectrum from the shortest wavelength (violet) to the longest (red). The lowermost band illustrates the position of visible light in the electromagnetic spectrum.

The color we perceive is determined by which wavelengths of light an object reflects to our eyes and which wavelengths it absorbs. If the object reflects all light, we see it as white; if it absorbs all light, we see it as black.

Notice that *rods* are most sensitive to light of shorter wavelengths, and *cones* are most sensitive in the yellow-red end of the spectrum about 560 nanometers. *Rods* are a hundred times more sensitive to *dim light* than *cones,* but they do not perceive color. We use them for night vision, and images appear as shades of gray. The photoreceptive pigment, called rhodopsin, is optimally activated by light of the blue-green wavelengths around 500 nanometers.

The *cones* of Old World monkeys, apes, and humans contain one of three photoreceptive pigments. Each pigment is optimally sensitive to a different wavelength of light, corresponding to the colors blue, yellow-green, and red. Their concerted photochemical responses to visible light are responsible for our ability to perceive the world in color.

THE WORLD OF DAY AND COLOR★

EYE MUSCLEA
LENSB
RETINAC
OPTIC NERVED

LIGHT●
DIME
BRIGHTF

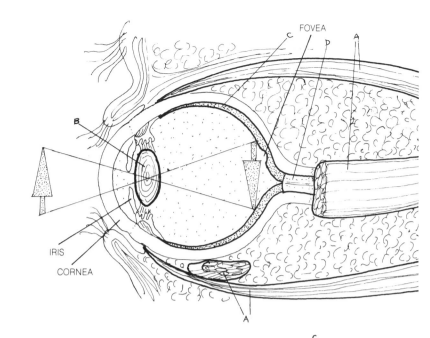

FOVEA
C
D
A
B
IRIS
CORNEA
A

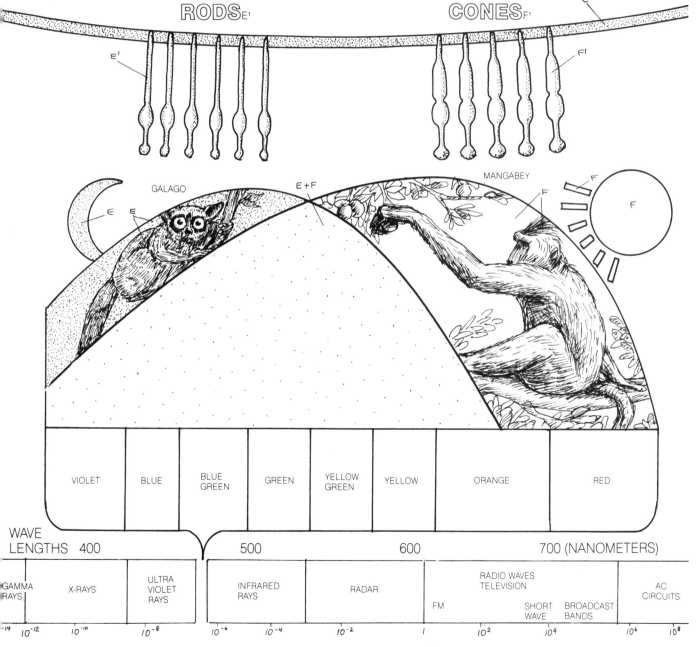

RODSE¹   CONESF¹

E¹   F¹

C

GALAGO   E+F   MANGABEY

E   E   F   F   F

VIOLET	BLUE	BLUE GREEN	GREEN	YELLOW GREEN	YELLOW	ORANGE	RED

WAVE LENGTHS 400   500   600   700 (NANOMETERS)

GAMMA RAYS	X-RAYS	ULTRA VIOLET RAYS	INFRARED RAYS	RADAR	RADIO WAVES TELEVISION			AC CIRCUITS
					FM	SHORT WAVE	BROADCAST BANDS	

10^{-14} 10^{-12} 10^{-10} 10^{-8} 10^{-6} 10^{-4} 10^{-2} 1 10^{2} 10^{4} 10^{6} 10^{8}

44

EYES AND SEEING:
VISUAL FIELD AND DEPTH PERCEPTION

Primates have evolved *stereoscopic vision* — the use of both eyes to view the same visual field with depth perception — as well as specializations in the retina that increase acuity and color discrimination. Stereoscopy also has evolved in predatory cats and birds. But most vertebrates survey the world in all directions (even behind them) with eyes placed on opposite sides of the head. Each eye covers a different visual field, which is advantageous in watching out for predators but is not good for depth perception.

The primate way of life depends upon stereoscopic vision for locating food and bringing it to the mouth and for skillful locomotion in trees. Stereoscopy, also called binocular ("two eyes") vision, underlies the primate ability to grasp branches when jumping from one tree to the next, though it is equally important for seizing, examining, and holding food items and objects. It is the basis for hand-eye coordination, which is critical in the evolution of early human tool use.

This plate illustrates the two anatomical bases of stereoscopic vision in primates.

Begin by coloring the visual fields at the bottom of the plate. Use two contrasting light colors to color the right, left, and overlapping visual fields of the lemur, human, and rabbit. As you color, notice that orientation of eye sockets results in different degrees of overlap between right and left visual fields.

We can see how the eyes move forward during evolution from primitive mammals, represented here by the rabbit, to prosimians and to higher primates. The zone of overlap, called the binocular field, is least in rabbits (about 30 degrees), intermediate in lemurs (about 90 degrees), and greatest in human (about 120 degrees).

When the visual fields overlap, each retina collects a slightly different set of visual information from nearby objects. Visual centers in the brain automatically analyze the data to produce a three-dimensional picture with accurate distances between objects and the observer.

The diagram in the upper left illustrates the second anatomical requirement for stereoscopy: a "rewiring" of the nerve fibers running from the retina to visual processing centers in the brain.

Color the visual field in the top left. The optic structures are labeled in this diagram but are not to be colored. Note the overlapping visual fields (A, A + B, B); these need not be colored. Use contrasting colors for H, I, J, K, and L to trace the flow of information from the visual field to the visual cortex. Note that there are no titles for H-L; they simply represent the visual information. Now, as each of the brain structures is discussed in the text, color the labels and the structures in the smaller diagram at the top right.

In nonbinocular animals like the rabbit, the *optic nerve* coming from the retina completely crosses over (decussates) at the *optic chiasma;* so all information from the left retina (and left visual field) is processed by the right side of the brain and vice versa. In primates, both eyes cover much of the same visual field, and only half the *optic nerve* fibers decussate. Again, the result is that the left visual field is processed by the right cerebrum and vice versa. The inner half of the retina sees the world as the rabbit's retina does, and these optic fibers cross; but the outer side of each retina sees the world on the opposite side of the head, and these fibers do not cross.

Leaving the *optic chiasma,* the *optic tracts* travel to the lateral geniculate nucleus (not shown), from which some fibers go to the superior colliculi (not shown), remnants of the optic lobes in mammals that localize objects in space. Most of the *optic tract* continues from the lateral geniculate nucleus as the *optic radiations* and projects to the *visual cortex* in the occipital lobes at the back of the cerebrum. Compare the colors from the visual field with those on the *visual cortex* to see how the sensory cortex is essentially an orderly "map" of the visual field. From the *visual cortex,* neurons send electrical impulses to many other parts of the brain to integrate incoming visual information with sensory inputs, memories, motivations, and motor activity patterns.

The evolution of vision in primates for color, depth perception, and memories is fundamental to their adaptive pattern and helps account for their acrobatic skill in the trees, facility for locating brightly colored fruits, ability to observe and remember the location of distant food sources, and heavy reliance on facial expressions and gestures for communicating with other members of their social group.

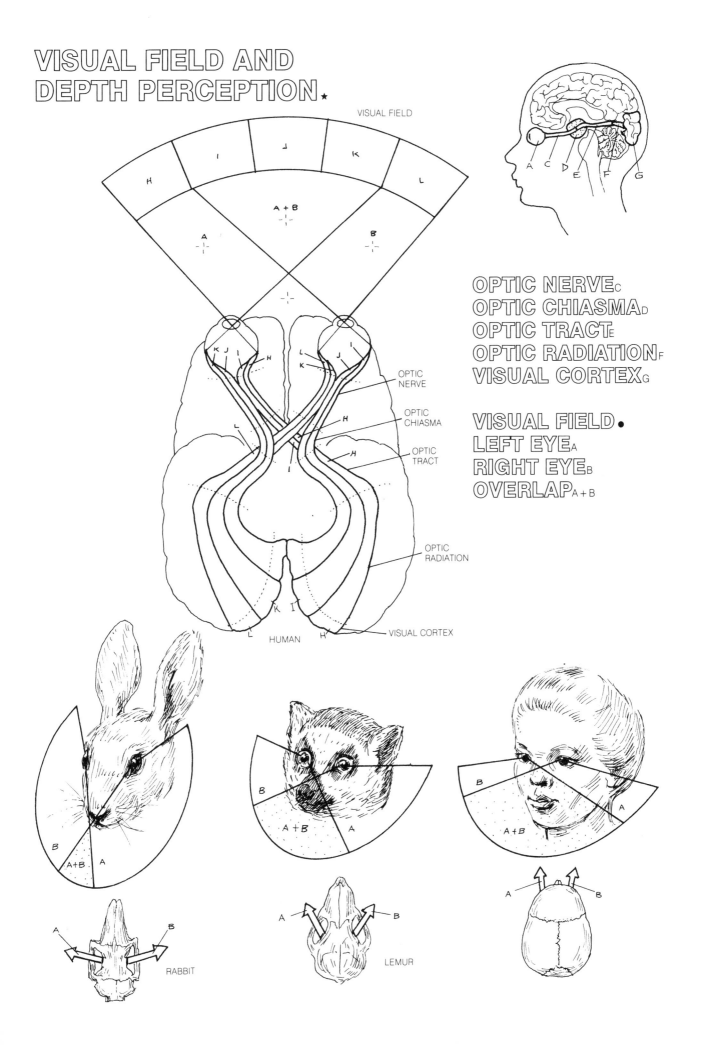

VISUAL FIELD AND DEPTH PERCEPTION.★

VISUAL FIELD

OPTIC NERVE
OPTIC CHIASMA
OPTIC TRACT
OPTIC RADIATION
VISUAL CORTEX

HUMAN

OPTIC NERVEc
OPTIC CHIASMAd
OPTIC TRACTe
OPTIC RADIATIONf
VISUAL CORTEXg

VISUAL FIELD●
LEFT EYEa
RIGHT EYEb
OVERLAPa+b

RABBIT

LEMUR

45
OLFACTION: NOSES AND SMELLING

As the abilities for color discrimination, visual acuity, and depth perception increased in the primates, their sense of smell began to play a less important role in their lives. We can see this by comparing the anatomical basis of smell in prosimians and humans.

Color the structures as they are discussed in the text. Choose light colors for the larger areas, and dark colors for the small areas.

Smooth and hairless, the *rhinarium* surrounds the lemur's nostrils and is underlain by many mucous-secreting glands that give the lemur a "wet muzzle." Within the mouth, the *rhinarium* comes in contact with *Jacobsen's organ* via a pair of tiny canals in the roof of the mouth. Inherited from our reptilian forebearers, this organ lies embedded in the *nasal septum,* which divides the *nasal cavity* in half. When a lemur touches an object with its moist nose, chemical substances from the object are dissolved in mucous and carried to *Jacobsen's organ* for analysis. Inside the pouch of *Jacobsen's organ* lies an isolated patch of olfactory epithelium. This specialized sensory tissue contains nerve cells sensitive to many kinds of dissolved molecules. The *olfactory receptors* respond by sending electrical impulses to the *accessory olfactory bulb* in the brain, a sensory structure also inherited from reptilian ancestors but lost in the higher primates. The openings of *Jacobsen's organ* into the mouth allow the prosimian to analyze the chemical substances in its food. Since human beings do not have a *Jacobsen's organ,* it is difficult to imagine the sensations the lemur experiences. Perhaps the organ aids the animal in deciding which potential foodstuffs are edible and desirable.

Finish coloring the frontal sections and side views of the lemur and human noses. Notice the longer nasal cavity in the lemur.

In the human, the *eye orbits* and the *sinuses* (cavities in the flat bones of skull) lie in the same plane as the nose. The *turbinates,* or conchae, are tiny bony scrolls within the *nasal cavity,* covered by a mucous-secreting epithelium; their main function is to increase the surface area of the overlying tissue so as to humidify and warm the incoming air as it passes through the *nasal cavity.* Toward the back of the nose the *turbinates* are covered with olfactory epithelium like that in a *Jacobsen's organ* (shown in the side view). The *olfactory receptors* in this tissue analyze dissolved chemicals on the basis of the size and shape of the molecules. Each *olfactory receptor* is specialized to respond to only one specific molecular shape. When this shape contacts the odor receptor, chemical information is translated into electrical impulses and carried by olfactory neurons through holes in the *cribiform plates* to the brain. The *olfactory bulb* is the first neural processing center for smell information; then it is relayed via the *olfactory tract* to many other parts of the brain for further analysis and association.

The number of *turbinates,* the extent of olfactory epithelium (represented by the number of *olfactory receptors*), the size of the perforated *cribiform plates,* and the relative size of the *olfactory bulbs* are all much greater in the lemur than in the human. These anatomical differences, along with the loss of the *rhinarium, Jacobsen's organ,* and the long snout, indicate a decreasing reliance on smell in higher primates.

Remember that mammals evolved originally as nocturnal creatures dependent on the sense of smell for detecting food and predators and for organizing social behavior through territorial markers and olfactory sexual signals. Prosimians like the lemur have retained these characteristics; whereas the anatomy of high primates reflects their more diurnal, visual way of life.

NOSES AND SMELLING ★

RHINARIUM_A
JACOBSEN'S ORGAN_B
NASAL CAVITY_C
 SEPTUM_D
  TURBINATES_E
EYE ORBIT_F
SINUS CAVITIES_G
BRAIN CASE_H
TEETH_I
CRIBIFORM PLATE_J
OLFACTORY
 RECEPTORS_K
  BULB_{K1}
   ACCESSORY BULB_{K2}
    TRACT_{K3}

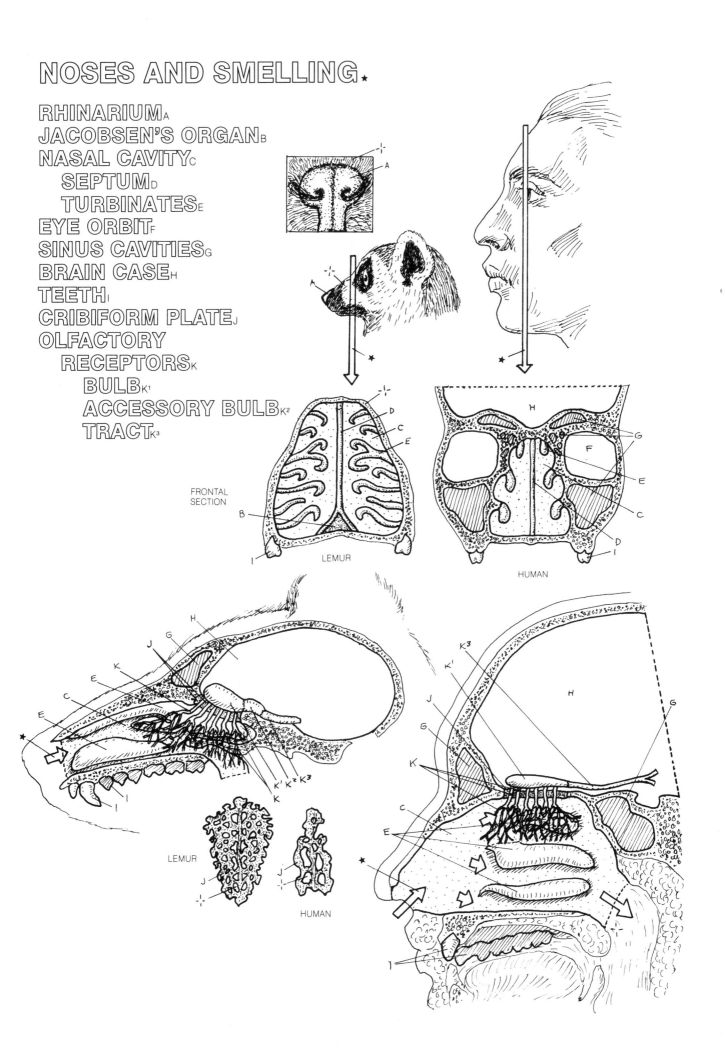

FRONTAL
SECTION

LEMUR

HUMAN

LEMUR

HUMAN

Humans, unlike other primates, use language for communication, for structuring their social systems, and for accumulating traditional knowledge. Sometime in the past 3 to 5 million years, hominids began using vocal signals to represent objects, to identify each other, and to coordinate group activities. As *humans* became more linguistic and cultural creatures, the sense of hearing became more essential for living successfully in the social group.

Begin by coloring the outer ear, canal, and eardrum, using related colors.

Humans have a typical mammalian ear. Nocturnal mammals of the Mesozoic evolved an *outer ear* to channel sound waves into the *ear canal* toward the *tympanum*, protected inside the bony skull. Most mammals optimize sound collection by moving their *ears* around, an ability that has diminished in higher primates more dependent on sight than hearing. Can you wiggle your *ears*? Most *human ears* have entirely lost this ancestral function.

Next color the bones of the middle ear.

As you color, imagine the path of sound waves through the *human ear* as a succession of high- and low-density packets of air. The number of these pressure waves reaching the ear each second is the frequency. *Low frequencies* are perceived as deep pitch, *high frequencies*, as shrill. When pressure waves strike it, the *tympanum*, stretched across the entrance to the middle ear, vibrates like the drum for which it is named. This sets the three middle ear ossicles (little bones) into motion: The *malleus* (hammer) strikes the *incus* (anvil), which in turn strikes the *stapes* (stirrup). The three mammalian ossicles are better for sound transmission than the one possessed by our reptilian ancestors (Plate 12).

Finish coloring the top two-thirds of the plate.

The inner ear, a legacy from our fish forebearers, consists of the *vestibule* and *semicircular canals*, fluid filled chambers sensitive to motion, necessary to maintain balance. The *cochlea*, a snail-like, triple-barreled tube, is vital for sound perception. Beneath the *cochlear canal* lies the thick *basilar membrane*, and on this *membrane* lies the *Organ of Corti*, with its fine, hairlike projections.

The *stapes* transmits pressure waves via the flexible *oval window* to the *cochlea*, where they spiral through the *vestibular* and *cochlear canals*, set up vibrations in the *basilar membrane*, and stimulate the hairs on the *Organ of Corti* to send nerve impulses to the brain. Different hairs respond to different sound frequencies and are perceived in the brain as different pitches.

The auditory fibers decussate (cross) completely before reaching the sensory *auditory cortex* in the temporal lobe.

Sound waves go down the *vestibular canal* to the apex, then travel back through the *tympanic canal* to the flexible *round window*, where sound energy passes out of the inner ear.

Finish the plate by coloring the sound wave frequencies, using dark, contrasting colors. Color the hearing range of the galago and human with light colors.

High frequencies excite nerve fibers at the base of the cochlea, *low frequencies*, at the apex. The range of frequencies an animal can perceive depends on the length of the *cochlea*. Mammals with a longer *cochlea* can hear higher frequencies than reptiles can hear. The *human Organ of Corti* contains more hairs than that of the other mammals (1500), reflecting the importance of auditory discrimination for understanding speech.

Prosimians can hear frequencies up to 60,000 Hertz (cycles per second), sounds beyond the perceptual range of monkey, apes, and *humans*, whose upper limit is 20,000 to 25,000 Hertz. *Galagos* are most sensitive at 8,000 Hertz, *humans* between 2,000 and 4,000.

In everyday life, we are bombarded with a bewildering array of overlapping sounds, differing in intensity, frequency, and rhythm. Our capacity to sort these out and make sense of them, to locate their different sources simultaneously, even to listen to more than one conversation at a time is truly remarkable. The basic structure of our *ears* is the same as that of our mammalian ancestors, which share with us an excellent sense of hearing. Our capacity to comprehend *human* speech depends not on the distinctiveness of *human ears* but on the way our brains analyze auditory information.

EARS AND HEARING.★

OUTER EAR_A
CANAL_B
EARDRUM
(TYMPANUM)_C
MIDDLE EAR ★
MALLEUS_D
INCUS_E
STAPES_F
INNER EAR ★
VESTIBULE_G
SEMICIRCULAR
CANALS_H
COCHLEA_I
AUDITORY NERVE_J

OVAL WINDOW_L
ROUND WINDOW_M
COCHLEAR CANALS_I ( )
VESTIBULAR_N
TYMPANIC_O
COCHLEAR_P
BASILAR MEMBRANE_Q
ORGAN OF CORTI_R

AUDITORY
CORTEX_K

SOUND WAVE FREQUENCIES.●
HIGH_S
MIDDLE_T
LOW_U

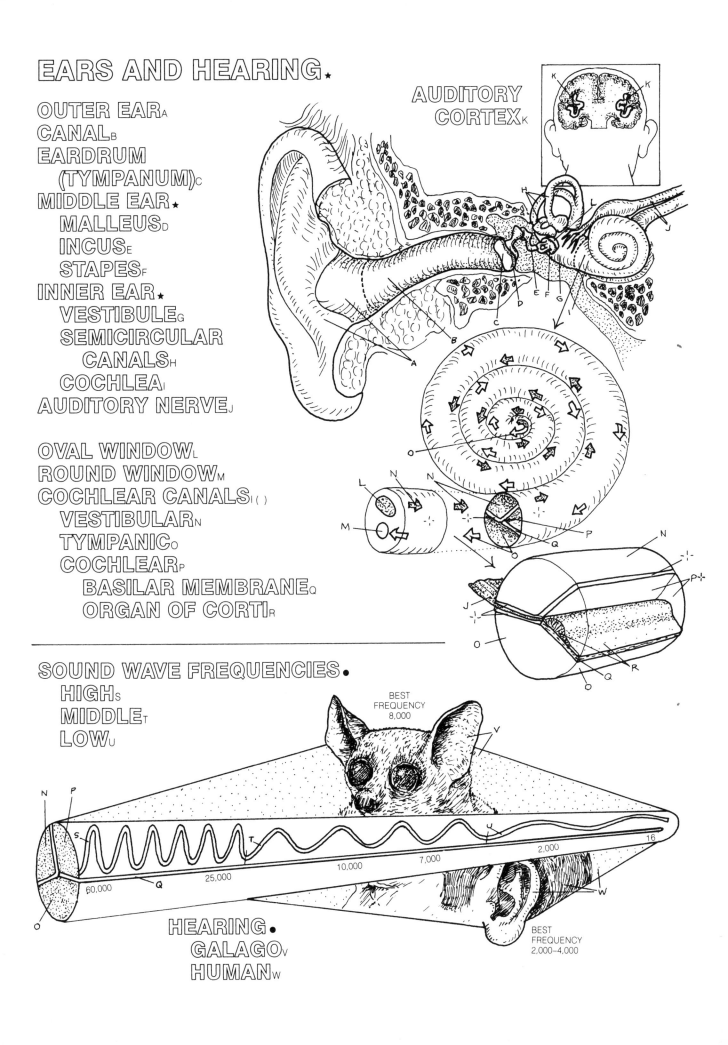

BEST
FREQUENCY
8,000

60,000   25,000   10,000   7,000   2,000   16

BEST
FREQUENCY
2,000–4,000

HEARING.●
GALAGO_V
HUMAN_W

CEREBRAL CORTEX: COORDINATION OF SENSORY AND MOTOR FUNCTIONS

A large brain relative to body size is one of the most distinguishing characteristics of primates. However, the primate brain is not merely a larger version of the brain of primitive mammals; it has been reorganized with greatest elaboration in the neocortex. The neocortex which first appeared in reptiles (Plate 13), has expanded in primates much more than the rest of the brain and occupies about half the brain's volume compared to about one-third in most other mammals. While retaining the general organization of other mammals, the neocortex has expanded to accommodate improvements in primate special senses: skilled movements of the hands and complex social behavior.

Color the primary sensory areas of the neocortex (the olfactory, visual, auditory, and somatosensory areas) and the motor cortex in the five primate species in the center of the plate.

The sensory areas receive and process stimuli from the nose, eyes, and ears as well as temperature, pressure, and tactile information from the musculoskeletal areas. The *motor cortex* controls movement of the voluntary muscles.

The outside layer of the brain, the gray matter, has become convoluted with distinctive folding patterns. This is an effective way of increasing surface area — or brain power — without enlarging the brain case. The tree shrew brain has almost no folds; the human brain has many. The two most distinctive are the central sulcus, separating the sensory and motor areas, and the sylvian fissure. Gray matter has expanded mostly in areas not directly concerned with sensory perception and motor control.

Color the remaining areas of the cortex gray.

These areas, the association areas, integrate information from several sense organs with knowledge, stored up from past experiences. They enable the higher primates to devise novel solutions to complex problems and are primarily responsible for the process of human "thinking". As the nerve tracts in the brain mature, they become myelinated, that is, covered with an insulating sheath that greatly increases the speed of electrical impulses traveling along them. Most of the association areas do not mature until after birth. Such slow neurological development would be fatal to most mammalian newborns, which are essentially on their own soon after birth. But in higher primates, it reflects their reliance on maternal dependence and social learning.

Obtaining detailed "maps" of brain organization is difficult. A variety of experimental techniques is combined in order to discover how the brain is organized. Postmortem examinations of patients with brain tumors or hemorrhages have pinpointed the locations of lesions causing specific neurological defects. Lesions may be experimentally induced in animals, and specific parts of the brain may be electrically stimulated to provide functional maps of the brain. The neurological underpinnings of general thought processes like memory or foresight cannot be located in any one part of the brain, but researchers have been successful in elucidating the organization of some parts of the brain, particularly of the cerebral, sensory/motor regions.

Complete the plate by coloring each segment of the human body and its corresponding neural processing region in the motor and sensory cortex. Note that there are no titles for F-M.

Notice the upside-down arrangement of body parts on these two neocortical strips — how, for example, the foot is represented in the top of the sensory and motor strips, and the uppermost parts of the body are represented in the lower sections of the neocortex. Notice also that the extent of cortical representation does *not* correspond to the size of the body part. The head, mouth, and tongue, for instance, occupy a much larger area than their relative sizes would suggest because of the important flow of information from and to them in the process of tasting, eating, talking, and communicating via facial expressions. Likewise, in primates, the areas devoted to the hand, particularly the thumb, are large to facilitate their high degree of manual dexterity.

Increase in brain size and complexity has played a critical role in all primate evolution, but particularly in hominid evolution. We are less able to document the evolution of behavior than the evolution of the special senses by examining brain anatomy. Many motor and sensory adaptations can be studied only by looking at living species and asking what they do.

CEREBRAL CORTEX: COORDINATOR OF SENSORY AND MOTOR FUNCTIONS.★

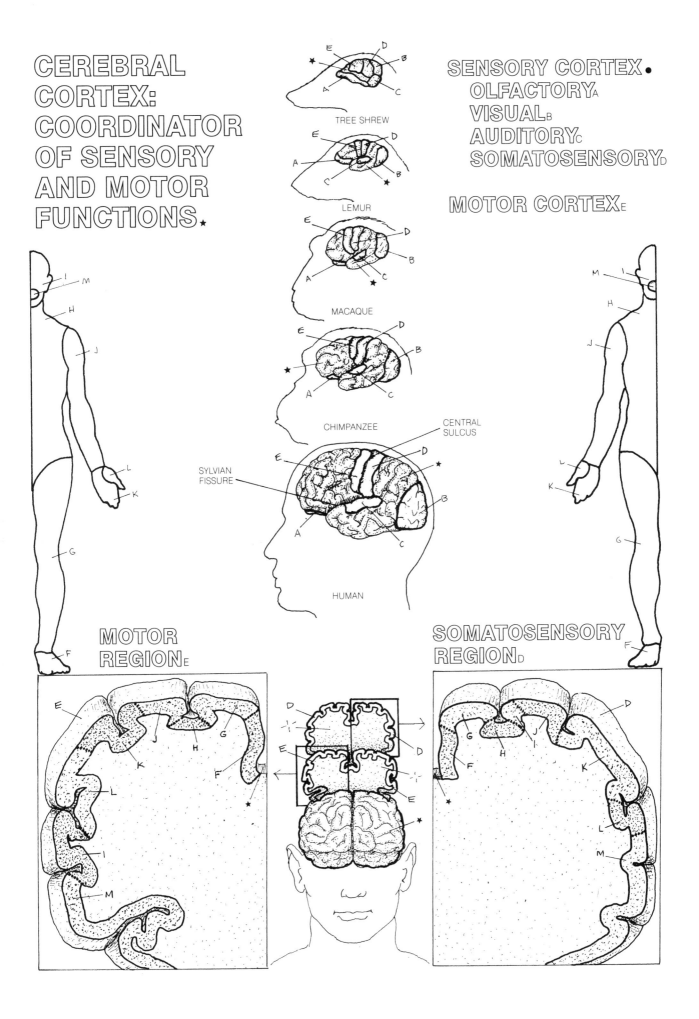

SENSORY CORTEX ●
OLFACTORY A
VISUAL B
AUDITORY C
SOMATOSENSORY D

MOTOR CORTEX E

TREE SHREW

LEMUR

MACAQUE

CHIMPANZEE

CENTRAL SULCUS

SYLVIAN FISSURE

HUMAN

MOTOR REGION E

SOMATOSENSORY REGION D

48
HANDS AND FEET:
SENSORY AND MOTOR FUNCTIONS

The complex functions of primate hands and feet illustrate why so much of the cerebral cortex is devoted to processing the sensory input from them and the motor output to them. Instead of paws, hooves, or flippers, primates have hands and feet that can grasp and manipulate objects. Unique characteristics are an opposable *thumb* (pollex) and *big toe* (hallux); long, straight *digits* with flattened nails; and thick friction skin on the *palms* and *soles* rich with sensory nerves, blood vessels and sweat glands.

Color the structures of the claw and nail, shown in side and bottom view in the diagram at the top of the plate.

Notice that the *claw* has two layers: The *deep stratum* fits tightly over the laterally compressed *terminal phalanx* and the *superficial stratum* provides a protective cover. The joint between the *terminal* and *middle phalanx* is sharply bent (flexed).

The *nail* of primates contrasts with the sharp curved *claws*. The *nail* lacks a *deep stratum*. The *superficial stratum* provides structural support for the weight-bearing *tactile pad*. The *terminal phalanx* is broad and flat, and the joint with the *middle phalanx* is straight (unflexed). The *tactile pads* on the ends of *fingers* and *toes* are richly supplied with sensory touch receptors, which discriminate temperature, texture, pressure, and shape.

The skin of the *digits, palms,* and *soles* forms a hairless surface. This friction skin has abundant sweat glands to keep it moist and pliable, essential for fine tactile discrimination.

Color the hand and foot of the tree shrew, our model of the common primate ancestor.

This primitive primate still has *claws,* but it also exhibits the beginning of a divergent (opposable) *thumb.* Tree shrews, like other primates, are hand feeders, and their hands are manipulative.

Continue coloring the hands and feet of the other primates as they are discussed.

The loris uses its hands and feet as though they were clamps. The second *digits* are reduced in size to maximize the loris' reach, the widest of any primate's. One *claw* — used for scratching — is retained on the second *toe;* all other *digits* have *nails* (not visible in this view).

The tarsier's hand is splayed in order to grasp its prey securely and also to grasp branches at whatever angle it may find them at the end of a leap. The tarsier has distinctively long, extended *digits* and well-developed *tactile pads*. The *thumb* lies in the same plane as the other *digits* and is not very opposable.

Long, slim hands and feet characterize the howler monkey. This New World primate can move its index *finger* and *thumb* independently, a departure from the "whole-hand" motor control characteristic of the prosimians. The howler uses its *thumb* and second *digit* in a sideways "scissors" action against the remaining *digits* when manipulating objects and, peculiarly, grasps branches between its second and third *digits* while moving through the trees.

The *fingers* and *toes* of baboons are somewhat shorter than those of the more arboreal monkeys because they spend so much time walking on the ground. The skin on the *palm* and *sole* is thick. The *thumb* can be opposed to the index *finger,* and fine control of the hand is further enhanced by independent motor control of the other *fingers.*

The orangutan, an acrobatic climber, has particularly long *palms, soles,* and *digits,* except for relatively short *thumbs* and *big toes.* Hands and feet serve as hooks to support the weight of a large body, usually distributed over several branches.

The human hand is similar in shape to other primate hands, with a noticeably well-developed *thumb.* An easy *thumb*-to-*finger* grip underlies our capacity for fine manual dexterity; the *fingers* are individually controlled, and their large projection areas in the sensory and motor cortex attest to their importance as tactile as well as manipulative organs.

Our foot differs remarkably from all other primate feet. Adapted for weight bearing, it has a broad heel and a thick-skinned ball cushioned by a fat pad. The *toes* are even shorter than on the land-dwelling baboon.

HANDS AND FEET.★

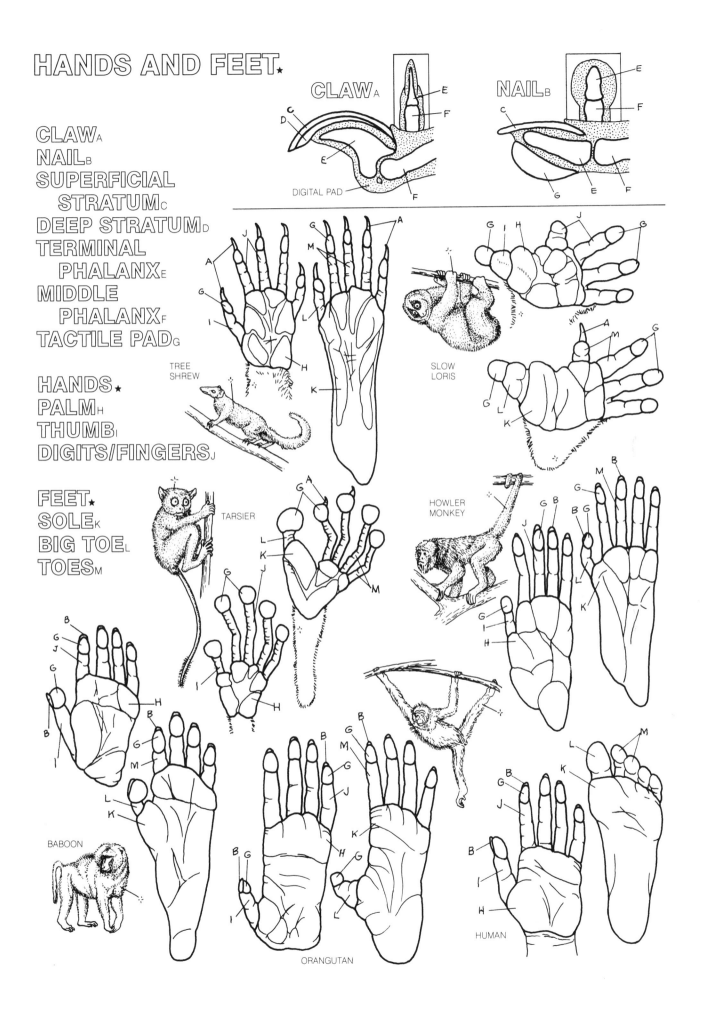

CLAWA
NAILB

SUPERFICIAL
    STRATUMC
DEEP STRATUMD
TERMINAL
    PHALANXE
MIDDLE
    PHALANXF
TACTILE PADG

HANDS ★
PALMH
THUMBI
DIGITS/FINGERSJ

FEET ★
SOLEK
BIG TOEL
TOESM

CLAWA

E

F

D
C

E

DIGITAL PAD

F

NAILB

E

F

C

G

E

F

TREE
SHREW

SLOW
LORIS

TARSIER

HOWLER
MONKEY

BABOON

ORANGUTAN

HUMAN

49
HANDS AND DOING

Primates use their hands not only for locomotion but also to manipulate and explore their environment, especially in feeding. Many of these activities require hand-eye coordination, which is more refined in anthropoids than in prosimians. Here we explore some of the ways primates use their hands.

Color the titles on the top left. Then find the matching letter in one or more of the pictures and color it as the activity is discussed.

For the first few weeks of life, a baby primate instinctively *clings* to its mother's hair; the mother nurses and carries the infant until it is weaned. The hands and feet of primate babies have relatively more muscle than those of adults (Plate 52), an anatomical feature that confirms the importance of the primate hand for the infant's early survival.

Primates are characteristically hand *feeders:* They bring food to their mouths rather than their mouths to the food. Hand *feeding* entails a complex of functions: finding, reaching, plucking, seizing, and holding, none of which would be possible without hand-eye coordination.

Searching beneath the ground for roots or tubers is another way of getting food. The baboon is *digging* for roots of savanna scrub plants. Gorillas and pygmy chimpanzees also dig for roots in the forests. Primates may have to *explore* the environment, move aside leaves or rocks, as the baboon is doing, to find lizards and luscious insects.

Some primates "process" their food a bit before eating. The Japanese macaque has learned to *wash sweet potatoes* given to it by human researchers. This behavior has become a tradition in some groups of Japanese monkeys, but not in others living in the same areas. It all began when a young female monkey, Imo, ran down to the ocean to *wash* the sand off her sweet potato. She was copied first by peers in her play group, then by the mothers who watched them, then by other adult females, and finally by most, but not all, the adult males. The spread of the *potato-washing* tradition provides a fascinating lesson for anthropologists, showing them how a "cultural practice" originates through the innovation of a single young individual and which members of the group pay attention to which other members.

An essential part of the early primate adaptation was the ability to *grasp* branches for moving around in the trees. The tarsier's long prehensile fingers and toes encircle a skinny vertical branch. The young monkey in the center holds on to the trunk of a fallen tree while its *grooming* partner removes dirt from parts of its body it cannot clean itself.

The gibbon *hangs and swings* (brachiates) as it swishes through the air beneath horizontal limbs. *Weight-bearing* on the hands is a typical monkey or ape posture, in trees or on the ground, as illustrated by the langur.

Primates are generally curious animals, continually focusing all their senses on new stimuli in the environment. Fine manual dexterity allows primates to *examine* objects in great detail. Here the langur *examining* a leaf is perhaps noticing its shape and color or watching an insect crawl across it.

Another example of manipulation is *tool using,* illustrated by the human hand holding a crayon. Hand-eye coordination is clearly necessary for using this and other human tools. We are unique among primates in not using our hands for locomotion, but we are not unique as tool users. Our close relatives, the chimpanzees, use tools in several innovative ways.

Primate hands also have social functions. Here one monkey is *grooming* another. The practical function of this activity is to remove particles and to cleanse wounds so they will heal more quickly. More important, *grooming* expresses affectional bonds. By observing who *grooms* whom and how often, field researchers discover the social ties that bind a group. For example, primatologist Donald Sade reported that in a macaque troop, a mother and her offspring remain in close physical proximity, and members of this mother-centered group *groom* each other more often than they *groom* nonrelated individuals. *Grooming* may express friendship and often precedes mating.

HANDS AND DOING.★

CLINGINGA
FEEDINGB
DIGGINGC
EXPLORINGD
POTATO WASHINGE
GRASPINGF
HANGING/SWINGINGG
WEIGHT BEARINGH
EXAMININGI
TOOL USINGJ
GROOMINGK

50
PRIMATE LOCOMOTION: IN THE TREES AND ON THE GROUND

Primates spend much of their time in trees—feeding, sleeping, and socializing. They are not the only animals that live in trees, but they are the only large mammals that do so. Trees form discontinuous pathways with branches oriented at many angles to each other; so a primate moving through the tangled network must be flexible, particularly in its forelimbs, for grabbing supports as it moves. Agility in the trees demands excellent hand-eye and muscular coordination and a good memory for maneuvering through established pathways and locating food.

Primates vary in size, and an animal's size limits where in the trees it can safely travel to obtain food and escape predators, as well as which methods of locomotion it can use. Body size is much less of a limitation for primates than for other mammals living in trees because their grasping hands and feet and flexible limbs enable them to distribute their body weight over two or three branches at the same time. Locomotor pattern relates not only to body size, but also to the length of limbs and their relative weight, length of the back, movement at the joints between segments, and shape of hands and feet.

Color the top half of the plate as the various primates are discussed in the text.

The little *tree shrew* scampers about the trees, low shrubs, and bushes. The *lemur* is a generalized climber and jumper. The *potto* is a slow climber with hands and feet that clamp about branches and unclamp, holding on to two or three supports at once as it moves carefully through the trees. The *galago* specializes in hopping and jumping. It has light upper limbs; long, heavy lower limbs; and a long foot.

New World monkeys, unlike their Old World cousins, live in the trees and seldom come down to the ground. The *marmoset,* small like the *tree shrew,* runs on the tops of branches. The *squirrel monkey,* like the *lemur,* also walks and runs on branches, leaps and jumps, stirring up flying insects, which it catches and consumes. The aptly named *howler monkey* has a prehensile tail, like the spider and woolly monkeys. These primates can wrap their tails around branches and hang suspended. But they generally move on top of branches, using the tail as a fifth limb for balance and security.

Color the lower part of the plate as the various primates are discussed.

Most of these are Old World monkeys and apes. Their pattern of motion is extremely varied; so it is oversimplifying to designate a species as "arboreal" or "terrestrial," as has often been done. *Langurs* run along the ground, climb trees, and may make spectacular flying leaps from one tree to another, catching the leafy branches with hands and feet as they land. The *baboon* lives and moves mostly on the ground but sleeps in the trees. Mothers carry their babies on their bodies, as shown here, and despite the extra weight remain as mobile as males. The *macaque,* a close relative of *baboons* and *guenons,* demonstrates here a typical primate method of tree climbing; the *guenon* walks quadrupedally along a branch.

Apes also move on the ground and in the trees. A very mobile shoulder and short back allow them to climb and to hang beneath and swing from branches. The *gibbon,* smallest of the apes, is our star brachiator: It whips through the trees by a gymnastic combination of bipedal running, rapid climbing, and mid-air leaps. The *orangutan* is a much larger ape (adult females weigh over 25 kilograms, adult males, over 75 kilograms). Yet, owing to its large, hooklike, grasping hands and feet and its flexible shoulder and hip joints, the *orangutan* is still quite acrobatic.

Gorillas and *chimpanzees* spend more time out of the trees than do either of the Asian apes. They move on the ground with their long arms and hooklike fingers, walking on their knuckles, like the *gorilla* shown here. Despite their great size, *gorillas* sometimes climb into trees and sleep there; young *gorillas* swing and climb more than their weightier elders. *Chimpanzees* also knuckle walk on the ground and climb and feed in the trees. It was from a long-armed, chimpanzee-like ape capable of occasional bipedal (two-legged) locomotion that we humans evolved our upright locomotor patterns, a repertoire that includes standing, walking, running, and dancing.

LOCOMOTION ★

TREE HABITAT ●

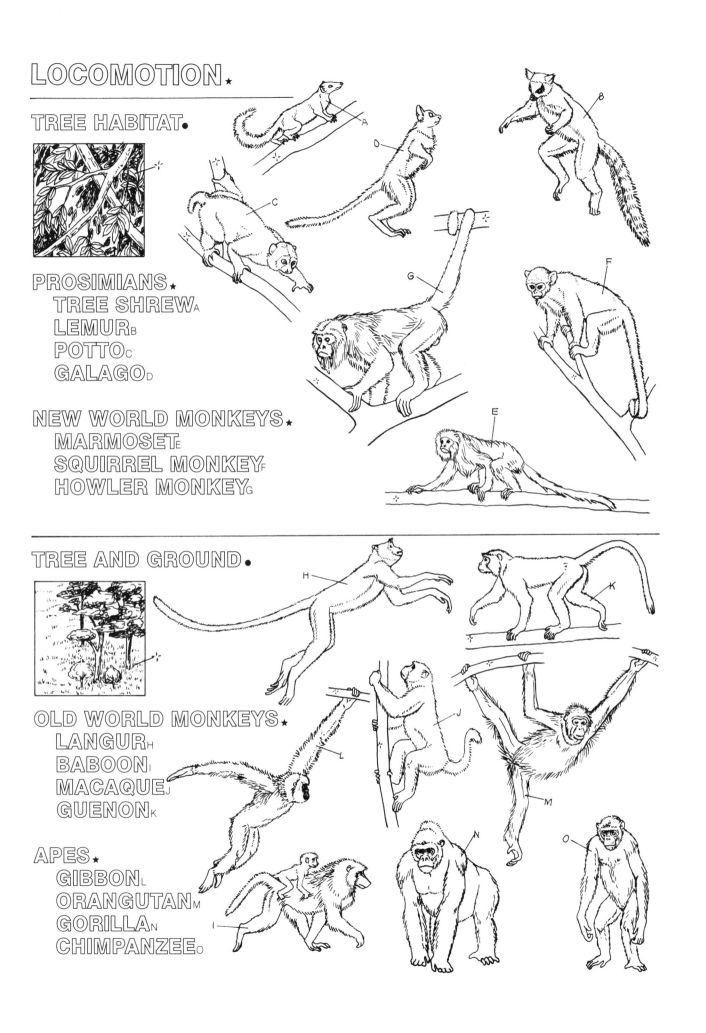

PROSIMIANS ★
  TREE SHREW A
  LEMUR B
  POTTO C
  GALAGO D

NEW WORLD MONKEYS ★
  MARMOSET E
  SQUIRREL MONKEY F
  HOWLER MONKEY G

TREE AND GROUND ●

OLD WORLD MONKEYS ★
  LANGUR H
  BABOON I
  MACAQUE J
  GUENON K

APES ★
  GIBBON L
  ORANGUTAN M
  GORILLA N
  CHIMPANZEE O

51

Most of an animal's body weight is devoted to moving it around its environment. Comparing the body build of different animals reveals how each is adapted for its specific mode of locomotion. The distinctiveness of primates, as fundamentally arboreal creatures, can best be illustrated by comparing their body build with that of a ground-dwelling mammal.

Comparative anatomist Ted Grand has developed a method for analyzing the bodies of animals in a way that helps to define the relationship between their anatomy and locomotor behavior. Grand's method consists of two steps: First, each segment of the body (the *trunk, head,* and *tail; arm, forearm,* and *hand; thigh, leg,* and *foot*) is weighed, and the percentage of whole body weight of each body part is determined. This calculation reveals the relative weight of each body segment. Next the relative weights of each tissue type (*brain, muscle, skin,* and *bone*) are measured for each body segment.

Drawing on Grand's research, this plate compares the segment weight and tissue composition of a macaque and a greyhound. The greyhound is adapted for running at high speeds along the ground; the macaque is suited for walking and climbing along branches in the trees and for walking and running along the ground. No one would mistake a dog for a monkey because they look so different, but have you ever considered exactly which anatomical features are responsible for these two varieties of body build?

Color relative segment weight. Color part of the body on the monkey and the corresponding percentage of body weight on the adjoining pie chart. Then color the same anatomical part and its relative percentage weight on the dog. In this way you can compare a particular segment and its corresponding weight in both animals. Color all parts of segment weight in this way.

A number of interesting comparisons emerge: In both animals, the *thighs* are the heaviest segments other than the *trunk* and *head.* The heavy hindlimb in both species is responsible for providing the power to propel the animal forward. The relative weight of the monkey's forelimbs is one and one-half times greater than that of the dog.

Although the fore *paws* of the dog are about the same relative weight as the monkey's *hands,* they are much longer. The monkey's *foot* is relatively two times heavier than the dog's hind *paw.* The monkey's *hands* and *feet* grasp arboreal supports and consist of about one-third *muscle.* The dog's *paws* are used for bearing weight on a flat terrestrial surface only and are *skin* and *bone,* with practically no *muscle* at all.

Adapted for life in the trees, the monkey has a much greater range of movement in the upper and lower limb joints. A macaque uses its *hands* and *feet* in such ways that it needs more *muscle* in its limbs. Monkeys can move well on the ground, but their heavier forelimbs and *hands* and hindlimbs and *feet* mean that they cannot run as fast as greyhounds.

The greater weight in the dog's *trunk* and less in the *legs* and *forearms* directly reflect its adaptation for speed. *Muscles* providing the power for locomotion are concentrated higher in the body (in the back—especially at the hip and shoulder joints). The lighter *paws* at the ends of the dog's limbs mean less weight to be swung forward with each stride. Finally, the longer the *paws* and *limbs,* the longer the stride and the greater the efficiency in running fast.

Complete the plate by coloring the tissue composition diagrams for both animals. (The category "other" includes the organs, such as heart, liver, lungs, and intestines.)

Tissue composition analysis reveals the greatest difference is in *brain* weight: The monkey's *brain* is relatively twice as heavy as the dog's. Notice also that the greyhound's body has more *muscle* than the monkey, an observation that extends to all comparisons between arboreal and ground-dwelling animals of similar body size. (Body segment analysis can also reveal differences in the locomotor adaptation of closely related species of the same body size. For example, a loris' forelimbs comprise 12 percent of body weight, its hindlimbs 14 percent, whereas a galago's forelimbs are 9 percent and hindlimbs, 22.6 percent.) From the monkey-greyhound comparison, we see how segmental analysis provides quantitative data on the relationship between form and function.

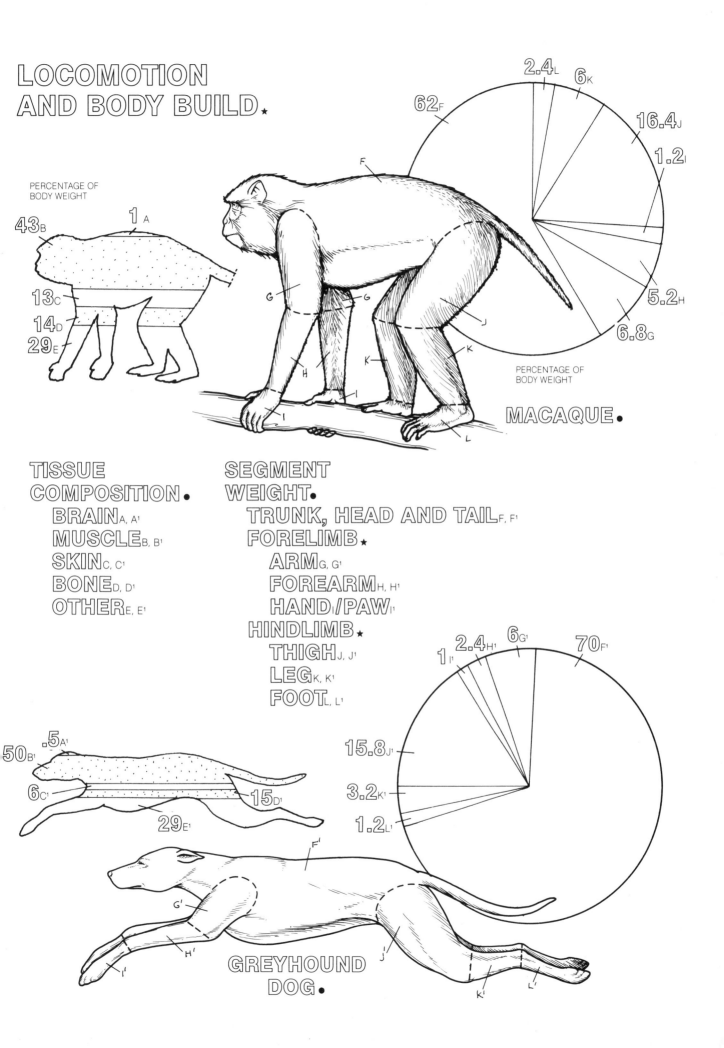

LOCOMOTION AND BODY BUILD. ★

PERCENTAGE OF BODY WEIGHT

43_B 1_A 13_C 14_D 29_E

62_F 2.4_L 6_K 16.4_J 1.2_I 5.2_H 6.8_G

PERCENTAGE OF BODY WEIGHT

MACAQUE.●

TISSUE COMPOSITION.●
BRAIN$_{A, A^1}$
MUSCLE$_{B, B^1}$
SKIN$_{C, C^1}$
BONE$_{D, D^1}$
OTHER$_{E, E^1}$

SEGMENT WEIGHT.●
TRUNK, HEAD AND TAIL$_{F, F^1}$
FORELIMB ★
  ARM$_{G, G^1}$
  FOREARM$_{H, H^1}$
  HAND/PAW$_{I^1}$
HINDLIMB ★
  THIGH$_{J, J^1}$
  LEG$_{K, K^1}$
  FOOT$_{L, L^1}$

1_{I^1} 2.4_{H^1} 6_{G^1} 70_{F^1} 15.8_{J^1} 3.2_{K^1} 1.2_{L^1}

5_{A^1} 50_{B^1} 6_{C^1} 15_{D^1} 29_{E^1}

GREYHOUND DOG.●

52

BODY BUILD AND LOCOMOTOR DEVELOPMENT

A baby primate is not born with an adult body build. As it grows from a small newborn into an adult, important changes take place in the types of tissue of which the body is composed. The distribution of *muscles, skin,* and *bone* over the *limbs, trunk,* and *head* changes according to a pattern which reflects the development of locomotor behavior. Using the methods described on the previous plate, Ted Grand explains how macaques undergo an anatomical "transformation" during development from their 400 gram weight at birth to their 8 kilogram adult stature.

Begin at the top of the plate with segment weight. Color the head and its percentage of body weight, first in the infant, then in the adult. Proceed to the trunk and its percentage in the infant, then in the adult. Continue until you complete all segment weights.

A large *head* at birth is typical of primates and other mammals. The newborn's *head* is relatively over three times heavier than that of the adult.

Another important comparison is that relative to total body weight, the *lower limb* is almost twice as heavy in the adult as in the newborn. However, there is less of a discrepancy in the newborn and adult *upper limbs.* Perhaps the most significant differences are in the distal ends of the extremities: The newborn's *hands* are more than twice as heavy as the adult's and the *feet* one and one-half times as heavy. Differential growth of the various body segments occurs: The weight proportions of certain segments—namely, the thigh, leg, upper arm, and shoulder—become greater; the relative weights of the *head, hands,* and *feet* become smaller.

These observations reflect differences in the anatomical requirements of an adult, as opposed to an infant, way of life. The body build of the newborn equips it for early survival. Its heavy *hands* and *feet* provide a grip strong enough to remain attached to its active, mobile mother. After about two weeks, coordinated muscle movements of the *limbs, hands,* and *feet* replace the clinging reflex; so the infant can break and regain its contact with its mother voluntarily.

For the first three months of life, the infant is totally dependent upon the mother for getting around, for unlike the babies of other mammals left behind in a nest or den, an infant primate travels with the social group. By tightly clinging to her coat, an infant leaves its mother's arms free for foraging and for outdistancing predators at a moment's notice. After several months, the baby occasionally and awkwardly begins to venture out on its own to play with other young monkeys. Soon it shifts from riding on its mother's belly to riding on her back. For at least six months, the young macaque depends on its mother for protection, nourishment, and locomotion. Then, as the infant loses its natal coat and begins to acquire adult coat color, its mother gradually begins to ignore its pleas for nourishment and locomotor assistance. As the infant is slowly weaned from its mother's milk, it practices the skillful movements of adult locomotion through play. By one year of age, a macaque moves well on its own.

Finish the plate by coloring the tissue composition diagrams.

As expected from findings on the relative weight of the *head, brain* tissue is relatively ten times heavier in the infant than in the adult. Notice that at birth the infant's body is composed of almost equal amounts of *skin, bone,* and *muscle.* During development, the amount of *skin* and *bone* decreases while the relative amount of *muscle* almost doubles. These changes in tissue composition document a trend toward increasing muscularity, a trend that correlates with the shift to locomotor independence during the first year of an infant's life.

Grand's method allows us to see how the center of gravity changes position, "migrating" from a higher position in the *trunk* of the clinging newborn toward the heavier hindlimbs in the adult. As the hindlimbs increase in relative size and muscularity, they provide the greatest propulsive thrust in climbing and running. The anatomical "transformation" of the newborn into the adult occurs with the transition from locomotor dependence to locomotor independence. At each stage, the growing primate has the locomotor equipment, reflected in its body segments, that provide for its survival to the next stage.

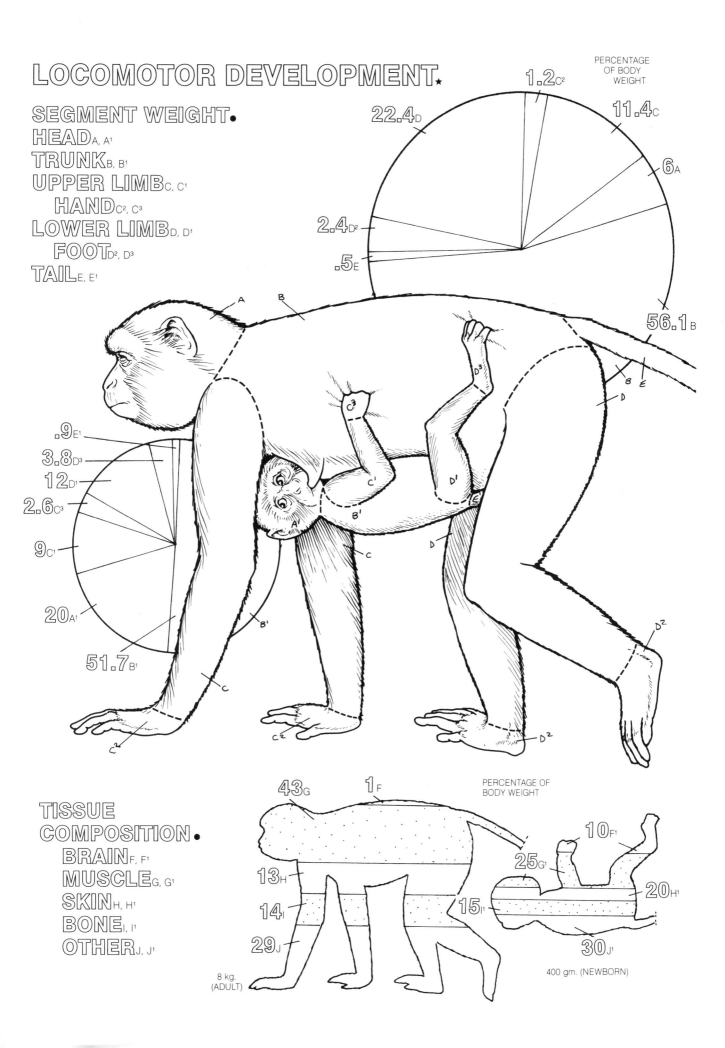

LOCOMOTOR DEVELOPMENT.★

SEGMENT WEIGHT.
HEAD_{A, A¹}
TRUNK_{B, B¹}
UPPER LIMB_{C, C¹}
   HAND_{C², C³}
LOWER LIMB_{D, D¹}
   FOOT_{D², D³}
TAIL_{E, E¹}

PERCENTAGE OF BODY WEIGHT

1.2_{C²}
22.4_D
11.4_C
6_A
2.4_{D²}
.5_E
56.1_B

.9_{E¹}
3.8_{D³}
12_{D¹}
2.6_{C³}
9_{C¹}
20_{A¹}
51.7_{B¹}

TISSUE COMPOSITION.
BRAIN_{F, F¹}
MUSCLE_{G, G¹}
SKIN_{H, H¹}
BONE_{I, I¹}
OTHER_{J, J¹}

PERCENTAGE OF BODY WEIGHT

43_G
1_F
10_{F¹}
13_H
25_{G¹}
14_I
15_I
20_{H¹}
29_J
30_{J¹}

8 kg. (ADULT)

400 gm. (NEWBORN)

53

THE TRUNK AND LOCOMOTION

Early in human evolution, our ancestors no longer climbed into trees for sleeping and feeding; yet we carry in our anatomy and behavior the legacy of our arboreal past. This plate draws on data collected in the 1930s by primatologist Adolph Schultz.

Color the titles—macaque, chimpanzee, and human. Then color the top and front views of the rib cage and vertebral column in contrasting colors. Use two shades of the same color for the scapula (shoulder blade) and clavicle (collar bone). Color the pelvis, sacrum, and ischial callosities.

Ape shoulder joints have a greater range of motion than monkeys', and humans have retained an apelike shoulder adapted for swinging or hanging from a branch. This functional difference between monkeys and hominoids (a term that includes both apes and humans) is reflected in the orientation of the *rib cage* and pelvic girdle. The *macaque's rib cage* is compressed on the sides, whereas the hominoid *rib cages* are relatively broader and flattened from front to back. On the top view notice that the hominoid *scapulae* lie against the back. In the monkey, the *scapula* lies on the side of the chest, and the *clavicle* is shorter than in hominoids. Critical for locomotor function is the orientation of the shoulder joints: out to the side in hominoids and toward the front of the body in monkeys.

The wider *rib cage* is complemented with wider *pelvises* on the *chimpanzee* and *human*. The upper part of the *pelvis* serves as an "anchor" for the wider splay of trunk muscles on the hominoids, a pattern that maintains the cylindrical shape of the primate trunk.

The *macaque's pelvis* has prominent tail vertebrae that contrast with the tiny vestigial ones called the coccyx in *chimpanzee* and *humans*. Monkeys use their tails for balance as they walk along branches. The Old World monkey's ischium, the part of the *pelvis* on which it sits, is flat and padded by thick calluses, *ischial callosities,* for sitting and sleeping in

trees. The shorter *human pelvis* curves forward at the upper edge and forms sort of a bowl.

Look at the *vertebral column*. Refer back to Plate 8 if you need to refresh your memory on its parts. The lumbar region of the *macaque* is much longer and there is more space between its *ribs* and *pelvis*. Its lumbar region is capable of motions that increase stride length and speed in quadrupedal running. The *chimpanzee's* lumbar region is so short, and its bottom *ribs* are so close to the top of its broad flaring *pelvis* that it lacks the *human* "waist."

In apes, and to a greater extent in *humans,* the *vertebral column* becomes a central pillar around which the body weight is distributed. Notice that *humans* have the thickest lumbar vertebrae and the widest *sacrum*.

Now color the bottom of the plate, including the graphs showing the relative percentages of back muscles to all of the other body muscles, and the relative weights of the hands and feet.

The back muscles run from the *pelvis* to the neck and comprise more of the total body muscle in the *macaque* than in the hominoids. The back muscle is not evenly distributed through the spinal column. In the *macaque* most of it lies in the lumbar region. Over one half of the *chimpanzee's* back muscles lie in the shoulder and neck region indicating reliance on the forelimbs for suspending body weight while swinging and climbing in trees or knuckle-walking. *Human* back muscles are similar in relative weight to *chimpanzees',* but most muscle is in the lumbar region and functions to maintain upright posture.

Even with different locomotor patterns, the *macaque* and *human* have hands similar in relative weight; all three species have feet of similar relative weight. Primate hands and feet bespeak our common arboreal heritage, as do the similar upper limbs and trunk of *apes* and *humans*. Though we can no longer brachiate or climb trees as adeptly as they, our common ancestry expresses itself in our ability to throw, lift, and become accomplished gymnasts.

THE TRUNK AND LOCOMOTION.★

RIB CAGE_A CLAVICLE_B SCAPULA_C
VERTEBRAL COLUMN_D PELVIS_E SACRUM_F

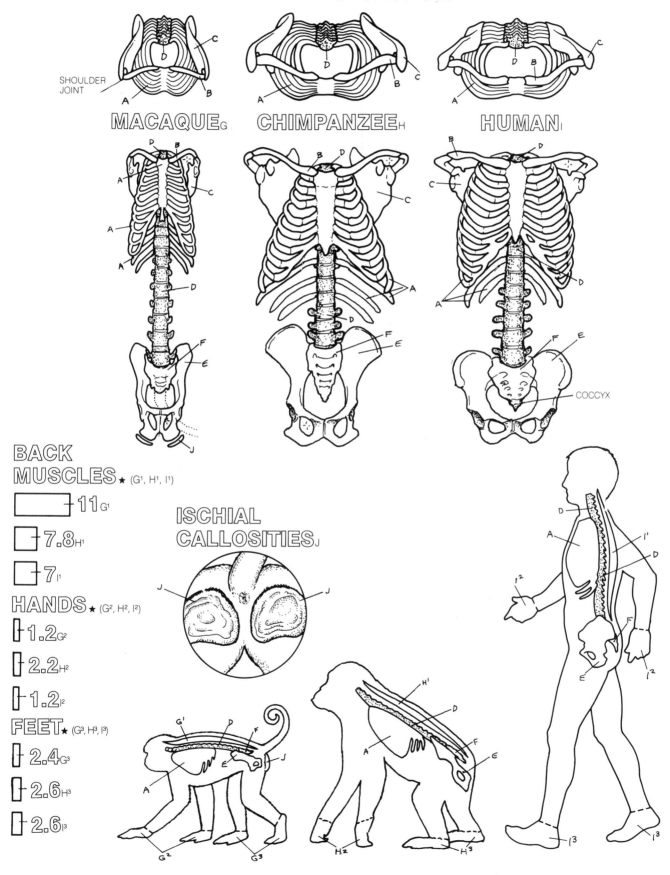

MACAQUE_G CHIMPANZEE_H HUMAN_I

SHOULDER JOINT

COCCYX

BACK
MUSCLES★ (G¹, H¹, I¹)

11_G¹
7.8_H¹
7_I¹

HANDS★ (G², H², I²)

1.2_G²
2.2_H²
1.2_I²

FEET★ (G³, H³, I³)

2.4_G³
2.6_H³
2.6_I³

ISCHIAL
CALLOSITIES_J

Walking upright all the time distinguishes humans from other primates, and this distinction is expressed anatomically in the unique human pelvis and foot. Other primates and other mammals, such as the kangaroo, sometimes move bipedally. This plate explores the characteristics of primate bipedalism that form the evolutionary backdrop of the human gait and stance.

Apes and monkeys are primarily quadrupedal but may become bipedal under certain circumstances. We have already looked at the many things primates do with their hands (Plate 44) and have observed that they are hand feeders. A sitting primate using its hands to feed or manipulate an object is strikingly different—especially more vertical and more "human" in appearance than a sitting dog or cat. The human resemblance is even more pronounced when the activities of the hands require a primate to stand upright, for example, holding on to a branch (like a commuter in a subway).

Color the primates as they are discussed in the text.

The patas monkey lives in savanna woodlands of west, central, and eastern Africa. It frequently stands up to *look* over the tall grass in order to spot food sources, predators, and other monkeys. Similar behavior has been observed in the closely related vervet monkeys, savanna baboons, chimpanzees, and other primates. Our ancestors were also a savanna species for whom standing and *looking* over the grasstops for food, enemies, and kin would have been an adaptive behavior.

Young gorillas *playing* and tussling with each other occasionally stand up like boxers, as do young chimpanzees and orangutans. Jane Goodall has observed even more aggressive bipedal action in her Gombe Stream chimpanzees fighting with baboons. The taller chimpanzees definitely have the advantage when they stand and swing or kick at the baboons, who cannot use their arms and legs as effectively when bipedal (see the previous plate for the anatomical basis of their back and shoulder). The baboon tries to bite with its long, knifelike canines, but the chimpanzee's long arms expertly keep the teeth out of range. For our human ancestors, as for young humans today, bipedal *play* was probably an important stage in the development of muscular skill, strength, and balance for a bipedal way of life.

Monkeys and apes perform *displays,* many of them bipedal, for attracting attention, asserting dominance, or scaring off predators. The spider monkey, a New World species with a long, prehensile tail, is standing and waving its arms at nearby (bipedal!) birds. When the gorilla stands up and beats its chest (Plate 65), the formidable sight and sound is enough to drive away almost any potential predator. The chimpanzee makes up in energy what it lacks in size by running around, waving its arms, and screaming when excited or frightened.

The talent for *throwing* is not limited to humans. This young chimpanzee is standing and throwing a stone at a baboon; its aim may be poor, but the *display* is effective in threatening an intruder, whether of a different or of the same species. *Throwing* as part of a *display* complete with vocalizations was probably one way our early ancestors scared off predators. Much later in our history, *throwing* probably became perfected for spearing animals.

Carrying things is not exclusive to humans. The Japanese monkey is carrying a sweet potato to a nearby stream to wash the sand from it. Obviously, an animal cannot *carry* things and still walk on four legs. Gombe chimpanzees are notorious for the incredible number of bananas they can manage to *carry* off at one time, wedging them under their arms and in the folds of the groin. One important incentive to bipedalism in our early ancestors would have been the invention of containers, such as slings, gourds, animal skins, or crude baskets for *carrying* gathered goods back to a home base. (A visit to any market anywhere in the world will reveal that baskets remain an important feature of the human adaptation.) A sling for *carrying* babies as well as food may have been one of the earliest human tools, necessitated by a human bipedal form of locomotion. Our primate relatives can go bipedally in many situations, but only humans have the anatomical specializations for being bipedal all the time.

BIPEDALISM IN PRIMATES.★

LOOKING A
PLAYING B
DISPLAYING C
THROWING D
CARRYING E

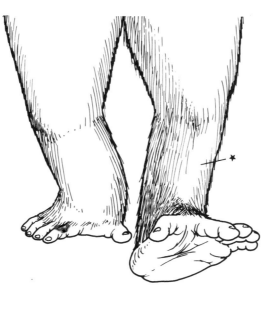

A

PATAS MONKEY

B

GORILLAS

C

SPIDER
MONKEY

D

CHIMPANZEE

E

JAPANESE
MACAQUE

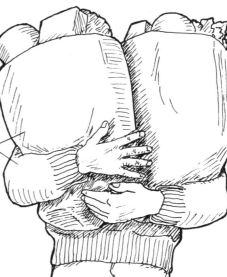

55

Chimpanzees are "jacks of all trades" in locomotion; whereas humans are "masters of one." Besides climbing, hanging, walking, and running on all fours, chimpanzees occasionally walk bipedally for short distances. Humans are specialized for walking long distances even while carrying food, tools, or babies. Endurance was selected for in early human ancestors.

This plate compares the major muscles involved in bipedalism.

Begin by coloring in gray the limb on the chimpanzee and human figures at the bottom and top of the plate and notice the body build of each.

Chimpanzee arms are much longer and much heavier than those of humans. Humans' lower limbs are not only longer but heavier than their arms.

Color the circle marking the center of gravity (the point where a suspended body would balance) in the chimpanzee and human.

The *center of gravity* is located higher in the chimpanzee's body than in the human body because more of the chimpanzee's weight lies in the upper trunk. In a bipedal animal, a lower *center of gravity* due to greater body weight in the lower limbs means greater stability. Human knees are straight rather than bent, as the chimpanzee's are, so that human muscles use less energy to maintain a vertical posture.

Color the lumbar region, which includes the vertebrae and the lumbar curve.

The chimpanzee's lower back is straight; it lacks the lumbar curve of humans, and human lumbar vertebrae are larger to support the weight of the trunk.

Color the sacrum and femur in the insets to the left. Notice the large human sacrum and the smaller chimpanzee one. Using light and contrasting colors for the major muscle groups, begin by coloring gluteus maximus in the large drawings.

Gluteus maximus is indeed "greatest" and gives the buttocks its unmistakable human shape; as the largest single muscle in the human body, it comprises over six percent of all the body's muscles. It is also large in chimpanzees but is much longer, and the heavy part of the muscle is low on the thigh, lying near the *hamstrings*. For humans, the large size of *gluteus maximus* reflects its important function: It straightens and

supports the hip joint and is active during much of the time in walking. The change in shape of this muscle was an important part of developing effective bipedalism.

Color gluteus medius and minimus. (G. minimus lies hidden under g. medius.)

These large muscles cover the surface of the ilium, cross the hip joint, and attach to the top of the *femur*. They are important in rotation and balance.

These muscles are also important when chimpanzees walk quadrupedally along the top of the branches; but in that position the trunk and ilium are in front of, rather than on top of, the hip joint. When the chimpanzee is upright, these muscles lie on top of rather than in front of the hip joint and cannot act as rotators. The ape ilium had to be redesigned for human use into one that curves in front of the hip joint. Notice in the inset to the left that the human ilium curves around, whereas the chimpanzee's is relatively straight across. This modification makes *gluteus medius* and *minimus* effective rotators in humans.

Color the quadriceps femoris (quad = four; ceps = head; femoris = of the femur) on the front of the thigh. Color the hamstrings, which lie on the back of the thigh, gray.

In typical quadrupedal animals like the greyhound, the *hamstrings* are twice the weight of the *quadriceps* because they propel the dog forward. In contrast, the *quadriceps* are twice the weight of the *hamstrings* in humans because in bipedal walking the straight knee acts as a "brake." *Hamstrings* are less important than the *quadriceps* in walking, but are important in running. In chimpanzees the *quadriceps* and *hamstrings* are nearly equal in weight. When chimpanzees climb, they straighten the hip and knee together, a motion similar to human walking, and so differ from typical quadrupeds in the propulsive action of the *hamstrings*.

Finish the plate by coloring the calf muscles.

These are well developed in humans and push the foot against the ground. Most of the muscles lie high on the leg, creating a slender ankle and inserting into the foot via the Achilles tendon, named for the Greek hero whose mother dipped him in a magic protective solution but left him vulnerable in the heel by which she held him.

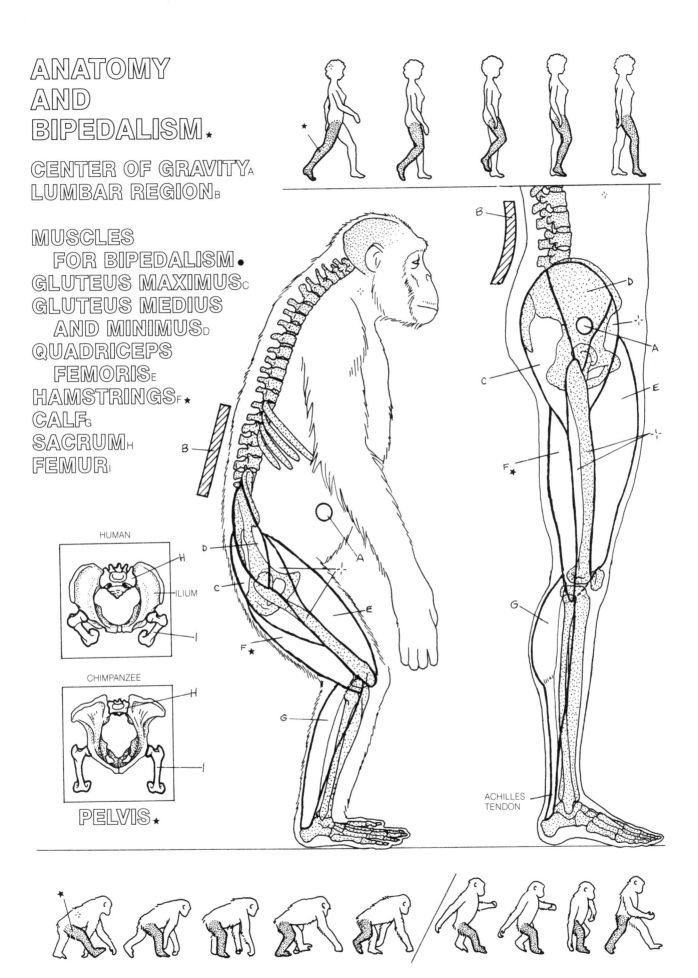

ANATOMY AND BIPEDALISM ★

CENTER OF GRAVITY A
LUMBAR REGION B

MUSCLES
 FOR BIPEDALISM ●
GLUTEUS MAXIMUS C
GLUTEUS MEDIUS
 AND MINIMUS D
QUADRICEPS
 FEMORIS E
HAMSTRINGS F ★
CALF G
SACRUM H
FEMUR I

HUMAN

H
ILIUM
I

CHIMPANZEE

H
I

PELVIS ★

ACHILLES
TENDON

56

HUMAN DEVELOPMENT: BODY SIZE AND PROPORTIONS

Human adult body proportions develop over a period of years. A 3.4 kilogram human infant gradually transforms into a 65 kilogram adult in 15–20 years. This plate traces body proportions from the large-brained newborn to the locomotor-independent adult.

Color each body part in all the figures at the top. In this series, height is kept constant and other body parts are shown relative to that.

The human fetus and newborn have a relatively large *head, hands,* and *feet.* The large *head* dominates in the fetus, but diminishes in relative size as other parts of the body begin to grow. The *brain* grows early in fetal life and has completed 25 percent of its growth at birth. The lower limbs grow a great deal, whereas the human trunk remains similar in relative size throughout development.

Color the human tibia and foot in the middle panel at three stages of development: two month fetal, newborn, and adult.

The early fetal tibia and *foot* are equal in length with a slightly divergent great toe, reminiscent of a typical primate grasping foot. By birth the *foot* is relatively shorter; the adult *foot* is still shorter, and its length is surpassed by the growth of the tibia.

Color the upper and lower limbs in the human and chimpanzee at three stages of development in the figures below. In these diagrams, trunk length is held constant so you can see the ontogenetic changes in limb length.

In humans the lower limbs grow disproportionately and later in development. In chimpanzees, they change much less in relative length; their upper limbs grow more.

Humans grow much slower than apes and for a much longer time. For example, although a male gorilla full grown may weight over 180 kilograms, it will have its full growth by 15 years of age; a male human weighing 65 kilograms may not reach his full size until about 18 years of age.

Age-related changes reflect the development of the locomotor system. In human children, the disproportionate size of the *head* and thorax make the center of gravity higher; therefore they are less stable when standing and walking. By two years of age, children are walking; compare the two-year-old body build to the fetus and newborn. Unlike adults, small children spread their *feet* apart and stretch out their *arms.* Such motions increase stability and compensate for short *legs* and an underdeveloped lumbar curve. As their *legs* lengthen, children can take longer steps and learn to rotate their thorax and pelvis in the proper rhythm. Changes in oxygen consumption accompany body build changes, and adults consume relatively less oxygen than children. Adult body build and metabolism are reached by about 15 years of age.

The sexes exhibit interesting differences in growth. The influence of female sex hormones (estrogens) causes the long bones to fuse, so that a girl's height is reached soon after puberty. However, female hormones stimulate the growth of the pubic bone in the pelvis, which results in broadening of the bony ring that serves as the birth canal. The female hip joints then are further apart, requiring more body rotation during walking and running. Due to changes in the pelvis, muscles, and *leg* length after puberty, women generally cannot run as fast as men.

Males are stimulated to grow taller under the influence of male hormones, the androgens, produced in larger quantities at puberty. The clavicles especially grow, giving males broad shoulders.

At the other end of the locomotor spectrum, changes in human gait are a sign of aging. Loss of muscle tone and a less flexible skeletal system restrict movements at all joints; so the stride is shorter and *feet* are placed wider apart to maintain maximum stability, rather like that of the child when it begins to walk.

Locomotor abilities vary with age and sex due to the differences in body build, metabolism, and bones. These factors must be considered in human origins, for the long dependence time of the human young is partly related to a slowly maturing locomotor anatomy and metabolism.

CHANGES IN BODY SIZE AND PROPORTIONS ★

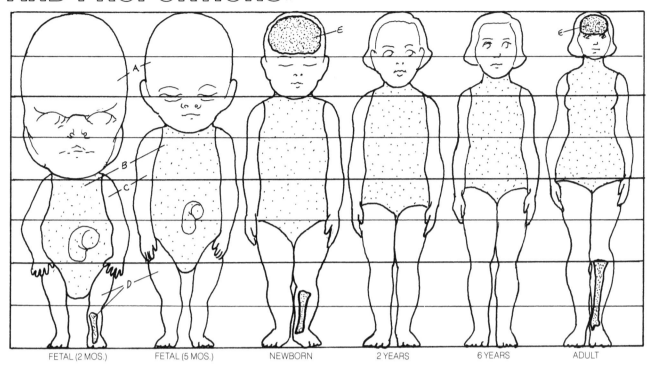

| FETAL (2 MOS.) | FETAL (5 MOS.) | NEWBORN | 2 YEARS | 6 YEARS | ADULT |

HEADA
TORSOB
ARMS AND HANDSC
LEGS AND FEETD
BRAINE

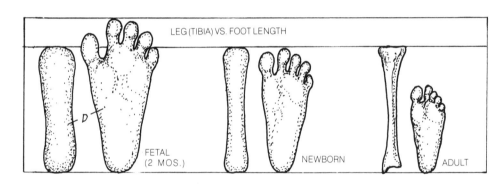

LEG (TIBIA) VS. FOOT LENGTH

FETAL (2 MOS.)   NEWBORN   ADULT

HUMAN ★

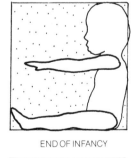
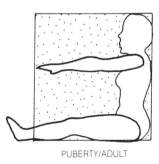

BIRTH   END OF INFANCY   PUBERTY/ADULT

CHIMPANZEE ★

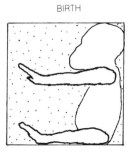
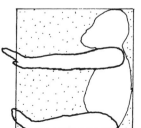
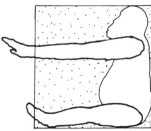

57

HUMAN DEVELOPMENT: FACE AND DENTITION

Just as head size changes relative to the rest of the body during development, there is differential growth of the head itself, particularly of the braincase and face, with the various parts developing at different rates. As with the locomotor features of the human body (Plates 55 and 56), form follows function. We now consider the evolving form and function of the human face and dentition.

Color the facial bones in the newborn, adult, and aged individual at the top of the plate.

The *nasal* passage, an airway and organ of smell, lies in the center of the face; above it are the orbits of the eyes; below it are the bones that support the *teeth*; lateral to it, are the bones to which the chewing muscles are attached, particularly the *zygomatic* arches. At birth, the orbits and braincase seem to dominate the face; in the adult, the *teeth* and lower jaw are more prominent; and in old age, with loss of *teeth* and supporting bone, there is somewhat of a reversion toward the child's proportions.

The bones of the newborn's skull are not fused and there are fontanelles between the bony plates, covered only by membranes and skin. In order to pass through the bony birth canal of the mother's pelvis, the baby's large head must be relatively malleable. The two halves of the newborn's lower jaw are also separate but fuse by adulthood. In newborns, who suck milk from their mothers' breasts with the help of fat, strong cheeks, the *teeth* are unerupted, and the chewing muscles are as yet small and weak.

As the braincase expands around the growing brain, erupting *teeth* stimulate the formation of bone in the jaws that support them. If *teeth* are removed before they erupt, the jaws never form properly. After the *teeth* are gone, due to disease or age, the supporting bone is resorbed.

Now complete the plate.

In this X-ray view, we see the forming *tooth* buds. The lower incisors erupt first, at six months of age. Between ages five and six, the dental arches are completely developed and all the *milk dentition* are in and have good biting function due to development of the chewing muscles attached to the jaws and skull. The distinct pattern of *tooth* wear characteristic of each species has already begun. At this age, the brain has completed 85 percent of its growth.

The first *permanent* molars, followed by the incisors, start coming in at about six years of age — two to four months later in boys than in girls. By this age, limb proportions are near those of the adult (Plate 56).

In advanced age, with loss of *teeth* or extreme wear, the face shortens from nose to chin, giving one of the characteristic features of the "old" face.

The interrelations of *tooth* size and shape with the muscles and underlying bone can help us understand the different forms of the face in closely related species — like the two australopithecines or in males and females (Plate 63). When males have much larger canines than females, as in many primate species, there are large canine roots, requiring more bone in the upper and lower jaws, larger masseter and temporal muscles, and larger sites of bony attachments for these muscles than females have.

All our knowledge of developmental anatomy, male-female and species differences, and form-function relationships is used in interpreting the often fragmentary remains of *teeth,* jaws, and skulls found in the fossil record left behind by early humans.

HUMAN DEVELOPMENT: FACE AND DENTITION ★

FACIAL BONES ★
　NASALₐ
　MAXILLARYʙ
　ZYGOMATICᴄ
　DENTARY
　　(MANDIBLE)ᴅ

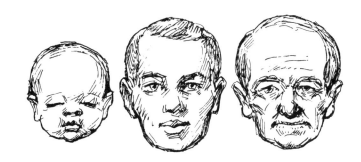

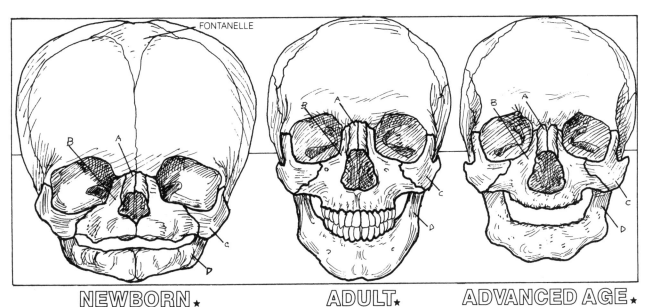

NEWBORN ★　　　ADULT ★　　　ADVANCED AGE ★

DENTITION ★
　MILK DENTITIONᴇ
　PERMANENT TEETHꜰ

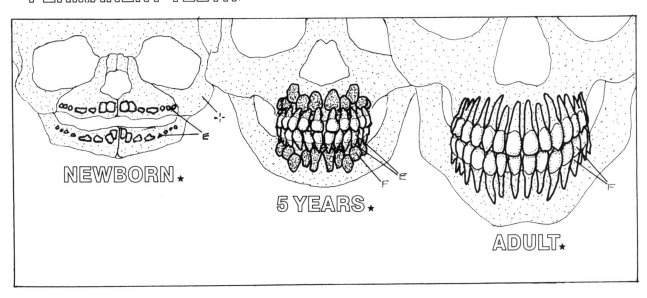

NEWBORN ★　　5 YEARS ★　　ADULT ★

58

DENTITION AND DIET

Primates eat *insects, fruits, leaves,* and occasionally lizards, tree frogs, birds, their eggs, and smaller mammals. Humans retain this primate "omnivory," (omni = all; vore = eat) though we use tools to obtain and process food and cultural and economic practices determine what, how, and when we eat.

The earliest primates were small bodied, fed at night on the forest floor, and depended upon *insects.* As diet expanded to include plant parts, *fruits,* flowers, and *leaves,* specializations of primate teeth and digestive systems evolved. The shape of the teeth relates to the manner in which different kinds of food are procured, prepared, and processed.

This plate discusses diet as it relates to food type, dentition, locomotion, and social organization. We compare three species: the slender loris, which eats *insects;* the mangabey, which relies heavily on *fruit;* and the howler monkey, which depends upon *leaves* as much as on *fruit.*

Color all parts of the dentition and diet of the slender loris as they are discussed in the text.

The total number of teeth, 36, can be represented in the dental formula 2•1•3•3 (2 *incisors,* 1 *canine,* 3 *premolars,* and 3 *molars*), the teeth in one quarter of the mouth.

The slender loris, a small (400 grams), nocturnal, slow climbing prosimian, lives in high forests of southern India and Sri Lanka and eats sluggish *insects.* These *insects* are widely scattered, so many lorises cannot feed in one place. Therefore, the loris is relatively solitary as it moves about the trees at night catching *insects* with its hands. As the food requires no preparation, the upper *incisors* are small. The sharp pointed cusps of the *premolars* and square-shaped *molars* all "interlock" for cutting and pulping *insects.* The dental comb, or tooth scraper, formed by the 4 *incisors* and 2 *canines* of the lower jaw is found in most prosimians and has two functions: one dietary and one social. It can be used for scraping gum off the bark of trees and is also used for combing the fur.

Color all parts of the dentition and diet in the middle picture.

Mangabeys are African monkeys with cheek pouches.

They eat *fruit* and seeds primarily but also flowers, *leaf* buds, bark, and *insects.* Their tooth number, 32, is common to all Old World monkeys, apes, and humans (the catarrhine group). A striking feature of their *incisors* is their breadth. Anatomist William Hylander showed that biting, peeling, or tearing *fruit* wears down the *incisors* — hence their larger size. The square-shaped *molars* have two pairs of cusps joined to form ridges, a characteristic of Old World monkeys called "bilophodonty," for pulping and grinding *fruit* before swallowing.

Notice the large projecting *canines,* larger in males than in females, used in social displays and threats as well as for obtaining food. The upper *canine* is sharpened against the lower *premolar* (note side view) by the "honing mechanism." There is a space, or "diastema," between the upper *canine* and *incisor* and between the *canine* and the specialized *premolar* below.

Fruits are available seasonally. Often patches of *fruit* are widely separated and the intervals between them are frequently unpredictable. Mangabeys live in groups of about 15 individuals but forage in smaller subgroups; their home range is several times larger than those of other forest monkeys.

Complete the plate by coloring the howler monkey's dentition and diet.

Howler monkeys, at 5–7 kilograms, are among the largest New World monkeys and eat *leaves* and *fruits.* Mature *leaves* require a lot of chewing and time to digest. Their dentition (2•1•3•3) includes well-developed *incisors* (not as large as the mangabey's) for biting off food and stripping *leaves* from branches. The square-shaped *premolars* and *molars* provide large shearing and cutting surfaces; the prominent cusps facilitate the cutting action required for *leaves.* The *canines* are large with a diastema. The specialized type of mangabey sectorial ("sharpening") lower *premolar* is not present. Howler monkeys may use their prehensile tail to assist them in reaching for *leaves.* Because *leaves* are abundant and continuously available, howler monkeys have a relatively small home range and live in stable groups of about 18 individuals.

DENTITION AND DIET★

UPPER LOWER

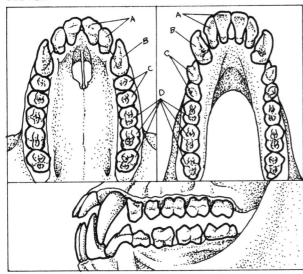

SIDE VIEW

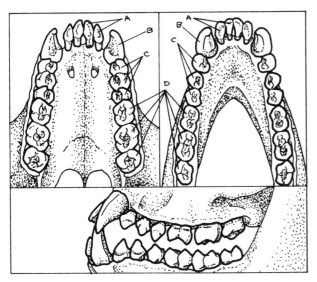

TEETH★
INCISOR A
CANINE B
PREMOLAR C
MOLAR D

DIET★
INSECTS E
FRUIT F
LEAVES G

SLENDER LORIS★

MANGABEY★

HOWLER MONKEY★

59

USE OF SPACE: HOME RANGE AND TERRITORY

The way primates use space in their daily search for food over a year's seasonal cycle is reflected in their locomotor system, dietary preferences, and social organization. Year after year a particular primate group can be located in the same area. Most primate groups live in an undefended *home range* (a large defined area in which the group roams during the course of a year), but a few primate species are territorial and defend a particular space and attempt to maintain exclusive access to it. To illustrate variability in the use of space, this plate contrasts the anubis, or savanna *baboon (Papio anubis)* and the *gibbon (Hylobates lar).*

Color the baboon in the middle of the plate. Use light colors throughout.

Baboons are quadrupedal and live in a variety of habitats, from tropical forests to woodland savanna. The savanna-dwelling *baboons* spend more time on the ground than do most other primates and have one of the largest *home ranges*: 20 square kilometers.

Baboons are omnivorous; they eat fruit and foliage from available trees but also feed on food items from the ground or low bushes. On the open savanna predators such as leopards are a great threat to the *baboon*. The *baboon's* protection on the open savanna lies in its large cohesive social group, called a troop, where all the members stay together all the time, and in the ability of the large adult male *baboons* to protect the group. The size of the troop can range from about 20 to 80 individuals.

With seasonal plants distributed over large areas on the savanna, *baboons* may travel up to 4 kilometers a day. Within each *home range* there is at least one *core area* where most of the troop's activity is concentrated. These *core areas* contain not only the best food sources, but more important, water holes and trees. Although anubis *baboons* spend most or all of the day on the ground, they always return to the safety of the trees before dusk to sleep.

Color the picture at the top of the plate showing the savanna mosaic environment, several baboon troops' home ranges, and their core area.

Baboon troop *home ranges* often *overlap; core*

areas, however, *overlap* much less frequently. Troops generally avoid one another in *overlap areas* so there is rarely open conflict. When they do meet, sometimes at streams or water holes, troops may intermingle and the young may play, but most often members of different troops simply ignore one another. The possibility for conflict increases when *core areas* are shared, as has been observed in the tension which endures when two troops must share the same sleeping trees.

Color the picture of the gibbon in the middle of the plate.

Gibbons live in the tropical rainforests of Southeast Asia and eat fruit. They live in small social groups that consist minimally of an adult male and female and their young offspring. Because *gibbons* are small-bodied, their nutritional requirements can be met within a small *home range,* averaging about one-seventh of a square kilometer. Also in contrast to the *baboons,* the *home range* is exclusive. No other *gibbon* groups share it; the area is defended and therefore is called a *territory. Gibbons* are among the few primates that are truly territorial.

Color the tropical rainforest at the bottom of the plate, remembering that each range represents a three-dimensional place to live about one-seventh of a square kilometer.

Note that the *territories* of *gibbon* groups are exclusive with some *areas of overlap*. These are *conflict areas,* and territorial battles take place within them. *Gibbons* have an elaborate system of long distance calls which signal their location to neighboring groups. This is a means for groups to avoid each other and helps to keep groups from encountering each other at the edges of their *territories*. About every other day ritual "territorial battles" do occur; they include loud vocalizations and charges and chases, crashing around through the trees. Occasionally the males of neighboring groups may actually engage in combat, but females do not. When these combats occur, the "winning" pair gains the right to feed in that area.

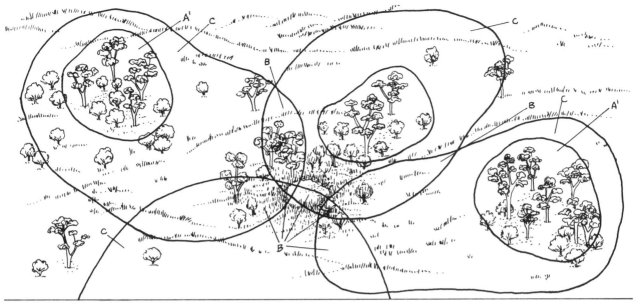

BABOON_A
TROOP 20–80 ★
CORE AREAS_{A¹}
AREA OF OVERLAP_B
HOME RANGE_C

USE OF SPACE: HOME RANGE AND TERRITORY ★

GIBBON_D
GROUP 2–5 ★
TERRITORY_{D¹}
CONFLICT AREA_E

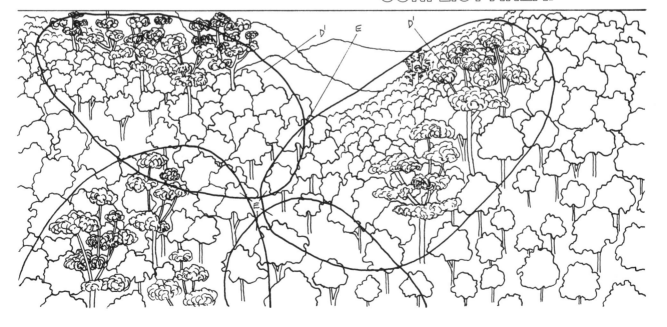

60
PRIMATE BEHAVIOR: GROUP STRUCTURE AND SOCIAL BONDS

An important characteristic of primates is that they live in year-round social groups consisting of both sexes and all ages. Among other mammals, the sexes may live apart for most of the year and come together during the mating season only; the young may be driven from the group before they are full adults. Primates are born into a social group whose membership is relatively stable throughout their lifetimes. The size and organization of the social group differ tremendously among primate species, but they usually include individuals of different age-sex classes.

Color the illustration at the top of the plate showing a baboon troop as each age-sex class is discussed in the text.

As the troop moves, *dependent young* are carried on the *mother's* belly or, when older, on her back. The *juveniles* move independently and often play together but still stay close to their *mothers*. There are usually many *adult females*; most are either carrying an *infant* or pregnant. *Adult males* are fewer in number than *adult females* and are twice their size. They play an important role in defending the troop against predators on the open savanna.

Because primates stay together as a group for the duration of their lives, important social bonds develop between individuals of each of the age-sex classes.

Color the infant-mother pair.

From the moment of birth, a baby primate clings to its *mother* with its grasping hands and feet, though she may assist it slightly with a touch or body posture adjustment. A primate's *mother* is the sole food source for several months. The close physical proximity of the immature *infant* to its *mother* enables the formation of the affectional primary bond, a bond that is necessary for learning to occur. The *infant-mother* bond is long lasting and in many species, lasts throughout life.

Now color infant and juvenile peer bonds.

As an *infant* grows up and becomes more independent of its *mother*, it spends time in peer play groups. In play, young animals learn how to control their movements and behave properly in social encounters.

During a period of rest in a baboon troop, one can observe mischieviousness, running, chasing and wrestling among agemates. The peer interactions become the basis for the formation of adult social bonds.

Color the pictures illustrating adult male-male bonds, female-female and male-female bonds.

Adult females have strong, long-lasting friendships with other *females*, which may be expressed by physically being together: sitting, moving and feeding together, and grooming. They may also "babysit" one another's *infants* and support one another in conflicts. Primatologist Shirley Strum recognizes that *adult female* relationships play an important part in influencing the course of social interaction within the group.

Adult males form bonds with one another and express them in the same way as *females*. There are social ranks among both *females* and *males* in a baboon troop, but for the *males*, rank is usually determined by the outcome of fights; for *females*, the outcome of conflicts is often determined by the social rank of the *females* involved.

Bonds between *adult males* and *adult females* are for reproduction, but *male-female* bonds are not always sexual. Companionships may form between siblings or nonrelated individuals. When *females* are in estrous, a *male* may interact with her exclusively: grooming and resting only with her.

Color the caretaking bonds.

These bonds include the primary one of a *mother* and her young; a *mother* and her offspring form a *mother*-centered subunit within many primate societies. Caretaking bonds also include the social ties that *adult males* and *adult* or *juvenile females* form with *infants* and *juveniles*. In primates, there is no concept of social fatherhood; *males* are generally protective of all *infants* and *mothers*.

Thus the integrity of the primate social group is maintained through social bonds developed over the lifetime of individuals. Group-living has important adaptive value in allowing a long time for the young to grow up and learn within a protective social environment.

SOCIAL GROUP STRUCTURE★

DEPENDENT YOUNG_A
JUVENILE_B
ADULT FEMALE_C
ADULT MALE_D

SOCIAL BONDS★
INFANT_A & MOTHER_C

INFANT_A & INFANT_A
JUVENILE_B & JUVENILE_B

MALE_D & MALE_D   FEMALE_C & FEMALE_C

MALE_D & FEMALE_C

MOTHER_C &
INFANT_A &
JUVENILE_B

MALE_D &
JUVENILE_B

61

In comparison to other mammals, primates have evolved an extended period of development after birth. This plate demonstrates the evolutionary changes in proportion of time spent in each developmental stage, as well as the general increase in life span and period of female fertility. The five primates depicted are a prosimian, the ringtailed lemur; an Old World monkey, the macaque; two apes, the gibbon and chimpanzee; and humans. This primate trend to increase rearing time allows the developing young the opportunity to increase their behavioral repertoire through a long period of learning.

Color the gestation periods. Color the proportion of the life cycle taken up by each stage for all the species before moving on to the next stage. Note on the scale to the left that gestation is measured in weeks, whereas all the other periods are measured in years.

The lemur has a *gestation* of 18 weeks, which, among mammals, is relatively long, given its small size. (A cat, similar in body weight, has a *gestation* of 6 weeks.) Human *gestation* is 40 weeks. The increased *gestation* time is necessary for the development of the complex brain and nervous system. Even with a longer *gestation,* primates are born less mature than most other mammals.

Color the infancy period for all species.

During *infancy* the mother nurses, carries, and protects her offspring, providing security and warmth as well as being the focus of learning. The close physical ties between mother and infant are necessary for infant survival. Human infants, unlike all other primates, cannot cling to their mothers. The babies have no grasping foot and mothers do not have body hair to cling to. Consequently, human mothers carry their infants in some type of sling or container, which varies from culture to culture.

Color the juvenile/subadult period, which has increased even more. Color the sexual maturity period.

This period is primarily one of socialization; that is, of learning to behave as adults do in social interactions. The *juvenile* is weaned from the mother but often maintains close ties with her and her other offspring. Youngsters learn what foods are safe to eat by observing their mother. In peer play groups they learn the physical and social skills they will need as adults. In *juvenile* play groups, one can observe object manipulation, play fighting, and gestures and actions that mimic adult communication. During this time, *juveniles* learn the behavioral characteristics of particular group members and learn to adjust their social responses accordingly. Human children must learn to walk bipedally and use language. The long *juvenile/subadult* stage in humans increases the responsibilities of adults who care for them, particularly the mother's. The long period of dependency in all primate groups is important in the formations of social bonds (Plate 60).

The length of the adult life varies primarily with body size. The figures given are estimates, as there are few data from animals living in a natural habitat. Animals living in captivity live longer because they are not subjected to predation, food shortages, and disease. The very long human life span is relatively recent in history because of advances in culture, mainly modern medicine, rather than genetic change.

The final segment to be colored in each bar graph represents the period of adult life in which females are able to reproduce.

Unlike all other primate species, human females have a life span that extends far beyond their childbearing years. When a woman stops her monthly ovulation cycle and can no longer become pregnant, she is said to go through menopause. Menopause usually occurs between ages 40 and 50. In modern society, the ''postmenopausal'' years are used by many women to pursue nonreproductive roles. In nonhuman primates, and presumably for much of hominid prehistory, females were capable of bearing offspring until their death. In zoo primates, a few cases of ''menopause'' have been reported, probably because their life span has been extended beyond its usual limit through the protection of captivity.

The increasing length of developmental schedules has important implications for mechanisms by which the anthropoid (higher primate) adaptive pattern is maintained, that is, by learning. As generation time increases, the possibility of an older generation accumulating ''knowledge'' and ''wisdom'' increases, and with human language, it becomes possible to pass more of this knowledge and wisdom on to subsequent generations.

LIFE CYCLES OF PRIMATES ★

GESTATION_A
INFANCY_B
JUVENILE/SUBADULT_C
ADULT ★
    SEXUAL MATURITY_D
    FEMALE REPRODUCTIVE PERIOD_E

HUMAN

GIBBON   CHIMPANZEE

MACAQUE

LEMUR

YEARS
BIRTH
WEEKS — 18
      — 40

62

FEMALE REPRODUCTIVE CYCLES: RHESUS MONKEY

Most female mammals ovulate once or twice per year during a breeding season which may be the only time when males and females of the species come together. When researchers first began observing primates in the wild, they were surprised to see sexual activity all year around. We now know that female higher primates differ from other mammals in ovulating monthly. In this plate the rhesus monkey *(Macaca mulatta)* illustrates female primates' monthly, yearly, and lifetime reproductive cycles.

Color the top section on the monthly cycle.

The ovulatory cycle is under the control of *estrogen* and *progesterone,* which in turn are regulated by hormones released from the pituitary gland in the brain. During each cycle, one ovarian follicle matures and releases *estrogen* into the bloodstream. Around day 14, *ovulation* occurs: the follicle releases the ovum into the oviducts, where it may be fertilized. The ruptured follicle develops into a secretory organ releasing *progesterone.* This hormone stimulates the uterus to build up a vascular, nutritive lining in preparation for implantation. If the ripe ovum is not fertilized, *progesterone* release declines and the uterine lining sloughs off in a menstrual flow by day 28.

In many primates, *ovulation* is accompanied by *estrus,* marked physical and behavioral changes in the female. In the rhesus monkey, specialized *sexual skin* on the rump, base of the tail, thighs, and face enlarges and reddens. Females in *estrus* increase their grooming and foraging activities, become more aggressive, and actively solicit sexual activity from males. Chemical changes in their vaginas yield olfactory cues that inform males of their willingness to mate. *Estrus* lasts approximately seven days in rhesus monkeys and ensures that *mating* occurs when a female is most fertile. Estrous cycles continue until pregnancy and do not resume until the infant is weaned.

Next color the middle diagram showing the annual cycle: a mating season and a birth season. Use light colors.

Many species of primates may be seasonal in their reproductive behavior, with a "birth peak" in one season, which corresponds to seasonality of food, moisture, and temperature. Among Japanese macaques, for instance, the young are born in the spring with warmer temperatures and available food so that by winter they are larger and more likely to survive the cold. In the macaques at the field station at Cayo Santiago, on which this diagram is based, there is a *mating season* during half the year and a *birth season* during the other half. Notice that 75 percent of the births occur in March and April.

Complete the plate by coloring the lifetime cycle.

A significant proportion of a female's reproductive life is spent *pregnant* or *lactating.* Notice that only one-fifth of an adult female's life is spent sexually active. This information was important in disproving a hypothesis of Solly Zuckerman, who, in the 1930s, proposed that sexual activity was the "social glue" keeping males and females together.

There is a great deal of variability in monkey and ape sexual behavior. Some species, like the gibbon, mate infrequently, even though they live in pairs. Others, like pygmy chimpanzees, which live in groups, engage in sexual activity often and not always during *estrus.*

Many theories of human evolution have stressed the "uniqueness" of human females in having lost the behavioral changes associated with *estrus,* resulting in continual sexual receptivity. This is misleading, for recent studies confirm that human females are cyclic in their initiation of sexual activity on a time schedule that correlates with *ovulation.* However, in many cultures, the attitudes and practices that surround sexual behavior mask these hormonal effects. Human females are not as unique in this regard as was once thought. Thus, any theory about the origin of human mating systems based solely upon sexual activity—or upon any single feature—must be seriously questioned.

REPRODUCTIVE CYCLES: RHESUS MONKEY ★

MONTHLY CYCLE (28 DAYS) ●
ESTROGEN$_A$
PROGESTERONE$_B$
OVULATION$_C$
MENSES$_D$
   SEXUAL SKIN$_{D'}$
ESTRUS$_E$
   SEXUAL SKIN$_{E'}$

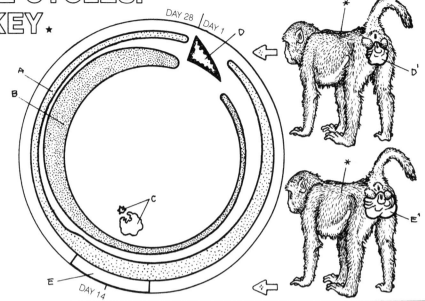

ANNUAL CYCLE (12 MONTHS) ●
MATING SEASON$_F$
BIRTH SEASON$_G$

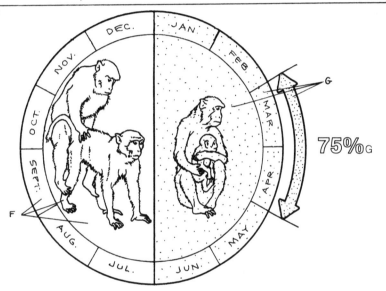

75%$_G$

LIFETIME CYCLE (25 YEARS) ●
INFANT/JUVENILE$_H$
CYCLING ADULT$_I$
PREGNANT$_J$
LACTATING$_K$

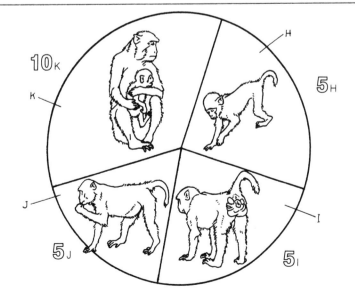

63

SEXUAL DIMORPHISM: TELLING FEMALES FROM MALES

Aside from reproductive anatomy, female and male primates may differ in *body weight* and proportions, muscular development, *cranial capacity, canine tooth* size, tail length, and coat color and markings. Recently, Walter Leutenegger has shown that in many nonhuman primate species as well as in humans, the female pelvis and birth canal are larger than that of males.

Although it is relatively easy to tell females from males when both are alive, it is often difficult in an extinct species that has left behind only fragmentary bones and *teeth* or skulls. My own research has revealed a distinct dimorphic pattern in each species of living ape and in humans, as a mosaic of features that includes *body weight* and size, head proportions and *cranial capacity, canine tooth* size, and pelvic dimensions. It may be possible to learn something of the social behavior of extinct species from observing anatomical and behavioral correlations in living species.

Here, the pattern of sexual dimorphism in *gibbons, chimpanzees,* and *gorillas,* with regard to *body weight, cranial capacity,* and *canine tooth* size is compared.

Begin by coloring the gibbons and their characteristics at the top of the plate, represented by averages for each sex. Body weight is given in kilograms, and cranial capacity in cubic centimeters.

Male and female *gibbons (Hylobates lar)* are similar in *body weight, cranial capacity,* and *canine tooth* size, which is large in both sexes. *Gibbons* live in pairs in tropical forests and defend a small territory. They are least social of the three species; they do not tolerate other adults, especially of the same sex, and the offspring are expelled from the parental territory as soon as they mature. The large *canines* of the adult female —unusual for primate females—keep other adult females out of the territory and enable her to compete with her mate for food. In the male, large *canines* serve in defending the territory from males in neighboring groups.

It is often claimed that our ancestors' social life was comparable to that of *gibbons* because *gibbons* live in "family" groups. But we can see how misleading the term "family" can be when applied to *gibbons.*

Humans do not live in isolated families, defending their territory against other families and expelling their young forever. Rather, there are extensive social networks between human families, and social ties among parents and children endure through life.

Color the chimpanzees and their characteristics.

Chimpanzees (Pan troglodytes) have moderate sexual dimorphism. Females have 80 percent of male *body weight,* slightly smaller *cranial capacity,* and moderately smaller *canines. Chimpanzees* have the most flexible social group, which varies in size and composition from day to day. There may be mixed groups of males and females, all male or all female groups, or solitary animals. Male *chimpanzees* seem to protect their community's boundaries against neighboring groups. The large *canines* in males have been attributed to threat displays between males competing for females. But both sexes are involved equally in threatening or attacking predators.

Color the gorillas and their characteristics.

Gorillas (Pan gorilla) are extremely dimorphic in all three features—males are much larger than females in *body size, cranial capacity,* and *canine tooth* size. They live in African rainforests and are vegetarians with enormous appetites. Eating occupies six to eight hours a day. Despite their fearsome appearance, aggressive behavior is rare. A group has about seventeen members, with one dominant silver-back (gray hair on rump and back) male, which is the oldest male in the group. Males display by beating their chests to intimidate intruders.

How does the human pattern of sexual dimorphism compare with these three? In *body weight,* and *cranial capacity* men and women differ about as much as *chimpanzees;* but in *canine tooth* size, there is no difference at all, which distinguishes the human pattern from that of the apes. In humans the *canines* are equally small, whereas in *gibbons* they are equally large. Human *canines* are so small that they often look like incisors (Plate 15).

The reduction in *canine* size apparently took place early in human evolution and implies greater sociability between males and females, but in particular, less competition and aggression between males.

TELLING FEMALES FROM MALES. ★

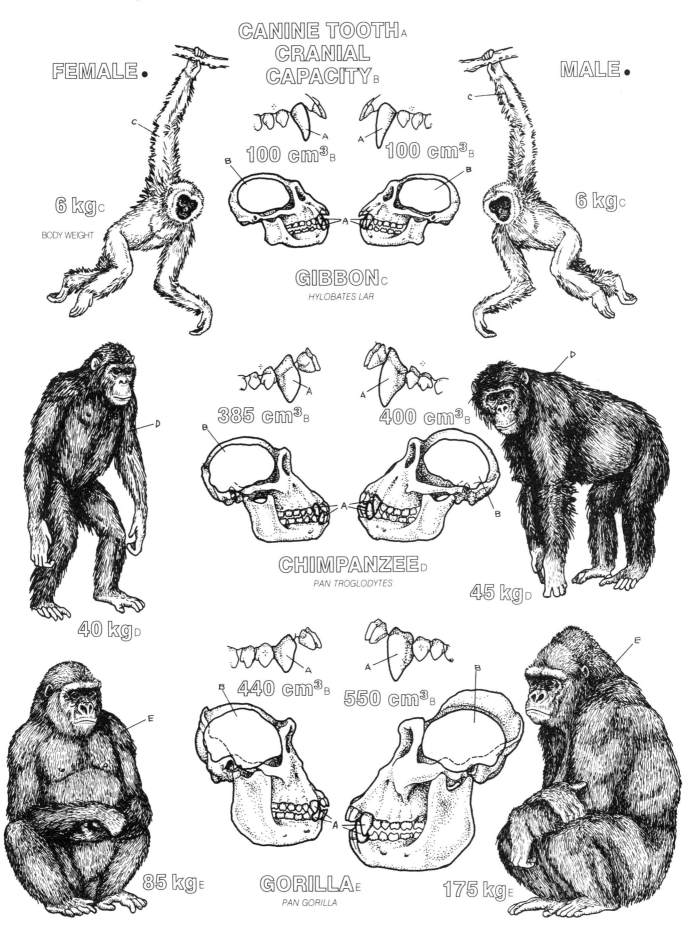

CANINE TOOTH A
CRANIAL
CAPACITY B

FEMALE. ●

MALE. ●

100 cm³ B   100 cm³ B

6 kg C   6 kg C

BODY WEIGHT

GIBBON C
HYLOBATES LAR

385 cm³ B   400 cm³ B

CHIMPANZEE D
PAN TROGLODYTES

45 kg D

40 kg D

440 cm³ B   550 cm³ B

85 kg E   GORILLA E   175 kg E
PAN GORILLA

64

OLFACTORY AND TACTILE COMMUNICATION

In order to coexist in a social group, primates have evolved complex signaling systems for the exchange of information. Like other mammals, they utilize all sensory channels to communicate information about their location, reproductive status, and level of arousal. The relative role of each sense as an avenue of expression varies between primate species. Prosimians and some monkeys have retained the primitive, nocturnal mammalian reliance on smell; the higher nonhuman primates utilize touch, visual, vocal, and other auditory displays to communicate their understanding of the prevailing social order. To some extent they can convey their intentions and even some factual information about the environment. Human language is not the only difference between human and nonhuman communication. Our species relies more heavily than any other primate on facial expression and gestures to relay subtle meanings.

This and the next two plates illustrate the major modalities of primate communication. As you color, remember that primates are always using several modalities—a visual flash of the eyes, and a vocal bark, for example—simultaneously. Furthermore, the message conveyed by primate signals must be interpreted in light of the social context in which it is conveyed.

Color the illustrations of olfactory communication.

Smell remains important for some New World monkeys and all the prosimians, especially those that are nocturnal. These animals have specialized *scent glands* or enlarged sebaceous glands on their chest, forearms, and anal-genital areas, as shown here on the ringtailed lemur. These glands emit odorous secretions when rubbed on branches and tree trunks. Males of territorial species *scentmark* to map out their exclusive area of the forest as a signal to others to keep out. Females emit a specialized *scent* indicating their willingness to mate. Many prosimians and monkeys regularly urinate over their hands and feet and use chemical markers around their range. In *sniffing* the sites where others have gone before and *scentmarked,* prosimians can gather information as to the whereabouts of group members and neighbors and about sexual identity and receptivity. Chemical signals may possibly be so subtle as to identify the

"scentmarker" as a specific individual whom the *"sniffer"* knows well.

Old World monkeys and the apes do not have *scent glands;* however, females emit a pheromone (volatile organic substance) from their vaginas when they are ovulating. These chemical signals are more essential than brightly colored sexual swellings for communicating to a male that a female is sexually receptive.

Human females may also emit pheromones which may be responsible for the synchrony of menstrual schedules often experienced by women living in the same social group.

Color the illustrations of tactile communication.

Monkeys, apes, and humans—with their high degree of manual dexterity—use their hands in a great deal of bodily contact *(holding, touching* and *grooming)* that expresses social bonds. Newborn primates instinctively desire the tactile sensations of the mother's body. Her *touches* are among the very first messages a baby receives concerning the social world she or he has just entered. Primatologist Harry Harlow has shown that if deprived of their mothers, baby monkeys prefer a terrycloth surrogate mother to a wire mother equipped to provide only milk. An infant will cling for hours to the terrycloth dummy, desiring the softness, comfort, and a bit of warmth it brings and will spend time on the wire mother only long enough to fill its stomach.

A *touch* may relay reassurance, calm, reconciliation, or express the relative social rank of individuals, depending on the context in which it is made. Chimpanzees will wrap their arms around one another, pat one another on the back, and embrace when they meet a kin or companion in order to communicate their friendly intentions. Here an adult male chimpanzee reaches out to assure a youngster who had been playing nearby.

Grooming is particularly expressive of the prevailing social order. Young infants, mothers, and high-ranking males are the recipients of *grooming* more often than others. Companions *groom* one another; males and females *groom* and otherwise *touch* one another before mating. Here a macaque mother is *groomed* by her eldest daughter in an expression of the affectional tie that lasts throughout their lifetimes.

SMELLING AND TOUCHING.★

OLFACTORY COMMUNICATION.
    SCENT GLANDS_A_
      MARKING_B_
    SNIFFING_C_

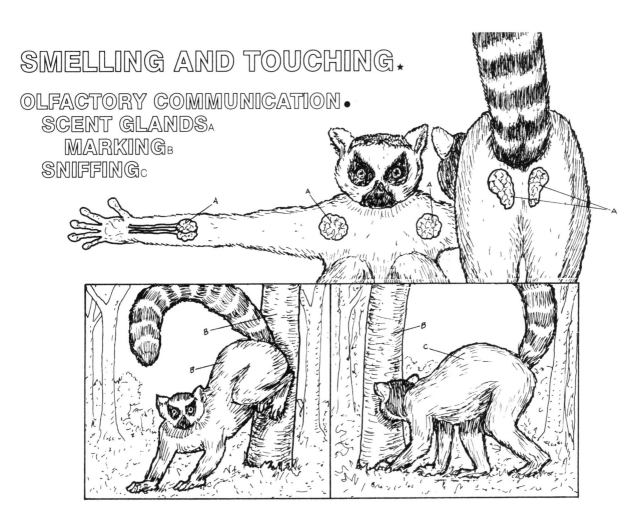

TACTILE COMMUNICATION.
    HOLDING_D_
    TOUCHING_E_
    GROOMING_F_

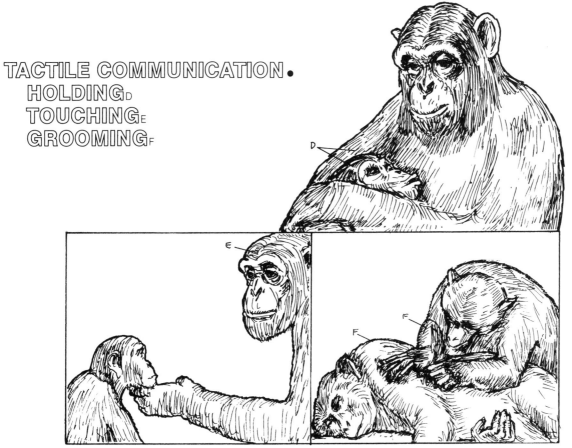

65

VISUAL COMMUNICATION: FACIAL EXPRESSIONS AND BODY POSTURES

Evolution of the mammalian facial musculature parallels increasing reliance on facial expressions and visual communication. This trend began when muscles from the neck of ancestral reptiles became mammalian face and scalp muscles. The new musculature on the front of the face was anchored to the skin and enabled mammals to use their faces in social communication. On the scalp, the newly evolved mammalian outer ears became mobile for hearing at night. The external ear muscles are reduced in higher primates because they rely less on hearing and more on vision for monitoring the environment. Their facial muscles, especially those around the eyes and *mouth,* are even more elaborated than in other mammals because they use facial expressions to convey complex and subtle meanings in everyday social interactions.

Color the muscles of the chimpanzee's face grey and then use contrasting colors for each of the components of the facial expressions.

Notice how different parts of the face can change position independently to create a large repertoire of subtle expressions. A primate's *mouth* and eyes are the most important components of a facial expression. The *eyebrows* can move up or down, to the middle, or outward. A direct, wide-eyed stare is a *threat* gesture for most primates; averting the eyes downward conveys *submission.* The highly mobile *lips* can be protruded forward, pushed tightly together, or retracted over the *teeth* while the jaws are open or closed.

The chimpanzee "hoot face" expresses excitement and affection, as when two individuals reunite after foraging separately for most of the day. The "play face" is observed most often among juveniles engaged in rough and tumble play and in infants when they are tickled. The "glare" is easily recognized because we use it when we are angry; chimpanzees do, too. The "silent-bared teeth" expresses *submission,* as when a younger chimpanzee wishes to express to one of the group elders that it intends no antagonism and does not wish to challenge the social authority of the older animal.

Similarities in the facial expressions of chimpanzees and humans are due to our nearly identical facial musculature. Some facial expressions even serve common functions for the two species. For example, the "play face" is thought to have evolved into human laughter. Our association of the smile with humor and enjoyment is not universal among humans, for in many cultures it is a sign of apprehension and discomfort. Many researchers think that the "silent-bared teeth"

expression—indicating *submission* in chimpanzees—is homologous to the human smile.

Facial expressions communicate an individual's internal emotional and motivational state to other members of the group. In specific social situations, these expressions influence the behavior of others.

Color the left square in the middle of the plate, illustrating a male baboon, his mouth wide open, exposing his long sharp canines.

This canine display functions to frighten off potential intruders or in other circumstances to notify group members that he can defend his social position if necessary. It signals other baboons to stay out of his way, to move off, or to give a *submissive* response meaning, "No harm meant, I'm harmless." Notice that the baboon's *eyelids* are lowered in its "*threat* face," but the chimpanzee's *threatening* expression is a wide-eyed stare.

Color the square illustrating a macaque giving a submissive gesture.

The characteristic crouching posture is supplemented by an averted gaze; *lips* are drawn back, as in the *submissive* face of the chimpanzee. The macaque is probably making soft "hoo" sounds to emphasize its feelings and intentions.

Color the squares on the bottom of the plate.

Gestures involving all or part of the body communicate messages to other animals. Here a chimpanzee is *begging* a morsel of food. She reaches out an arm and holds a hand palm upward under the possessor's chin.

As in the gibbon territorial rituals, other primates combine communicative body postures with noisy displays. Here a macaque is jumping up and down while *shaking* a tree limb. In such a display, the macaque jumps from branch to branch, *shaking* each limb a few times. Facial expressions are those of a *threat.* This display is often used in response to the sound of an airplane passing overhead, or to a strange human intruder below!

The gorilla *chest beating* display is especially loud and booming in adult males, who have inflatable air sacs in their chests. The display is part of a highly ritualized response to intruders, and can be heard from over a mile away. *Chest beating* is thought to release tension when a gorilla is unsure whether it should act aggressively or quickly round up its kin group and leave.

FACIAL EXPRESSIONS AND BODY POSTURES. ★

COMPONENTS.
FOREHEAD A
EYEBROW B
EYELID C
NOSE D
CHEEK E
LIPS/MOUTH F
TEETH G

"HOOT-FACE"

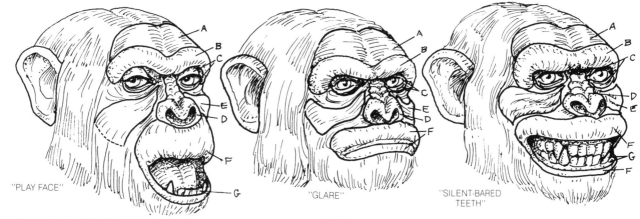

"PLAY FACE" "GLARE" "SILENT-BARED TEETH"

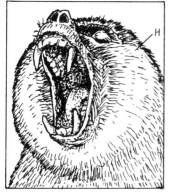

EXPRESSION.
THREAT H
SUBMISSION I

GESTURES.
BEGGING J
TREE SHAKING K
CHEST BEATING L

66

COMMUNICATION: PRIMATE VOCALIZATIONS AND THE BRAIN

Dependent as we are on language for thinking and communicating, a persistent problem in studying primate vocalizations is knowing what to look, and listen, for. We have learned that ear structure in various primates is nearly identical (Plate 46), but their brains may well be specialized for perceiving vocalizations with species-specific meaning, as ours are.

Begin by coloring the neocortex of the human and the limbic system of the monkey at the top of the plate.

Research in the neurosciences has established that human language and nonhuman primate vocalizations have different "control centers" in the brain. The limbic system is a group of forebrain structures comprising almost all parts of the forebrain except the *neocortex* (Plate 13). It is sometimes known as the "smell brain" because it includes the olfactory bulb and tract, a major brain system in primitive mammals. The angular *cingulate cortex* and the *orbitofrontal cortex* are parts of the limbic system which control the calls of nonhuman primates. Electrical stimulation of these parts of a squirrel monkey's limbic system elicits almost its entire range of vocalizations.

In humans, stimulation of many areas of the *neocortex,* the temporal association areas, will elicit words, phrases, and sentences, but stimulation of these areas in monkeys does not produce vocalizations. Conversely, limbic stimulation in humans may evoke emotionally charged sounds such as laughter, shrieks, or curses.

Pathological lesions in human brains and experimental ones in nonhuman primates further confirm that human language is largely controlled by the *neocortex,* whereas other primate vocalizations are controlled by the limbic system. Some parts of the limbic system regulate bodily processes, such as hormonal levels, and heart and respiration rate, in humans as well as other mammals; and it has been called the "visceral" or "emotional" brain because it seems to have a role in attention, motivation, and arousal. An intact limbic system is necessary to animals for various social behaviors—friendship, aggression, and submission and sucking, clinging, and affection seeking in infants—activities essential for surviving the first few weeks of life. Limbic stimulation may produce rage, alarm, fear, ecstasy, or surprise, and damage to limbic structures may produce deficits in facial expressions, sexual behavior, territorial defense, maternal protectiveness, or social communication.

These observations are consistent with primatologists' observations that nonhuman primate vocalizations invariably reflect emotional states such as pain, affection, and fear. So primate communication is said to be "self-oriented" and emotional, in contrast to human language, which is said to be "object-oriented," and symbolic. We can talk about danger without seeking cover, about affection when we are not embracing; but the detachment of language from emotional state is not complete. It is difficult, for example, to talk about our favorite foods without salivating! Even with nonhuman primates, the distinction between "emotional" and "referential" may not be clearcut. Primatologist Thomas Struhsaker discovered that vervet monkeys have three discrete warning calls for three different predators.

Finish coloring the plate.

Among vervets, short tonal *chirps* mean *"leopard,"* high-pitched *chutter* means *"python"*; and low-pitched, staccato grunts—a *r-raup* sound—warn *"eagle."* The vervet troop responds differently to the three signals: appropriately rushing up into the trees and out on the small branches to avoid the *leopard,* looking down and around to locate the *python,* and diving to hide near the center of the tree or into underbrush when the presence of an *eagle* is announced.

Struhsaker's colleagues, R. Seyfarth, D. Cheney, and P. Marler, in more recent studies, report on the development of these calls by young vervets. Infants at first give the calls indiscriminately to a wide range of objects, and the adults pay little attention to them. Gradually, infants learn to vocalize more accurately and thus elicit the appropriate reaction from other group members.

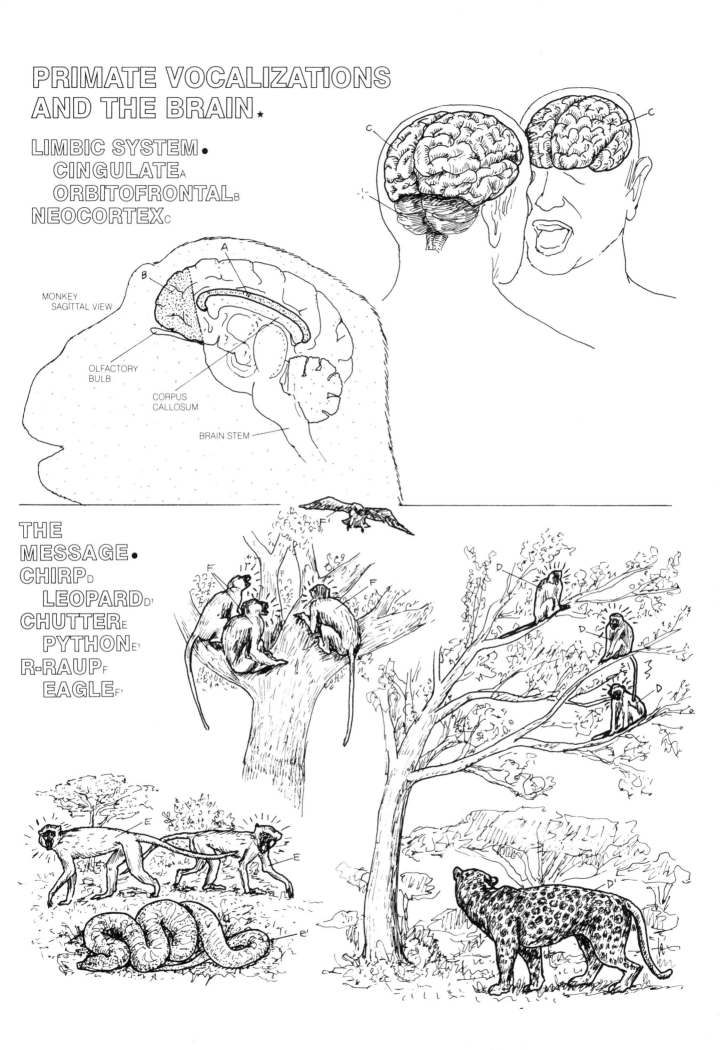

PRIMATE VOCALIZATIONS AND THE BRAIN ★

LIMBIC SYSTEM ●
CINGULATE A
ORBITOFRONTAL B
NEOCORTEX C

MONKEY
SAGITTAL VIEW

OLFACTORY
BULB

CORPUS
CALLOSUM

BRAIN STEM

THE
MESSAGE ●
CHIRP D
LEOPARD D'
CHUTTER E
PYTHON E'
R-RAUP F
EAGLE F'

67

EARLY PRIMATE EVOLUTION: PALEOCENE AND EOCENE PRIMATES

The oldest fossils identified as "primate" are dated to the beginning of the Cenozoic. New primate groups appear in each of the epochs of the Cenozoic, starting 67 mya (see Plate 4).

Color gray the time ranges of the Paleocene and Eocene epochs on the two scales at the top of the page.

In this plate we begin the discussion of primate evolution by comparing a Paleocene fossil primate, *Plesiadapis,* with an Eocene fossil primate, *Notharctus.* The differences between them reflect the trends that distinguish living primates from other mammalian orders.

Color the structures associated with vision: eye orbit orientation (noting the difference) and the postorbital bar, found only in *Notharctus.*

The *postorbital bar* first appears in the Eocene primates; it is related to increased reliance on vision and to the rotation of *eye orbits* toward the front for improved stereoscopic vision (Plate 44).

Color the muzzle up to the dotted line in the side views, and note that it is relatively shorter in the Eocene than Paleocene primates. Also color the diastema and the incisors.

The less-developed *muzzle* in the Eocene *Notharctus* reflects the primate trend toward decrease in olfactory acuity and increase in grinding efficiency and regular hand-to-mouth feeding.

The *diastema* or gap between the front teeth and the back teeth in *Plesiadapis* and the large protruding *incisors,* are reminiscent of modern rodents. Eocene primates are more recognizable as prosimians.

Color the claws and nails.

Plesiadapis had clawed hands and feet, as shown by the tips of fossil finger bones. There is evidence in Eocene *Notharctus* for grasping hands and feet equipped with *nails.* In locomotion, *Plesiadapis,* with short limbs and *claws,* resembled a squirrel more than a primate. *Notharctus* may have been similar to modern lemurs, but more heavily built. Some Eocene primates, like *Necrolemur,* show the beginning of a leaping adaptation.

Complete the plate by coloring the brain, which is relatively larger in size in the Eocene than in the Paleocene primates.

Difference in *brain* size is evident both from casts of the inside of the braincase (endocasts) and also from the relative proportion of the skull taken up by the facial skeleton versus the braincase. The olfactory lobes are not much reduced in *Notharctus,* but note that the visual area is relatively larger compared with *Plesiadapis.*

What do these differences between the Paleocene and Eocene fossil primates mean? The primate way of life has been defined traditionally as "climbing by grasping in an arboreal habitat." The dentition, *nails* on grasping hands, depth perception, and associated changes in the *brain* are adaptations to a locomotor-feeding mechanism that allowed primates to exploit a new adaptive zone: the seasonal flowering, fruiting trees that had begun replacing the coniferous forests in the late Cretaceous. In other words, primates shifted away from purely insect eating and developed omnivorous dietary habits.

The rise of the angiosperms resulted in an explosion of insects, which not only exploited this new source but helped in spreading its pollen. Anatomist Matt Cartmill maintains that tree-living per se does not account for larger *brains,* overlapping fields of vision, or the replacement of *claws* by *nails.* He suggests that the earliest primates may have been "visual predators" of insects. Increased intelligence and better vision are common characteristics of predators. These early primates may have lived not in the trees but in the low, bushy understory of the developing deciduous forests. In the next stage, they moved up into the trees.

In the above scheme, all the Paleocene fossils are excluded from the primate order. However, nature does not work in ways that permit easy classification. Rather, we might look at the Paleocene primates as experiments in "becoming a primate." These forms indicate a range of specializations from insectivory to omnivory—supporting the idea of numerous experiments. In the Eocene fossils, the majority show reliance on vegetation of some kind. The Paleocene primates then may be viewed as "becoming," and the Eocene primates as "arrivals."

PALEOCENE AND EOCENE PRIMATES ★

POSTORBITAL BAR(A)
EYE ORBIT ORIENTATION(B)
MUZZLE(C)
DIASTEMA(D)
INCISORS(E)
CLAWS(F)/NAILS(F1)
BRAIN(G)

F   PLESIADAPIS ●   NOTHARCTUS ●   F1   F'

F

B

A   OLFACTORY LOBE
A

G   G

VISUAL AREA

C   G   A   G

D
E
E   C
C   E
D   C

MODERN RODENT ★   MODERN LEMUR ★

68

PRIMATE EVOLUTION AND CONTINENTAL DRIFT: THE CHANGING GLOBE

The remarkable jigsaw puzzle fit of some continental margins has raised the question as to whether the continents might have once formed a single giant land mass. But as no mechanism was known for moving such huge and seemingly solid masses, speculations about continental drift were mostly ridiculed or dismissed.

We now know that the perception of the continents as heavy, immovable masses is erroneous; in fact, the continents of the earth's crust "float" on the underlying mantle. The continents have greatly altered their position, pattern, size, and number over millions of years. This fragmentation of the land masses has affected the world's environments, and thus its life forms, through changing ocean currents, rainfall patterns, and ocean temperatures.

Plate tectonics, the study of movements of the earth's crust, helps to explain mountain formation, the distribution of volcanoes, and earthquakes.

Begin on the lower left by coloring the continental positions as they are found in the world in the present day. Continue to use these colors for the rest of this plate and the next two plates. Next color the continents as they were in the Triassic, about 200 mya, when they were joined in the supercontinent Pangea ("all lands") surrounded by a single ocean, the Tethys.

Note that *India* was close to *Africa,* with *Madagascar* wedged between them. Plants and animals on this conglomerate territory constituted one continuous living system that could traverse the land masses without confronting oceans.

Color the land masses that had already begun to break up by the early Cretaceous, about 135 mya.

Two extensive rifts, which became the Atlantic and Indian Oceans, initiated the separation into two major land masses, *Laurasia* in the north and *Gondwanaland* (named for an Indian province) in the south, each moving away from the equator and the northernmost and southernmost parts becoming widely separated. *India* "floated" north, eventually collided with mainland *Asia,* and created the Himalayan mountains. *Madagascar* was still attached to eastern *Africa,* and *South America* was attached to western *Africa,* close enough for continuing exchange and mutual colonization of plants and animals.

The distribution of numerous living and fossil species that previously baffled zoologists, botanists, and paleontologists are now explained. Dinosaurs, for example, which were at their zenith in the Cretaceous, are found as fossils on all major land masses. And it used to be difficult to explain how the obviously related flightless birds—the ratites, which include ostriches in *Africa,* rheas in *South America,* emus and cassowaries in *Australia* and New Guinea, and kiwis and moas in New Zealand—managed to cross such enormous bodies of water to reach their present-day distribution. Now we know that they all evolved on the southern continent *Gondwanaland,* which, as it broke up, separated the different enclaves of ratites.

The impact of the changing globe on biological evolution may be illustrated by contrasting reptilian and mammalian evolution. The age of reptiles, lasting 200 million years, gave rise to 20 orders. The age of mammals, by contrast, lasting 65 million years (the Cenozoic), gave rise to 35 orders—more than one-half again more orders in one third of the time. Finnish paleontologist Björn Kurtén explains this seeming dilemma: the reptiles were evolving at a time when the land masses were less fragmented and climates, less diverse than they were 65 mya when the mammals, including the primates, began their radiation. The greater number of land masses and greater diversity in environments that developed with the breakup of Laurasia and Gondwana created new opportunities for independent evolution or convergence (for example, marsupial versus placental evolution—Plate 17).

In the present-day configuration of continents, the Mediterranean Sea, separating *Africa* and *Eurasia,* is a remnant of the once more widespread Tethys Sea. *South America* and *Africa* are widely separated by a huge fracture, the Mid-Atlantic ridge. *North America* has reconnected to *South America* through the Isthmus of Panama only within the last 3.5 million years.

Primate evolution must be interpreted within these constraints, and this will be done as we explore the primate fossil evidence.

THE CHANGING GLOBE.★

EURASIA A
NORTH AMERICA B
GREENLAND B'
SOUTH AMERICA C
AFRICA D
MADAGASCAR E
INDIA F
AUSTRALIA G
ANTARCTICA H

200 MYA TRIASSIC.★

PANGEA.

EQUATOR

TETHYS SEA

135 MYA EARLY CRETACEOUS.★

LAURASIA.

TETHYS SEA

GONDWANA.

PRESENT DAY★

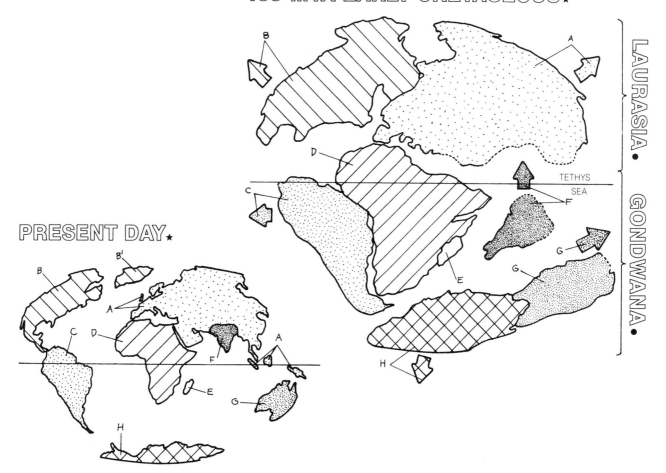

69

PRIMATE EVOLUTION: PALEOCENE FOSSILS

The initial primate radiation was almost exclusively a Paleocene phenomenon. The known fossil record is only North American and Eurasian: no primate fossils from this time period have been found in the southern hemisphere.

Color in gray the time range of the Paleocene. Color the continents using the same colors as in the previous plate.

Note that *North America* and *Eurasia* were still essentially a single continent during the Paleocene, separated from the "island" continents of the southern hemisphere. This may explain the "holarctic" distribution (entire northern hemisphere) of the Paleocene primates. During the Paleocene, the northern continents had a warmer and more tropical climate than now. The rise of the angiosperms occurred at the end of the Mesozoic era, providing new food sources for birds and mammals. Angiosperms have many edible parts — fruit, seeds, flowers, nectar, shoots, and leaves — and primates were quick to discover and exploit them.

The Paleocene primates are called Plesiadapiformes ("nearly like adapids," a widespread Eocene group) and consist of five families: Paromomyidae, Picrodontidae, Carpolestidae, Plesiadapidae, and Saxonellidae.

These Paleocene fossils are so primitive that they lack many of the features used to distinguish present-day primates from other mammals; therefore, some paleoprimatologists do not consider them primates at all (Plate 67). Even if the strange Paleocene "proto"-primates were not directly ancestral to the "true" primates first appearing in the Eocene, both groups had a common origin from a still earlier, more primitive, type of mammal.

Color the fossils and associated arrows from left to right, using light colors so as not to obscure details. The portion with shading indicates the actual fossil discovered; the remainder is reconstructed with the aid of other fossil material.

The genus *Purgatorius* (family Paromomyidae), named for Purgatory Hill, is the oldest fossil primate yet discovered from what is now the Rocky Mountain region in Montana. The climate was subtropical, much warmer than now, and the Rockies had not reached their present heights. This fossil is known only from a few teeth and the mandibular fragment shown here. How can we know that *Purgatorius* was "on the way" to being a primate? In animals evolving away from a diet of insects with hard exoskeletons to one of fruit or leaves, the molars tend to lose spiky cusps and become broader to provide more grinding surface (Plate 58). *Purgatorius'* molars show some of these changes.

Palaechthon, a member of the same family as *Purgatorius,* was also found in the Rocky Mountain region. A bit younger, 60 mya as compared to 65 mya, its remains include the earliest known fossil primate skull. The face, like that of other early mammals, is long and lies directly in front of the small braincase. Its molar tooth cusps are even lower than *Purgatorius',* suggesting that it, too, ate fruit and other vegetation.

Picrodus from western *North America* was first thought to be a bat because of the notch on the tongue side of the the first molar tooth. Overall tooth and jaw structure indicates that *Picrodus* is a primate with a paromomyid ancestor.

The carpolestids ("fruit stealers"), here represented by the genus *Carpodaptes,* have an unusual last premolar with a high, bladelike crest similar to that of some living marsupials, which use it to slice through the tough fibrous husk of an avocadolike fruit.

Plesiadapis skulls and skeletons have been found in *North America* and western *Europe.* At least fifteen species have been found — the most diverse and best known of early primate fossils. The large, rodentlike incisors protrude in front; the lost lower canine and premolar leave a large gap (diastema); and the upper canine is reduced.

A single species, *Saxonella,* of the family Saxonellidae, from a late Paleocene site in Saxony, Germany, has incisors like the plesiadapids but a bladelike premolar like the carpolestids.

For most Paleocene primate species, we have only teeth and so can say nothing about eyes, hands, or brains. But from the teeth, it seems clear that the Paleocene primates were changing from an exclusively insect diet to one that included fruits and other vegetation.

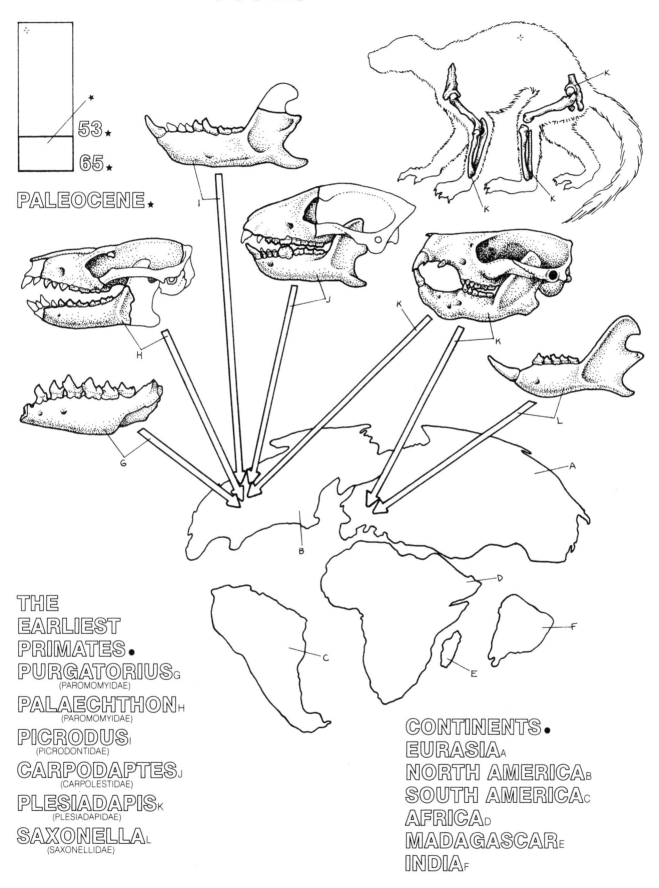

PALEOCENE FOSSILS ★

53 ★
65 ★

PALEOCENE ★

THE
EARLIEST
PRIMATES ●
PURGATORIUS G
(PAROMOMYIDAE)
PALAECHTHON H
(PAROMOMYIDAE)
PICRODUS I
(PICRODONTIDAE)
CARPODAPTES J
(CARPOLESTIDAE)
PLESIADAPIS K
(PLESIADAPIDAE)
SAXONELLA L
(SAXONELLIDAE)

CONTINENTS ●
EURASIA A
NORTH AMERICA B
SOUTH AMERICA C
AFRICA D
MADAGASCAR E
INDIA F

70

Most of the earliest primates had died out by the end of the Paleocene and beginning of the Eocene. Especially in the northern hemisphere, where most early primates have been found, a trend toward a more seasonal environment continued until the very end of the Paleocene, about 54 mya. Later, during the Eocene, the climate became even more tropically lush than it had been earlier in the Cretaceous. These changes brought about an environmental "crunch" that led to the extinction of some forms, but also to the subsequent radiation of other mammalian groups, including the highly successful rodents and the first modern-looking primates.

Color the continents as in the previous plate, and the time range 53-37 mya. The north Atlantic is widening further between North America and Eurasia, and the southern continents remain isolated. Using light colors, color the fossils and associated arrows. Leave *Amphipithecus* uncolored for now. Color the adapids (G, H, L, and M) first, then the omomyids (I, J, and K), so that the holarctic distribution of each will be apparent.

There are two families of Eocene primates, the lemurlike Adapidae and the tarsierlike Omomyidae. Each family had representatives in both *North America* and *Europe,* but *omomyids* were more numerous in *North America* and *adapids* in *Europe.*

Smilodectes and *Notharctus* represent the North American adapids, which tend to be earlier (54-47 mya) and slightly more primitive that the European ones (52-35 mya). *Smilodectes,* from the Rocky Mountain region, which at that time had a subtropical to tropical climate, is dated 48 mya and known from the complete skull shown here and other bones of the skeleton. Note its relatively short muzzle and post-orbital bar, characteristic of Eocene, in contrast to Paleocene, primates (Plate 67).

Notharctus, also from the Rockies (50-47 mya), is somewhat larger than *Smilodectes,* has a longer muzzle, and its diet was probably fruit. A complete skeleton is similar to that of living lemurs and suggests that these animals were active and agile in trees. The ends of their fingers and toes show evidence of nails rather than claws.

The Omomyidae are generally considered the ancestors not only of the modern prosimian *Tarsius,* but also of the monkeylike fossil primates of the Oligocene. *Tetonius* is one of the oldest omomyids and one of the few for which there is a skull. It is just over three centimeters, with a relatively short face and large eye orbits and brain, which differentiate the tarsierlike omomyids from the lemurlike adapids. *Tetonius* has a unique dentition for Eocene primates: large incisors and small molars, rather like those of omnivorous fruit eaters such as chimpanzees and some South American monkeys.

Omomys, also from the Rockies (from 53 to 47 mya), is one of about twelve genera in this subfamily.

Necrolemur belongs to another omomyid subgroup from western Europe (46-36 mya), is known from skulls and some limb bones, and is considered by some researchers to be the ancestor of the modern tarsier. *Necrolemur* had a fused tibia and fibula, an adaptation of the hind limbs for leaping.

Adapis of western Europe (40-35 mya) was the first fossil primate to be described, by Cuvier in 1821. Now there is a large collection of them. Like most other Eocene primates, the face is large compared to the braincase and the cheek teeth sharply crested in a trend toward what we see in living lemurs. However, as for all Eocene fossils, the front teeth lack the "dental comb/tooth scraper" of the pointed lower incisors and canines in nearly all living lemurs and lorises. In *Adapis* the lower canines are spadelike, not pointed; so they probably had a scraping function.

Pronycticebus, represented by a single specimen from the late Eocene of France (38 mya), has a relatively short face and very large eye orbits; this suggests that, like some modern prosimians with these features, it was nocturnal.

Complete the plate by coloring *Amphipithecus*. Known only by jaw remains from the late Eocene of Burma, *Amphipithecus* is considered an anthropoid primate and might be the earliest evidence of higher primates (Plate 74).

The Eocene primates are represented by two major families with numerous species, representing two major adaptive themes, each with many variations.

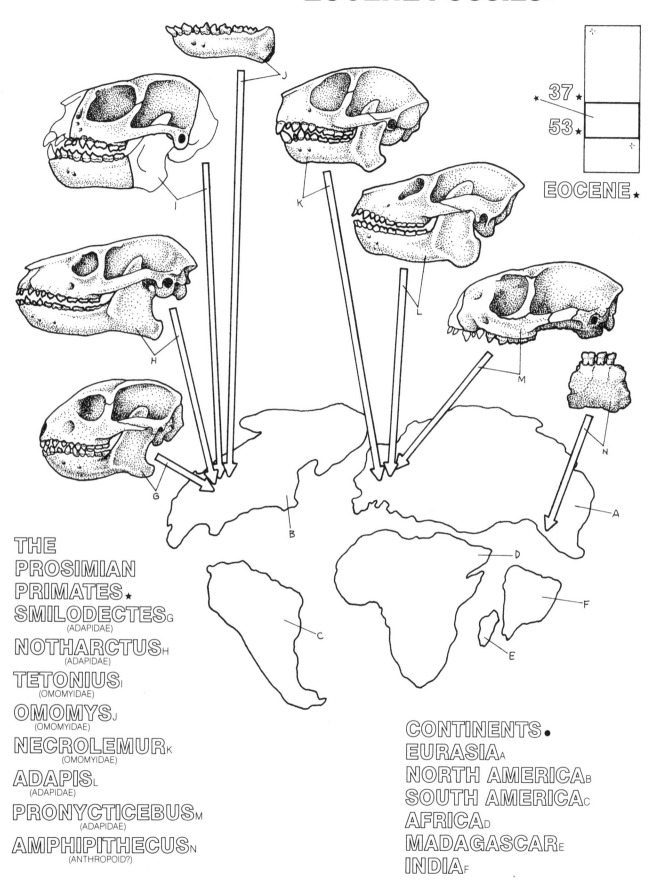

EOCENE FOSSILS.★

J

I

K

L

H

M

G

N

37★
53★

EOCENE.★

A

B

C

D

F

E

THE
PROSIMIAN
PRIMATES.★
SMILODECTES G
(ADAPIDAE)
NOTHARCTUS H
(ADAPIDAE)
TETONIUS I
(OMOMYIDAE)
OMOMYS J
(OMOMYIDAE)
NECROLEMUR K
(OMOMYIDAE)
ADAPIS L
(ADAPIDAE)
PRONYCTICEBUS M
(ADAPIDAE)
AMPHIPITHECUS N
(ANTHROPOID?)

CONTINENTS.●
EURASIA A
NORTH AMERICA B
SOUTH AMERICA C
AFRICA D
MADAGASCAR E
INDIA F

71

Prosimians *(tarsiers, lorises, and lemurs)* and *tree shrews* are diverse groups in which it is difficult to establish evolutionary relationships from anatomical or behavioral features alone. The family tree shown here is based on molecular evidence and, for the most part, agrees with inferences drawn from comparative anatomy and behavior.

Most prosimians are small bodied (mouse to cat-sized), nocturnal, forage alone, and rely primarily on insects for food. They use smell for location of food and other species members and generally do not form stable mixed social groups. Home ranges of adult females usually overlap, but those of males do not.

The family tree shows relationships of living primates and their approximate times of separation, based on molecular data (Plate 25). Begin by coloring the tree shrew on the left (its location is not indicated on the map).

Sixteen species of *tree shrews* are found in South and Southeast Asia and weigh from 40 to over 300 grams. Most species are diurnal and forage both in trees and among bushes. They are usually loners; even the young go out on their own at an early age.

The evolutionary status of *tree shrews* is disputed, with some specialists claiming they are not even primates. Arguments for their primate status include the muscular system, placenta, visual system, and above all, molecular similarities.

Now color the tarsier, whose primate status also is contested but at a different level — whether they are closer to the prosimians or to the anthropoids (monkeys and apes). Color the tarsiers' location on the map.

Tarsiers are the only prosimians possessing a fovea, the part of the retina found among anthropoids that confers visual acuity. Three *tarsier* species are found on the islands of Southeast Asia. *Tarsiers* are small (150 grams) and feed nocturnally on insects. Little is known of their social behavior, but they seem to live in pairs of adult male and female, have a single infant, scent mark, defend a territory, and communicate visually and vocally.

Color the lorises and their locations.

Lorises are subdivided into the "slow climbers" (African *pottos* and Asian *slow lorises*) and the "leapers and hoppers," seven species of *galagos (bushbabies)* of sub-Saharan Africa.

All the *lorises* are nocturnal and forage alone for insects, gums, and fruits. Single babies of slow climbers are "parked" hanging on branches while their mothers go off and forage; *galagos* build nests for two or more young.

Color the rest of the plate.

Lemurs reside only in Madagascar and the small Comoro Islands in the Mozambique Channel, near Madagascar. They have been there for about 55 million years, presumably having crossed from Africa when it was closer to Madagascar. There are 23 species, each with a distinct ecological niche, including diurnal ones that on the mainland have been preempted by monkeys and apes. Some species were ground dwellers but became extinct after humans arrived some 2000 years ago.

Of this remarkably diverse group, the *aye-aye* is perhaps the most peculiar. Its long middle finger, equipped with a sharp claw, spears insects living beneath the bark of trees — a strategy like that of the tool-using finch of the Galapagos.

The largest of the *Malagasy lemurs* are three species of indrids, two of them diurnal. *Indri,* pictured here, and the sifaka, have long powerful hindlimbs for leaping from one tree trunk to the other and ricocheting from ground to trees. The third, avahi, is nocturnal.

The genus *Lemur* (five species) and the *gentle lemur* (two species) are mostly diurnal, live in small groups of two to four and eat fruits, leaves, and insects. The cat-sized ringtailed lemur, here representing the genus *Lemur,* lives in groups of 20 or more.

The *sportive lemur* eats leaves, fruits, and flowers and is nocturnal and solitary.

The *dwarf* and *mouse lemurs* are nocturnal and solitary. They eat flowers, nectar, fruits, and insects. During the wet season, body weight increases by fat storage in their tails. The *dwarf lemur* hibernates during the dry season in hollow tree trunks; the *mouse lemur* only decreases its activity.

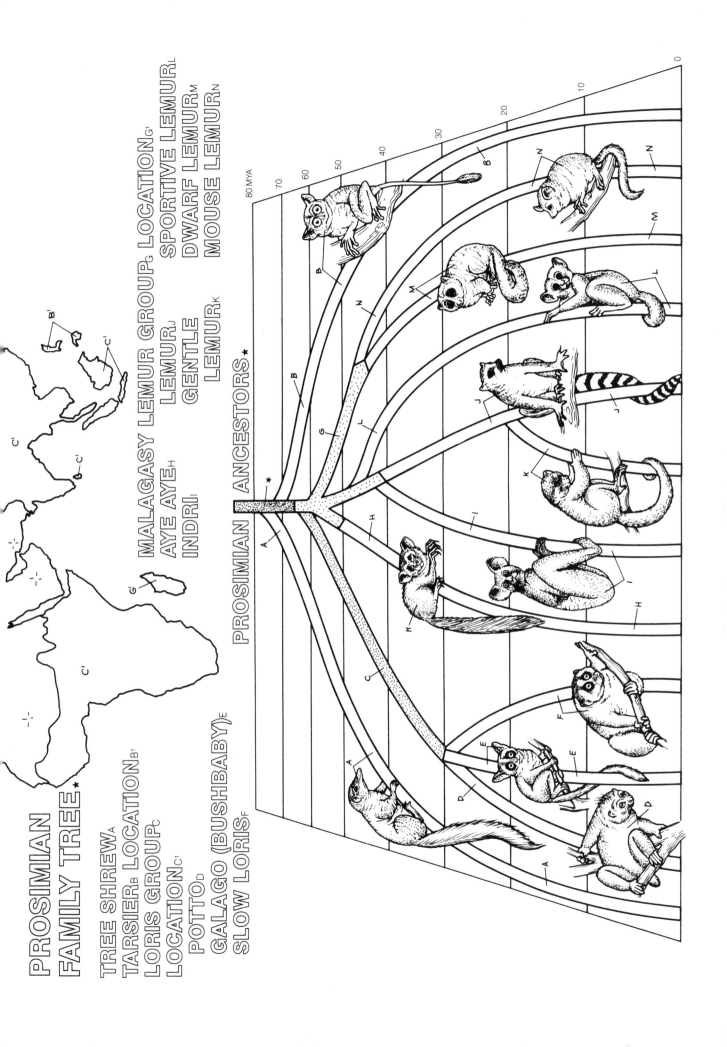

PROSIMIAN FAMILY TREE ★

TREE SHREW A
TARSIER B LOCATION B'
LORIS GROUP C
  LOCATION C'
  POTTO D
  GALAGO (BUSHBABY) E
  SLOW LORIS F

MALAGASY LEMUR GROUP G LOCATION G'
  AYE AYE H
  INDRI I
  LEMUR J
  GENTLE LEMUR K

SPORTIVE LEMUR L
DWARF LEMUR M
MOUSE LEMUR N

PROSIMIAN ANCESTORS ★

80 MYA 70 60 50 40 30 20 10 0

72

PROSIMIAN ECOLOGY AND NICHE SEPARATION

In some West African rainforests, as many as five species of prosimians live in the same trees in the same patch of forest. How can so many closely related species successfully survive and reproduce without directly competing for the same resources?

This plate is based on several years of research by French primatologist Pierre Charles-Dominique. In Gabon, there is a diversity of habits with various food types available throughout the year, a situation similar to what Darwin observed for the Galapagos finches. Charles-Dominique reports that in the rainforest there are 120 species of mammals; of these, 17 are primates. The five prosimians are nocturnal, each specialized for different food sources, which, by day, are utilized by birds and Old World monkeys.

The five sympatric prosimians (sym = same; patri = country) in Gabon are two lorises and three species of galagos. As shown on Plates 48 and 50, these two subfamilies have distinct patterns of locomotion: the lorises are slow climbers; the galagos are specialized for hopping, leaping, and running along branches. These differences are reflected in the ways they eat, avoid predators, and communicate.

Lorises detect their prey by smell, usually selecting *insects* that are ignored by other insect eaters: the slow-moving, repugnant, stinging caterpillers, ants, and centipedes. Galagos prey upon the speedier flying *insects:* orthoptera, (crickets and so forth), lepidoptera (butterflies), and coleoptera (moths), which they detect by the high-frequency sounds the *insects* make. They capture them by a stereotypic motion in which they lunge forward rapidly, with hind feet clasped to a branch and hands widely splayed to grab the flying bug with a single swipe.

Whereas galagos quickly flee from predators with leaps and bounds, the lorises move slowly and silently or "freeze" to avoid the attention of arboreal carnivores such as the palm civet. The cryptic lorisine strategy works only in areas where these small animals can effectively screen themselves in the dark shadows of dense, leafy forest coverage. The fast-moving galagos may utilize the more open forest areas because they may quickly leap away from potential danger. Contributing to their strategy of being discrete and unobserved, lorises rely more on olfactory than on vocal communication, in contrast to their noisier and more social galago cousins.

Choose five contrasting colors and color the five prosimians and their body weights, noticing that they are not drawn to scale. The potto is the largest and the demidoff galago, the smallest. Now color the pathways used by each animal as it moves through the forest.

The *potto* and the *demidoff* and *needle-clawed galagos* spend their time in the forest canopy between 5 and 40 meters from the forest floor; the *golden potto* and the *Allen's galago* stick to the understory on the ground and in the low bushes.

Color the components of diet for each prosimian.

The three species frequenting the forest canopy all eat *gums* (the thick sap beneath the bark), *fruits,* and *insects,* but the proportions of each food type varies. The *potto,* with the largest body, eats mainly *fruits;* whereas the tiny *demidoff galago* feeds primarily on *insects.* The *needle-clawed galago* eats large amounts of *gums.* Its needlelike "claws" are actually keeled nails, a specialization enabling this prosimian to move along smooth surfaces to find and extract *gums* without slipping and falling.

In the understory, the *golden potto* eats primarily *insects,* and *Allen's galago* eats large amounts of *fruits* as well. Neither species in the understory eats *gums,* available in the higher parts of the forest. All the prosimians eat *insects,* but as body size increases, the diet must be supplemented by bulkier foods, mainly *fruits.*

These five species of prosimians coexist by living on different "floors" of the same ecological edifice and by getting different foods in different, highly specialized ways.

ECOLOGY AND NICHE SEPARATION.★

LORISES.
POTTO_A
GOLDEN
POTTO_B

GALAGOS.
DEMIDOFF_C
NEEDLE-CLAWED_D
ALLEN'S_E

DIET.
GUMS_F
FRUITS_G
INSECTS_H

1100g_A

F 25%
G 65%
H 10%

60g_C

F 10%
G 15%
H 75%

300g_D

F 75%
G 5%
H 20%

CANOPY★

UNDERSTORY★

200g_B

G 15%
H 85%

260g_E

G 75%
H 25%

73
PROSIMIANS AND ANTHROPOIDS: SPECIAL SENSES AND DENTITION

Prosimians and *anthropoids* have followed different evolutionary pathways for 70 million years, according to the "molecular clock." The *anthropoid* fossil record does not exist before 35 mya, the earliest date of fossils found in north Africa and South America. If *Amphipithecus* from the late Eocene is an *anthropoid,* the record extends to 40 mya.

What distinguishes these two major groups of primates? What factors initiated the divergence into these different ways of life?

Color the lemur, a prosimian, and the vervet monkey, an anthropoid (specifically, an Old World monkey). Use light colors. Notice the difference in position of the eyes, length of the snouts, and the noses. Now color the structures associated with vision in each animal: the visual cortex, the orientation of the eye orbits, and the postorbital closure.

Notice the more frontal position of the *orbits* in the *anthropoid,* indicating better stereoscopic vision, and how they are totally enclosed within bone. *Anthropoids* also have color vision and are diurnal. *Prosimians* are nocturnal, except for a few Malagasy species, such as the *lemur* pictured here. The enlarged *anthropoid visual cortex* indicates more elaborate processing of visual information.

Color the structures associated with the sense of smell: the olfactory bulb, and the snout or nose.

The *olfactory bulb* and *snout* region are notably smaller in the *anthropoid,* and the wet muzzle (rhinarium) is lost entirely. Although vision is well developed in *prosimians* compared to other mammals, with some stereopsis, they rely on smell more than *anthropoids* do.

Color the dentition, noting the number of teeth and their shapes.

Anthropoids eat mainly fruits and foliage. Many *prosimians'* diet includes insects. Feeding on fruit and foliage requires extensive use of hands and of the front *teeth*—large incisors for biting, squarish molars for chewing. *Anthropoids* lack the *prosimian* tooth scraper/dental comb, composed of incisors and ca-

nines. Instead, they use dexterous hands in feeding and grooming. *Postorbital closure* in *anthropoids* may be related to an increase in size of the chewing muscles.

Prosimians are generally small, making it possible to survive on a diet of insects (and to hide or escape from predators, especially in a forest at night).

Anthropoids rely on more abundant foods; so it is possible for them to live in larger social groups. Some *anthropoids* have body and canine size dimorphism (Plate 63). *Anthropoids* generally have one baby at a time, and the mother carries it on her body at all times. *Prosimians* that have two or three young leave them in nests; a single young may be "parked" on a branch while the mother feeds.

Complete the plate by coloring the brain.

The advanced evolution of the *anthropoid* nervous system is linked to the manipulative hand and to visual acuity. The *anthropoid's* foraging techniques involve dexterity, learning, and memory.

What factors account for the evolutionary divergence of these two major groups of primates? Anthropologist Susan Cachel suggests it was due to changes from a more tropical warm and humid climate to one that was cooler and seasonal. Large bodies have more effective thermoregulation and need more food. A diet of fruit involved day living, color vision, a manipulative hand, and traveling over a larger range than is necessary for smaller-bodied, insect-eating species.

Day living and larger body size also made these emerging *anthropoids* more vulnerable to predators; so perhaps larger social groups, in which larger-bodied males with large canines could protect other group members, developed for defense.

Larger body size seems to correlate both with relatively bigger *brains* and with few babies. The *anthropoids* invest more time and energy in their babies, and each offspring needs an extended period of development (Plate 61). An increasingly learning-dependent way of life evolved as youngsters paid more attention to how adults did things, imitated them, at first in play, and then put their acquired learning to use. Primates are intelligent animals because it is so important for them to learn from one another and to communicate effectively with other group members.

SENSES AND DENTITION.★

BRAIN_C
VISUAL CORTEX_D
OLFACTORY BULB_E

PROSIMIAN_A

DENTITION ●
NUMBER_F
TOOTH SHAPE_{F¹}

ANTHROPOID_B

SKULL ●
SNOUT_G/NOSE_{G¹}
ORBIT ORIENTATION_H
POSTORBITAL CLOSURE_I

LEMUR_A

VERVET_B

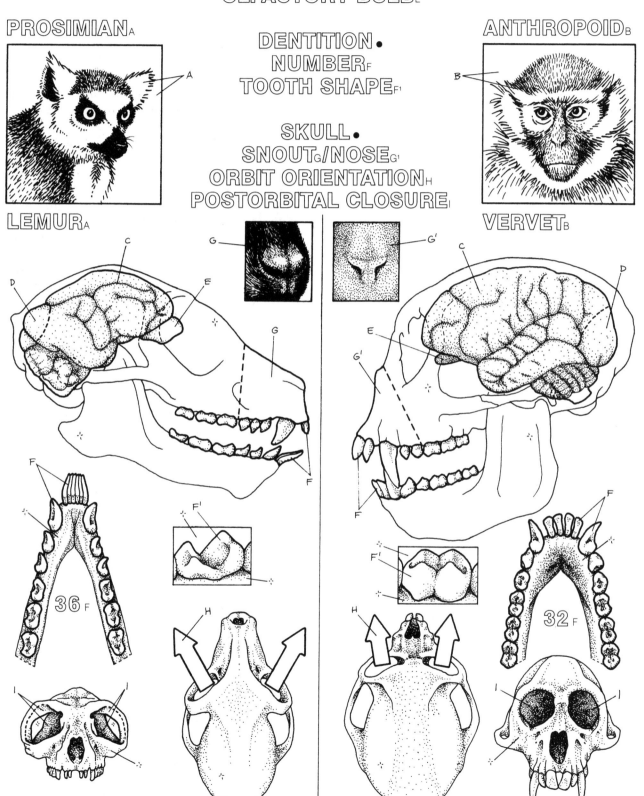

74

FIRST EVIDENCE OF ANTHROPOIDS FROM FAYUM EGYPT

The third primate radiation took place in the Oligocene, 37-22 mya, and is the first evidence for the anthropoid adaptation. Most of these monkeylike fossils come from northern Africa and a few from South America (Plate 75).

Color the time chart and the maps showing the Fayum area, past and present. Oligocene Africa was still an island separated from Eurasia by the Tethys Sea. The Red Sea had not yet formed between Africa and Arabia.

The Fayum today lies in a desert 60 miles southwest of Cairo, below the Nile delta. Its deposits were laid down within the deltaic floodplain. What is now desert was then gallery forest. Primate fossils were discovered during the 1906 expeditions of the American Museum of Natural History. Dozens more fossils were discovered from 1961 to 1967 by Yale University expeditions led by paleontologist Elwyn Simons, and the work has recently been resumed after being interrupted by the 1967 Arab-Israeli War.

Color the many layers that make up the thick Jebel el Qatrani formation of the Fayum. Then color the fossils found in each layer and the corresponding arrows.

The *basalt cap* at the top is dated at *25 mya* by potassium-argon method (Plate 93). The middle layer, *Upper Zone,* is 250 feet thick and has no primate fossils. The next layer is the *Upper Fossil Wood Zone,* named for huge logs, where most of the primate fossils have been found. These include the nearly complete skull of *Aegyptopithecus,* the fragmentary *Apidium* skull, and *Parapithecus* and *Propliopithecus* jaws.

The *Middle Zone* has yielded a smaller species of *Apidium* with the pictured limb bones and a *Propliopithecus* jaw. The *Lower Fossil Wood Zone* is early Oligocene; so all these formations span the time period *35-25 mya.* The oldest fossil, from the lowest layer, is an *Oligopithecus* jaw fragment. Although considered anthropoid, *Oligopithecus* is primitive in some respects and has been compared to Eocene omomyids (Plate 70); but its premolar is quite anthropoidlike.

Of the four remaining genera, *Apidium* and *Para-*

pithecus are "monkeylike," *Propliopithecus* and *Aegyptopithecus,* "apelike," but only in their teeth. It is doubtful that there were true monkeys and true apes in the Oligocene, for the monkey-ape divergence did not occur until the Miocene, about 20 mya, according to the molecular evidence and to the appearance of definite fossil apes based on dental evidence at that time. Pelvic bones from the *Upper Fossil Wood Zone,* found in 1977, the oldest known for anthropoid primates, are more monkeylike than apelike.

What anthropoid characteristics do the Fayum fossils share that distinguish them from prosimians? The frontal bone of the skull and the two halves of the lower jaw are not fused in prosimians; they are fused in anthropoids and in the Fayum fossils. The orbits are fully oriented toward the front; the olfactory area is smaller; the visual cortex area is larger than in the Eocene fossils. The posterior part of the eye orbit is not closed in prosimians; however, it is in *Aegyptopithecus,* to the same extent as in New World monkeys. Some prosimian traits remain: the long muzzle and the constricted "waist" behind the orbits (Plate 73). The endocast (a mold of the inside of a braincase) of *Aegyptopithecus* shows a brain pattern intermediate between prosimians and New World monkeys. *Apidium* and *Parapithecus* retain a third premolar, like prosimians and New World monkeys. *Aegyptopithecus* and *Propliopithecus* have only two premolars, like catarrhines (Old World monkeys and apes) and there appears to be sexual dimorphism in canine size, also as in catarrhines.

The locomotor anatomy of the Fayum fossils, the pelvic and limb bones, indicate that they were agile arboreal quadrupeds, certainly not like apes. They most resemble New World monkeys, the most primitive living anthropoids.

Some of the Fayum primates were thought to be "dental apes" because their molars are more like those of fruit-eating apes than those of Old World monkeys, but they are also like the molars of fruit-eating New World monkeys. So evidence from the teeth suggests that the Fayum primates were fruit eaters, not that they were apes. Their overall anatomy, including the dentition, suggests that at 30-35 mya the Fayum primates were near the common ancestry of both apes and Old World monkeys.

FIRST EVIDENCE OF ANTHROPOIDS FROM FAYUM EGYPT.★

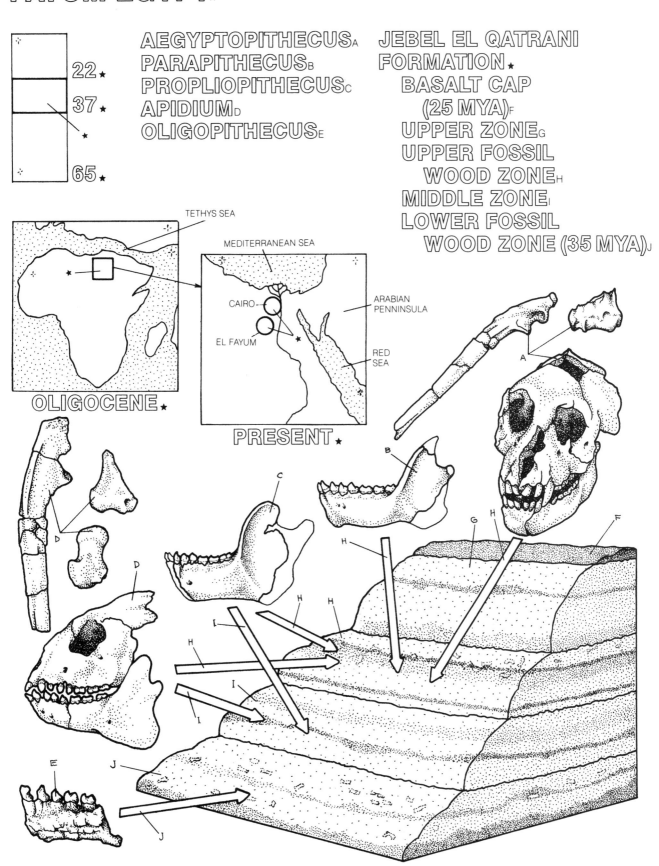

22★
37★
★
65★

AEGYPTOPITHECUS_A
PARAPITHECUS_B
PROPLIOPITHECUS_C
APIDIUM_D
OLIGOPITHECUS_E

JEBEL EL QATRANI
FORMATION.★
  BASALT CAP
  (25 MYA)_F
  UPPER ZONE_G
  UPPER FOSSIL
    WOOD ZONE_H
  MIDDLE ZONE_I
  LOWER FOSSIL
    WOOD ZONE (35 MYA)_J

TETHYS SEA

MEDITERRANEAN SEA

ARABIAN
PENNINSULA

CAIRO

EL FAYUM

RED
SEA

OLIGOCENE ★

PRESENT ★

75

NEW WORLD MONKEY ORIGINS AND THE RAFTING HYPOTHESIS

Prior to the 1960s, when continental drift theory began to gain acceptance, explanations for New World monkey origins relied on hypothesizing island hopping or an ancient land bridge connecting *South* to *North America*. *North American* primate fossils include only prosimians. The origin of New World monkeys then had to be explained through parallel evolution; that is, a *North American* Eocene omomyid evolved into *South American* monkeys and a related *Eurasian* omomyid evolved independently into Old World monkeys, probably in *Africa*.

The molecular data provided further evidence against parallel evolution: they showed that the New and Old World monkeys shared an extensive common evolutionary history prior to their divergence between 35 and 40 mya. These data suggested a probable Old World origin, but the *Atlantic Ocean* seemed so unlikely a barrier for monkeys to cross that parallel evolution and a *North American* origin for New World monkeys were generally accepted.

Growing information on plate tectonics made it clear that *South America* was an island for most of the last 65 million years. The recent direct connection (3.5 mya) between the *North* and *South American* land masses was much too late to have provided a route for a New World monkey ancestor. Wherever the origin, the ancestor had to cross a major water barrier.

Begin at the bottom of the plate and color the continental positions as they appeared in the early Oligocene. Now color the Atlantic Ocean, the continental shelf, and the mid-Atlantic ridge, using three shades of blue.

The *mid-Atlantic oceanic ridges* reach a height of about two miles above the sea floor; the deep valley or rift in the middle allowed the release of hot and molten material from the earth's mantle, causing the sea floor to spread. In the Oligocene, the sea level was at an all-time low; monkey ancestors could have "island-hopped" from one continent to another. *Africa* was no more distant from *South America* than *North America* was.

Proponents of the rafting hypothesis suggest that monkeys evolved from prosimians only once and in *Africa* and that it was a primitive monkey, and not a

prosimian, that made the water-logged trip to *South America*.

What is the fossil evidence? The Oligocene anthropoids in *Africa* had many New World monkey characteristics, the primitive ones for anthropoids.

Color the fossils according to chronological sequence, from oldest to youngest, beginning with the fossil primate, *Branisella*. Color the fossil representation and the time sequence on the scale to the left of each before moving on to the next fossil.

In the early Oligocene, a single described specimen from the Bolivian Andes is *Branisella*, with dental features primitive for an anthropoid but with some platyrrhine (New World monkey) similarities.

We then skip to the latest Oligocene in Argentina, to *Tremacebus*, comprised of a single, badly damaged cranium. *Tremacebus* had a short face and incomplete orbital closure.

Dolichocebus, also from the latest Oligocene of Argentina, consists of a distorted cranium with no teeth.

Homunculus, from the early Miocene of Argentina, is a nearly complete lower jaw, part of the face, and a femur.

Stirtonia, from the middle Miocene of central Colombia, is the largest of the fossil platyrrhines, about the size of a large male capuchin. It may be related to the living howler monkey.

Neosaimiri is very similar to and probably closely related to the living squirrel monkey.

Cebupithecia, from the middle Miocene of Colombia, is the best represented of the entire group, with dental and cranial fragments as well as several limb and tail bones.

On the basis of geology, molecules, comparative anatomy, and the fossil record, the split between New and Old World primates probably took place sometime in the late Eocene, after the two groups shared at least 20 million years of common evolutionary history. The early fossils in *South America* may not be directly ancestral to particular living species, but may represent "experiments" in becoming New World monkeys.

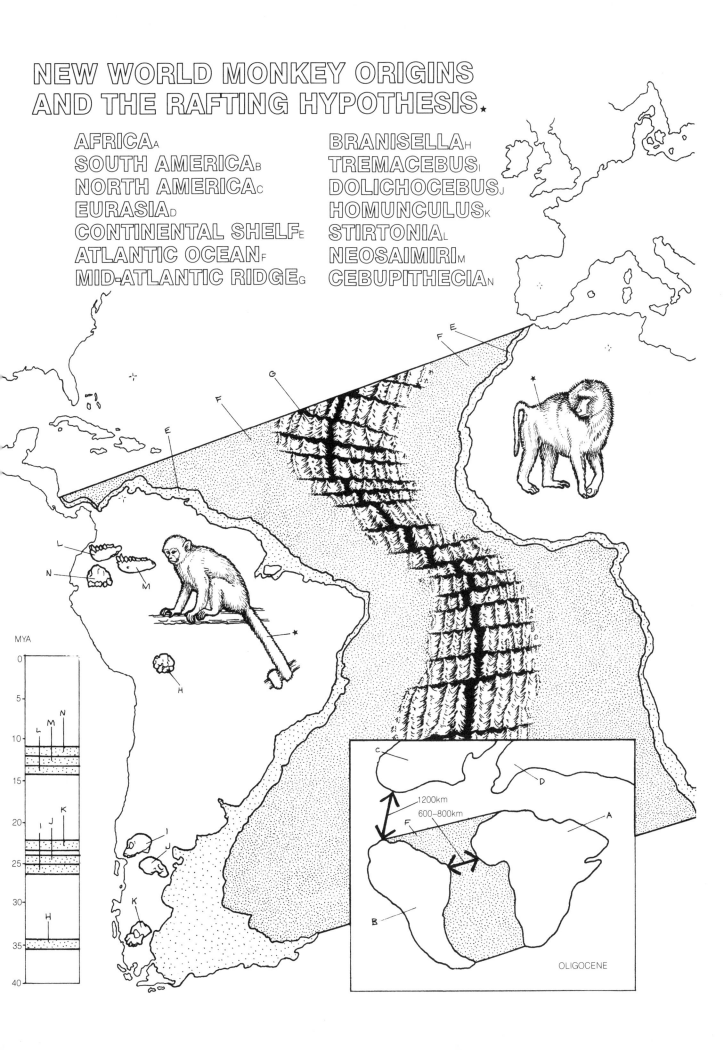

NEW WORLD MONKEY ORIGINS AND THE RAFTING HYPOTHESIS.★

AFRICA_A
SOUTH AMERICA_B
NORTH AMERICA_C
EURASIA_D
CONTINENTAL SHELF_E
ATLANTIC OCEAN_F
MID-ATLANTIC RIDGE_G

BRANISELLA_H
TREMACEBUS_I
DOLICHOCEBUS_J
HOMUNCULUS_K
STIRTONIA_L
NEOSAIMIRI_M
CEBUPITHECIA_N

MYA

1200km
600–800km

OLIGOCENE

76

NEW WORLD MONKEYS: TWO FAMILIES

New World monkeys consist of the marmosets (Callithricidae) and the cebids (Cebidae), distinguished by *body weight,* dentition, *hands,* method of feeding, and social behavior. This plate and the next draw on the work of primatologist Philip Hershkovitz.

Color all the features of the representative marmoset.

The smallest monkeys, marmosets, are all diurnal and eat insects, fruits, and gums. They forage in open forests, secondary growth areas, and riverine foliage. All their digits except the great toe (hallux) have clawlike nails. They have one less molar *tooth* than the cebids, and their triangular molars effectively crack open the hard, protective shell of arthropods.

Marmosets have so often been described as "squirrel-like" that primatologist Paul Garber compared the behavior of tree squirrels and marmosets in the same forest in Panama. He found that squirrels run up and down tree trunks, and marmosets take long, acrobatic leaps through the canopy from one thin terminal branch to another. Squirrels feed as well as travel on the thicker supports; marmosets use their grasping ability to forage on dense tangles of vines and fragile interlacing foliage. Unlike squirrels, marmosets feed on gums and may cling for hours to a vertical tree trunk while doing this. Their clawlike nails and small body size are well adapted for obtaining this food—an interesting convergence with the needle-clawed galago (Plate 72).

Marmosets' reproductive pattern sets them apart from all other primates. A female gives birth to two offspring each time and lives with an adult male, along with recent offspring. By doubling their output, so to speak, females increase their reproductive potential, but the combined weight of the young may exceed the mother's by 10–25 percent. This may be the evolutionary reason why males help care for the young. There is a male-female *pair* bond, unusual among primates. In some species, older offspring remain in the parental group and help with transporting more recent arrivals. In other species, the offspring leave at an early age and the adult male does most of the carrying. During an annual cycle, a female may have two sets of twins, a heavy burden physically and nutritionally.

Marmosets are unusually small for anthropoid primates. The evolutionary trend from prosimians to monkeys and apes has been toward larger body size, and this makes us wonder why marmosets are so small. Zoologist John Eisenberg believes that they evolved from a larger ancestor; their way of feeding and traveling favors a smaller form. They seem to rely more on the sense of smell than other anthropoids do: they possess a Jacobsen's organ (Plate 45), and both sexes have sternal and anogenital scent glands, resembling those of their prosimian uncles and aunts. Marmosets do not have a primate counterpart in the Old World; they are unique. Their body size and build is more like that of tree shrews, according to Ted Grand, than that of other monkeys.

Color the representative cebid and its features.

Cebids are generally heavier than marmosets; spider monkeys may reach 11 kilograms, and there is a gradient down to the smallest titi monkeys at .68 kilograms. Fingers and toes have nails used for grasping branches as they move, and their *hands* are effective manipulators in foraging and feeding. Cebids have three, rather than two, molar *teeth,* which are square rather than triangular, for grinding food in their varied diet consisting mostly of fruit, with supplements of insects and foliage. Their prehensile *tail* serves as an extra limb during feeding and locomotion and helps them to hang and reach for food or cross gaps in the pattern of branches in the arboreal habitat.

Their social group is more variable than that of marmosets. A couple of cebid species live in *pairs,* but most live in *larger groups* of males and females with stable composition and larger home ranges. In cebids, only one baby is born at a time, and it is always carried by its mother. Development may be slow, and in spider monkeys, births are spaced every three years. This longer period permits the infant to learn the complex motor and foraging patterns essential to survival in a large home range.

THE TWO FAMILIES.★

BODY WEIGHT_A
HAND_B
FOOT_C
TAIL_D
TEETH_E
SOCIAL GROUP.★
   PAIR_F
   LARGER GROUP_{F1}

MARMOSET.
.14–.5 kg_A

CEBID.
.68–11 kg_A

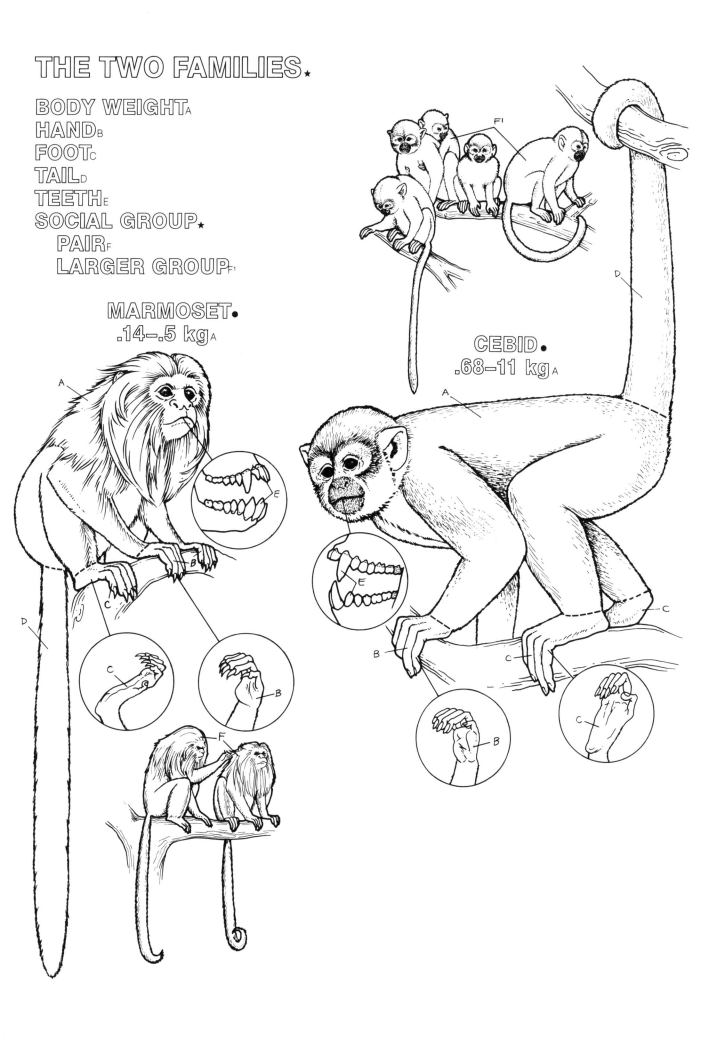

77

NEW WORLD MONKEY FAMILY TREE

From the molecular information, it appears that most living New World monkey groups diverged from a common ancestor between 15 and 20 mya.

Color the marmosets, which have diversified into twenty species during the past 10 million years.

The smallest (140 grams) is the *pygmy marmoset;* the most specialized, according to Philip Hershkovitz, is the *golden* or lion *marmoset,* named for its bright orange fur and lionlike mane. Its long, slender arm, narrow palm, and elongated middle fingers are used to pry insects from underneath bark or from crevices in trees.

Color the owl or night monkey, named for its large eyes and nocturnal habits. It is the only living anthropoid that is active at night.

The *owl monkey*'s diet is much like that of the sympatric *capuchin monkey*—largely fruits and insects. The two species avoid competition by foraging at different times, one at night, the other during the day. *Owl monkeys* live in small groups of an adult male, adult female, and their offspring.

Color the titi monkey, which has three species.

Like the *owl monkey* and *marmosets,* they live in groups of one adult male and adult female and offspring, and the males carry the young. Their diet is 70 percent fruit, supplemented by leaves and insects. They have an acute sense of smell and leave scent marks; they also communicate visually by postures and movements. *Titis* are territorial, like gibbons, and display aggressively, calling, rushing, and chasing around the edges of their home range.

Color the saki monkey.

Little is known about *saki monkeys,* seven species that include the uakari. They weigh between 1.5 and 3 kilograms and are mainly fruit eaters.

Color the well-known squirrel monkey, which includes two species.

Squirrel monkeys browse on fruit, averaging only 3 minutes at each tree. When not feeding or sleeping, they are restlessly inquisitive—foraging, unrolling dead leaves, peering here and there. Their social groups generally consist of about 20 individuals, and they scatter widely in a tree while foraging.

Color the capuchin monkey.

The *capuchin monkey* is the familiar "organ-grinder" monkey, known for its cleverness and manual dexterity in captivity and in the wild. There are four species; the average weight is about 3 kilograms; and group size is about 10. Their ecology is similar to that of *squirrel monkeys.*

Zoologist Richard Thorington observed *capuchin* and *squirrel monkeys* in Colombia. *Capuchins* linger longer in each tree, about 20 minutes, and so do not visit as many as the *squirrel monkeys* do. *Capuchins* also eat different insects from the *squirrel monkeys'* fare; they may spend hours extracting them from crannies in dead trees.

Complete the plate by coloring the spider and howler monkeys, with their prehensile tails.

This group contains the largest-bodied New World monkeys, 12 species of them, including the woolly and woolly-spider monkeys. Their specialized tails are equipped with friction skin, sweat glands, and sensory nerves leading to a large projection on the motor cortex of the brain like a fifth hand. By wrapping their prehensile tails around branches, they can hang suspended and reach otherwise inaccessible tidbits.

Spider and *howler monkeys* often live in the same forests. They represent dietary specializations in opposite directions; 80 percent of the *spider monkey* diet is fruit, the rest leaves and shoots. In general 20 animals range over an 8 kilometer range, but travel in groups of 3 or 4. They rarely feed in one tree for more than 15 minutes. *Howler monkeys* eat only 60 percent fruit and 40 percent leaves; they have a long gut, specialized for leaf eating. They live in stable social groups of about 18 animals and may feed all day in one tree.

New World monkeys are remarkably diverse in size, diet, and social behavior—testimony to the variety of ecological niches they have filled in South America during the past 20 million years or so.

NEW WORLD MONKEY FAMILY TREE ★

MARMOSETS_A
 TAMARIN_{A¹}
 PYGMY MARMOSET_{A²}
 TUFTED EAR
  MARMOSET_{A³}
 GOLDEN MARMOSET_{A⁴}

CEBIDS ●
 OWL MONKEY_B
 TITI MONKEY_C
 SAKI MONKEY_D
 SQUIRREL MONKEY_E
 CAPUCHIN MONKEY_F

PREHENSILE TAILED_G
 HOWLER_{G¹}
 SPIDER_{G²}

ANTHROPOID STEM ★

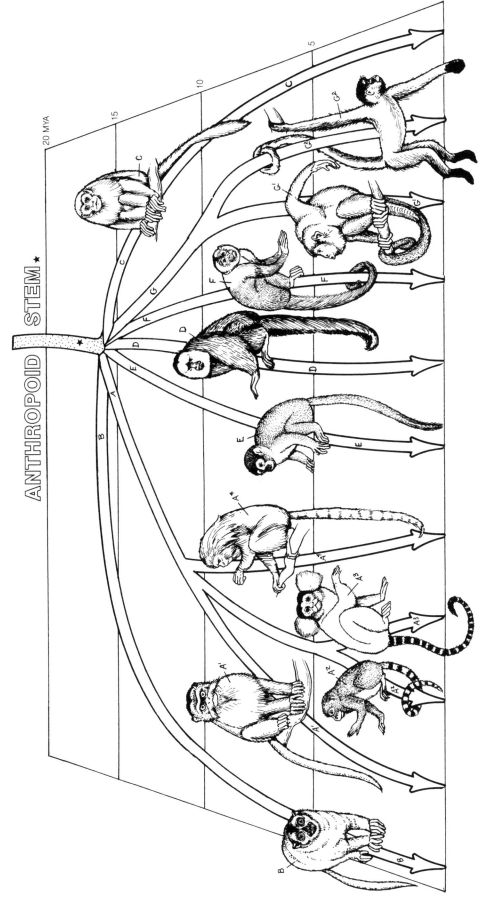

20 MYA   15   10   5   0

78
OLD WORLD MONKEYS: BANQUET AND SMORGASBORD FEEDERS

In contrast to the diversity of adaptive patterns found in New World monkeys, Old World monkeys as a group are more similar, but there are several variations on the theme. There are two groups of Old World monkeys: *leaf monkeys* and *cheek pouch monkeys*. The major distinction between the two lies in their dietary patterns, differences that show up in their anatomy, feeding style, food selection, and social behavior.

The so-called *leaf monkeys* — colobus monkeys in Africa and langurs in Asia — eat leaves, even dry, mature ones, as a major food item, and they feed mainly in trees. The *cheek pouch monkeys*, which include the baboons, guenons, and mangabeys in Africa, and macaques in Asia, eat a lot of fruit and feed in the trees and on the ground.

Color the monkeys in the center of the plate.

The *cheek pouch monkeys* derive their name from the pouch formed by one of the cheek muscles, which has stretched out to form a pocket. The monkeys stuff their pouches with fruits, leaves, or seeds, which they collect in the trees or on the ground. The pouches allow them to collect a lot of food quickly; eventually, they can move to a position of safety or comfort to chew and digest their food at leisure.

Leaf monkeys have no cheek pouches. Instead, their digestive anatomy is characterized by a large, chambered stomach analogous to that of a cow. They can store large quantities of food, especially leaves, in the stomach, where bacterial action helps in the digestion of plant cellulose. The process takes more time than does digesting fruits, fleshy seeds, nuts, and insects. The anatomy of the *leaf monkey* equips it for the types of food eaten and the style of feeding.

Color the leaf monkeys in the tree.

Leaf monkeys disperse in a tree, go to the ends of branches, and space themselves 1.2 to 1.8 meters from their nearest neighbor. They usually face away from the central core of the tree to avoid interacting socially with a newcomer. The number of animals in a tree is limited by the available branches and their three-dimensional arrangement.

The *leaf monkeys* spend on average 5 to 20 minutes, sometimes as long as 40–50 minutes, sitting and feeding in one place. Each mouthful of leaves must be chewed and swallowed before another bite is taken. Leaves are not very nutritious; so many of them must be eaten, which requires these monkeys to feed 3 to 6 hours a day. In contrast, monkeys with cheek pouches collect their food in a very different way.

Color the pathway of the cheek pouch monkey as it climbs into a tree and looks down a branch to see if there is food and to make sure that another monkey is not there.

A *cheek pouch monkey* will go down and quickly pick the fruits and young leaves from the ends of the branch, stuff them into its cheek pouches, then go down the next branch, until it has covered all available branches or has a full pouch. Next it will find a shady place, perhaps in company with other *cheek pouch monkeys*, and with its hand, push the food from the pouch into the mouth and chew it. This pattern enables some monkeys, like mangabeys, to continue to eat even after they have gone to bed. Cheek pouches are useful for feeding on the ground also, where possible danger might exist, but the animal can feed quickly and retire to the safety of trees.

Finish the plate by coloring the vignettes at the bottom of the page.

Primatologist Suzanne Ripley, who has studied feeding behavior of these monkeys, has dubbed the two kinds of monkeys as the "banquet" versus the "smorgasbord" styles. The leaf eaters, like the colobus and langur monkeys, banquet for a considerable time in one place, filling their large stomachs with a lot of vegetation. Unlike (most) human banqueters, they are too busy chewing to interact socially with their table partners, but they may socialize more during the rest-and-digest period after the meal. The *cheek pouch monkeys* move around, sampling the varied smorgasbord the forest offers, then take their "full plate" to a safe and comfortable spot, where they can enjoy the meal in the company of friends.

BANQUET AND SMORGASBORD FEEDERS ★

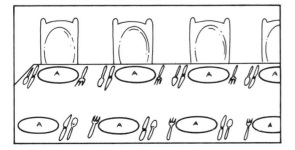

LEAF MONKEY A

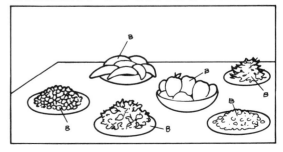

CHEEK POUCH MONKEY B

79

OLD WORLD MONKEY FAMILY TREE

Old World monkeys, the most successful and numerous species of nonhuman primates, inhabit a variety of habitats: tropical rainforests, high mountain areas, savannas, and even snow. The two kinds of Old World monkeys separated some 12 mya, each going its own separate course: the *leaf monkeys* specialize in the ability to digest leaves. Members of the other group possess cheek pouches and eat mostly fruit, but remain generalists.

Color the leaf monkeys.

Langur monkeys, widely distributed from India, Pakistan, and Sri Lanka into Southeast Asia and many offshore islands, consist of 14 species and dozens of subspecies. The Asian *leaf monkeys* inhabit a wider range of habitats than do the African *leaf monkeys,* and there is a greater range in body weight, from 4–17 kilograms. Their diet consists of at least 30 percent leaves, but the amount of leaves versus fruit eaten varies between species.

Colobus monkeys, widely distributed in Africa, consist of five species, weighing 5–7 kilograms. They often live in groups of one adult male and several adult females and young and usually do not come to the ground.

In several areas there exist sympatric species of *leaf monkey,* one eating more fruit, the other, more leaves: the gray *langur* and purple-faced *langur* in Sri Lanka; the banded and dusky *langurs* in Malaysia; and the red *colobus* and black and white *colobus* in Africa. In these pairs, the former eats more fruit and has a larger home range and a larger social group.

Color the guenons, or Cercopithecus group.

Widely distributed throughout sub-Saharan Africa, the *guenons* consist of 21 species and range from a little more than a kilogram (the talapoin monkey) to the large male *patas monkey* over 12 kilograms. The *red tail* monkey here represents the numerous species of *guenon.* It is not uncommon to find four *guenon* species in the same forest, each inhabiting different levels.

The *vervet* and *patas monkeys* are closely related to each other, but quite different in anatomy and behavior. *Vervet monkeys* are widely distributed throughout sub-Saharan Africa. They live on the ground and in trees in groups of several adult males and females having little sexual dimorphism. The terrestrial *patas monkeys* live in groups with one adult male and several adult females and young; males are twice the size of females. *Patas monkeys* converge with the greyhound in locomotor anatomy and are capable of sustained speeds of 64 km/h. Their main defense is running from predators.

The remaining groups of *cheek pouch monkeys* are all very closely related.

Color the mangabey monkey and the mandrill.

The *mangabeys* consist of four species, forest living and fruit eating. The *mandrills* consist of two species, inhabit the forests of west Africa, and are the largest monkeys.

Color the baboon group.

The *baboon* group consists of three species: the *savanna* or anubis *baboon,* which occupies forests in central and west Africa and the eastern and southern savannas, and the *gelada* and *hamadryas baboons,* found in Ethiopia.

Baboons are highly sexually dimorphic. The *savanna baboons* live in large groups with many adult males and adult females and their young. The *gelada* and *hamadryas baboons* forage in one-male multi-female groups, but come together in large groups for sleeping in the sides of cliffs above and below the grassy plains. *Gelada baboons,* which eat mainly seeds and grass obtained on the ground, may be found in areas with *savanna baboons,* which eat more fruit.

Color the representative macaque.

One *macaque* group is found in northern Africa along the Atlas Mountains and in Gibralter. The other 11 species are found throughout India, Pakistan, Sri Lanka, and into the islands of Southeast Asia and exhibit great diversity in body size, diet, habitat, and social behavior. *Macaques* seem to occupy those niches in Asia that *mangabeys, baboons,* and *vervets* occupy in Africa; the numerous species of *langurs* occupy the niches that the *colobus* and *guenons* do in Africa.

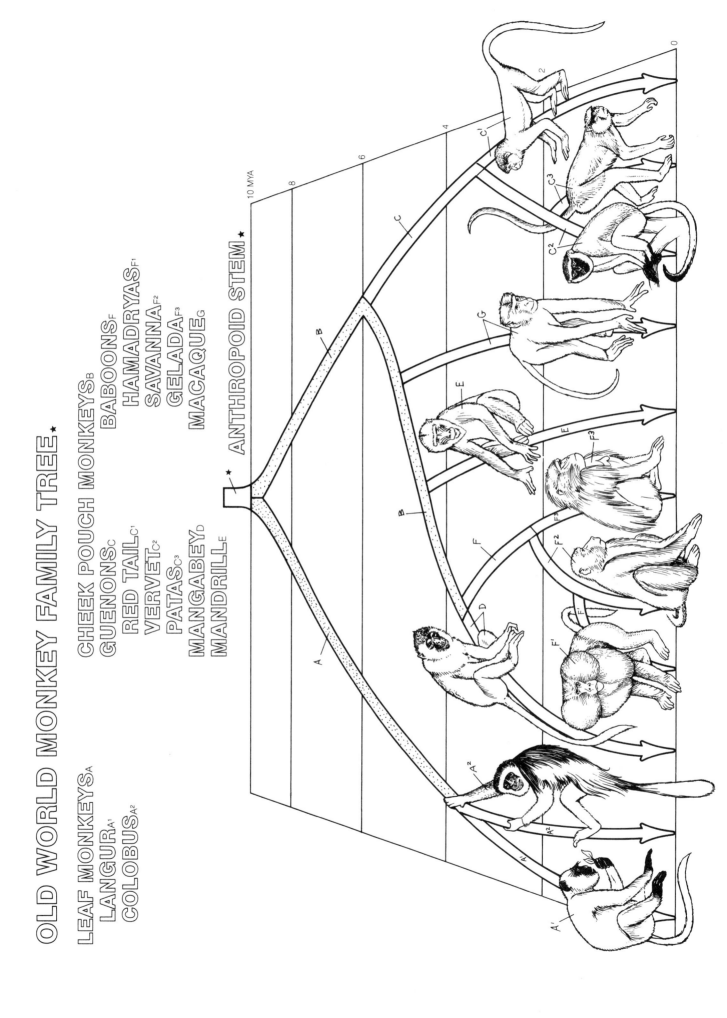

OLD WORLD MONKEY FAMILY TREE ★

LEAF MONKEYS_A
  LANGUR_{A'}
  COLOBUS_{A²}

CHEEK POUCH MONKEYS_B
  GUENONS_C
    RED TAIL_{C'}
    VERVET_{C²}
    PATAS_{C³}
  MANGABEY_D
  MANDRILL_E

BABOONS_F
  HAMADRYAS_{F1}
  SAVANNA_{F2}
  GELADA_{F3}
  MACAQUE_G

ANTHROPOID STEM ★

10 MYA

80
NEW AND OLD WORLD MONKEYS: MORE THAN A MATTER OF NOSES

New and *Old World monkeys* can be distinguished by the shape of their noses, but many other anatomical and behavioral features set the two groups apart.

Using two light colors, color the New World spider monkey and the Old World langur monkey, except their tails. Then color each characteristic as it is discussed in the text.

Platyrrhine means "flat nose"; *catarrhine* means "down-facing nose." In the *platyrrhine* noses of *New World monkeys,* the nostrils are far apart and open upward; in *Old World monkeys* (and apes and humans), the nostrils are closer together and open downward.

Primatologist Adolph Schultz notes that *Old World monkeys* vary less in *body weight* than *New World monkeys.* In the *Old World monkey* group, the large 30 kilogram mandrill is only 25 times heavier than the 1.2 kilogram talapoin. The 11 kilogram spider monkey is 79 times heavier than the smallest *New World monkey,* the 0.14 kilogram pygmy marmoset.

The *hands* of the two groups also differ, especially in the orientation of the thumb. In a typical *New World monkey hand,* shown in the box below the spider monkey, the thumb is in line with the other fingers and opposes the next digit in a scissorslike grip. The *Old World monkey* thumb is rotated and more opposable, more like ours, and provides the *hand* with greater dexterity and ability to manipulate objects. Note that the spider monkey illustrated has lost its thumb altogether.

The *New World monkeys'* well-developed *prehensile tails* are essential to them for feeding and locomotion. *Old World monkeys' tails* are enormously variable, from being absent to mere stumps to long ones, but none is prehensile or as well muscled as those of the New World cousins.

Old World monkeys have thick, calloused skin, called *ischial callosities,* on which they sit while feeding or sleeping in trees, and the ischial bones of the pelvis are flattened to support those pads (Plate 53). *New World monkeys* have no such feature.

The *ear region* of the skull and *dentition* also make it easy to distinguish *New* from *Old World monkeys.* *New World monkeys* have three rather than two *premolars,* which are relatively large—like the prosimians and Fayum anthropoids—and their last molar is

comparatively small or absent. The lower *premolar* specialized in *Old World monkeys* (and apes) for sharpening the upper canine (the honing mechanism) is not present in *New World monkeys* (Plate 58). In the *ear region* of *Old World monkeys* (and apes and humans), there is a bony tube from the tympanic membrane to the external *ear* that remains visible externally on the skull; whereas in *New World monkeys,* there is only a bony ring and no tube, also like Fayum anthropoids.

Finish the plate by coloring the habitats, using a light color.

New World monkeys live primarily in forest *habitats.* In contrast, *Old World monkeys* live in *habitats* as diverse as African and Asian rainforests, high mountain areas in Nepal and India, African savannas, semi-arid desert in Ethiopia, and areas in Japan with severe winters. Many *Old World monkeys* live on the ground during the day, feeding, moving around, and socializing, and seek the shelter of trees at night or for safety when threatened. No *New World monkey* species regularly moves and feeds on the ground.

In a number of *Old World monkeys,* especially ground dwellers like baboons and macaques, females have sexual skin that changes size and color during the estrous cycle (Plate 62). Marked sexual dimorphism (Plate 63) is common in *Old World monkeys,* but not in New World forms, and may have evolved in part for defense against large ground predators.

Dietary convergences occur in the two groups. Howler monkeys, in the New World, are leaf eaters; this niche is filled by the colobus and langur monkeys (the leaf monkeys) in the Old World. The small squirrel and talapoin monkeys are specialized insect eaters in the New and Old World, respectively. The greater size diversity of the *New World monkeys* means that they fill niches that in the Old World are occupied by nonmonkey primates. For example, the small New World marmosets are comparable to tree shrews and small galagos; whereas spider monkeys are similar in their diet and slow development of their infants to the fruit-eating gibbons, which are apes. Such convergences among the *New* and *Old World* primates reflect the separate evolutionary histories of the groups.

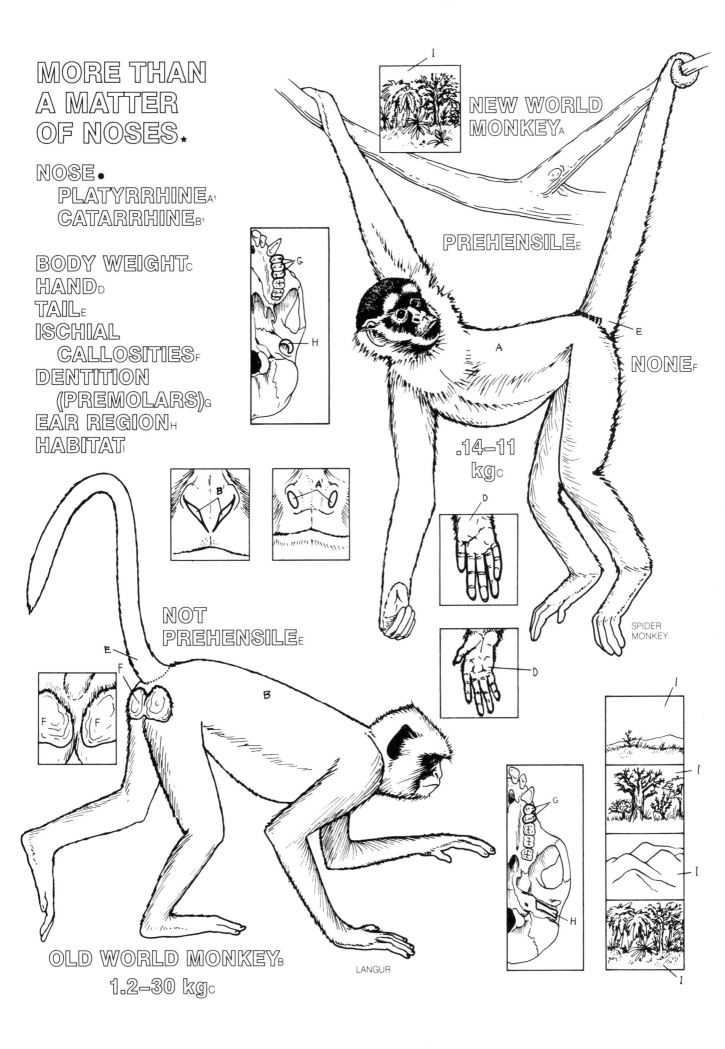

MORE THAN A MATTER OF NOSES. ★

NOSE •
    PLATYRRHINE A¹
    CATARRHINE B¹

BODY WEIGHT C
HAND D
TAIL E
ISCHIAL
    CALLOSITIES F
DENTITION
    (PREMOLARS) G
EAR REGION H
HABITAT I

NEW WORLD MONKEY A

PREHENSILE E

NONE F

.14–11 kg C

SPIDER MONKEY

NOT PREHENSILE E

OLD WORLD MONKEY B
1.2–30 kg C

LANGUR

81

MONKEYS AND APES: TWO LOCOMOTOR-FEEDING PATTERNS

Monkeys differ from *apes* in dentition and locomotor behavior and anatomy. The differences depend more on their ways of feeding than on their diet, as both groups eat fruits and foliage. As with the two kinds of Old World *monkeys* (Plate 78), feeding style strongly influences an animal's anatomy.

Color the molar tooth patterns.

Monkeys and *apes* can be distinguished by the cusp *pattern* on their lower *molars*. *Apes*, like humans, have five cusps on their *molar teeth* (called the Y-5 pattern), whereas *monkeys* have only four cusps, connected by two crests in a pattern called "bilophodont" (bi = two; loph = crest; dont = tooth). *Monkey molar* cusps are also higher (hypsodont), a specialization for eating more foliage rather than fruit, as *apes* do.

Color the apes and monkey, and their feeding ranges, using light colors.

Anatomist Ted Grand compared a gibbon, an *ape*, and a macaque, a *monkey*, of similar body weight to demonstrate feeding and locomotion.

Fruits and tender leaves, the most desirable morsels of food for many animals, usually lie at the ends of terminal branches, and animals have a variety of strategies for getting them. Insects, birds, and bats can fly there; squirrels and other small species easily scamper out and back, for their weight is no problem. Some New World *monkeys* hang from their prehensile tail while they reach for the goodies.

When a primate weighs more than about 3 kilograms, it may have trouble running on top of the branch, for the branch may bend sharply, creating an unstable footing, or break. *Apes* solve this problem by being able to hang beneath the branches as well as sitting on top of them. Hanging from a bent branch is a more stable position, and *apes* easily feed in this posture for several minutes. This form of hanging and feeding, as well as moving under branches, is a form of locomotion called brachiation. *Monkeys* do feed at the ends of branches, but in a more restricted way than do the *apes*.

The anatomy of the two groups reflects these differences in behavior. *Monkeys* have a long, muscular, flexible back for running and jumping quadrupedally and a tail for balance. Indicative of their more "vertical" locomotion, *apes* have a short back and no tail (Plate 53), but long arms. Vertical hanging and swinging requires specializations of the shoulder, elbow, and wrist joints.

Color the lower half of the plate. Begin with the shoulder joint elements and color each in the ape and monkey; proceed to the elbow joint and then the wrist joint.

The *ape's clavicle* is longer, its *humeral* head (top of the *humerus*) rounder, and the chest is shaped so that the shoulder joint is out to the side, permitting an *ape* to reach up easily over its head and suspend itself from its arms and to rotate the arm 360 degrees. This movement permits the *ape* to reach out in any direction while sitting on top of or suspended under a branch. The *monkey's* shoulder joint is less flexible than the *ape's* but more so than a dog's (Plate 51).

The *ape's* elbow joint can be extended out in a straight line, like our own, because the short olecranon process does not inhibit extension. The *monkey's* large olecranon process prevents full elbow extension. The stability of the elbow joint is maintained by the fit of the *ulna* against the *humerus*. But the elbow joint is also mobile. The *ape's* lower arm can rotate 180 degrees due to the round head (top) of the upper *radius*, from palm down to palm up (motions of pronation and supination); the *monkey's* rotates about 90 degrees.

The *ape's* carpals (wrist bones) articulate with the *radius* and little with the *ulna*; the *monkey's* articulate with both, producing less flexibility at that joint, too. Side-to-side motion at the wrist joint is called *ulnar-radial* deviation.

Improved shoulder, elbow, and wrist rotation, combined with long arms, gives *apes* their power and maneuverability in moving swiftly through the trees or leisurely hanging and feeding.

When we turn to the fossil record, all that is left for identifying fossil *monkeys* and *apes* are teeth and fragmentary bones. Evolution operates in a "mosaic fashion," that is, dental and locomotor adaptations do not evolve at the same rate. This makes it difficult to identify fossil anthropoids as either ancestral *monkeys* or ancestral *apes* on the basis of one anatomical feature alone.

TWO LOCOMOTOR-FEEDING PATTERNS.★

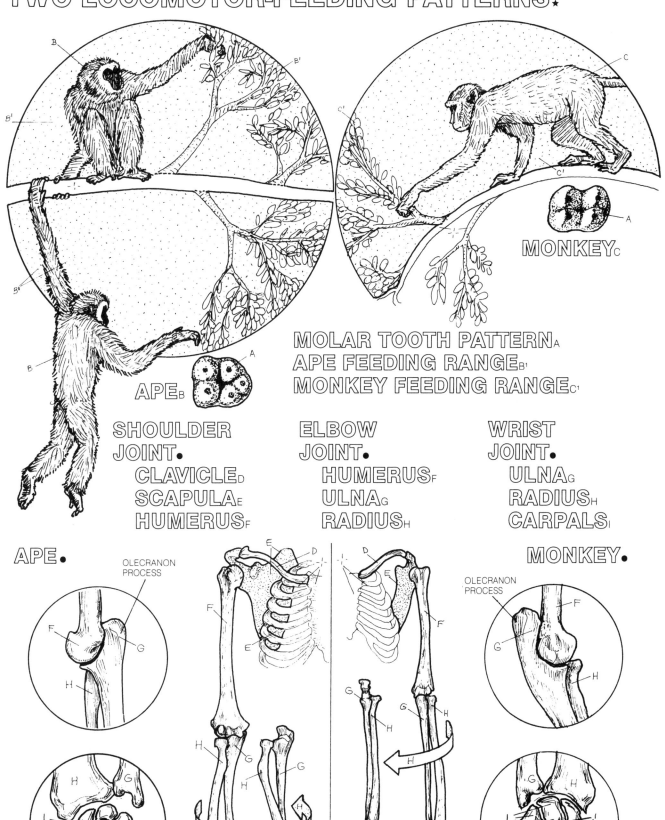

MOLAR TOOTH PATTERN_A
APE FEEDING RANGE_B'
MONKEY FEEDING RANGE_C'

MONKEY_C

APE_B

SHOULDER JOINT.●
CLAVICLE_D
SCAPULA_E
HUMERUS_F

ELBOW JOINT.●
HUMERUS_F
ULNA_G
RADIUS_H

WRIST JOINT.●
ULNA_G
RADIUS_H
CARPALS_I

APE.●

OLECRANON PROCESS

MONKEY.●

OLECRANON PROCESS

82

FAMILY TREE OF LIVING APES

Old World monkeys and *apes* separated about 20 mya, but the modern *apes* had their origin only in the last 10 million years or so.

Modern *apes* can be grouped by geographical region: the *Asian apes* are the *gibbons, siamangs,* and *orangutans.* The *African apes,* the *common chimpanzees, pygmy chimpanzees,* and *gorillas.* *Apes* are also grouped as "great *apes*," including the *orangutan* with the *African apes,* and the "lesser *apes*"—the *gibbon* and *siamang.*

Although there are relatively few species and fewer numbers of *apes* compared to *monkeys, apes* represent a greater range in body size, from the small 4 kilogram concolor *gibbon* to the 200 kilogram male *gorilla,* and in social behavior from the solitary *orangutan,* to the small "family" *gibbon* groups, to the large gregarious, social *chimpanzees.*

Color the Old World monkey. Next color the ape line (B) and the gibbon and siamang.

Gibbons range from 4 to 8 kilograms in body weight, with the larger *siamang* at 9 to 13 kilograms. The species of *gibbon* are allopatric (allo = different, patric = country) and replace each other geographically, but *siamangs* are sympatric with lar or agilis *gibbons* in Sumatra and Borneo. *Gibbons* and *siamangs* are primarily fruit-eaters, but compared to *gibbons, siamangs* eat more leaves, have a smaller home range, have shorter daily routes, and defend their territories less vigorously. All species are alike in their social behavior, living in small groups with one adult male, one adult female, and one to three young. They retain ischial callosities, an *Old World monkey* character.

Color the orangutan, an Indonesian word for "people of the forest."

Orangutans are highly arboreal; over half their diet is fruit, but they eat leaves and bark also. Female *orangutans* at 35 kilograms are half the size of males, showing extreme dimorphism like baboons and *gorillas.* Males and females do not travel or feed together, but mother and offspring may stay together for eight years before the young goes off on its own. *Orangutans* are difficult to study because they are rare, shy, stay high in the trees, and rarely come to the ground. Through the studies of primatologist Biruté

Galdikas, whose research spans 10 years, we are learning more about these rare *apes.*

Color the African apes, beginning with the gorilla.

Gorillas live in the tropical rainforests of Africa and eat mostly foliage, such as leaves, shoots, pith, and bark. They feed on the ground, live in stable social groups of about 18 animals with a home range of about 22 square kilometers. Several adult males in the group are "age-graded," with the oldest "silverback" having the highest rank. Zoologist George Schaller first studied these beasts intensively and dispelled the myth that they were fierce and dangerous. More recently, primatologist Dian Fossey has studied *gorillas* for more than a decade.

The *common chimpanzees* are perhaps best known due to primatologist Jane Goodall's continuing research at Gombe Stream Reserve in Tanzania, now spanning over 20 years. Her observations revolutionized our ideas about the uniqueness of human behavior by revealing that wild *chimpanzees* make and use tools, occasionally share food, engage in predatory behavior, and have an elaborate communication system and strong mother-offspring ties.

Until the 1970s, when Japanese primatologists began their field studies, we knew little of the social behavior of the *pygmy* or dwarf *chimpanzee,* sometimes called the bonobo. *Pygmy chimpanzees* eat less fruit than the *common* species and more foliage, like the *gorillas.*

There are some interesting differences between the two kinds of *chimpanzees.* All groups, whatever their size, contain both adult males and females. *Pygmy chimpanzees* are more often found in groups larger than 30 individuals than are *common chimpanzees,* and there are no observations of a mother with an infant alone. Social affinity among adults as measured by grooming, food sharing, and sleeping together is most marked between male and female, next between females, and least between males. In *common chimpanzees,* social bonds are most developed among adult males, next between males and females, and least between adult females.

We have a great deal more to learn about *pygmy chimpanzees,* and as long-term field research continues, about all of the *apes.*

FAMILY TREE OF LIVING APES ★

OLD
WORLD
MONKEYS_A

APE LINE_B
ASIAN APES ★
GIBBON_C
SIAMANG_C1
ORANGUTAN_D

AFRICAN APES_E
GORILLA_F
CHIMPANZEE_G
COMMON CHIMPANZEE_G1
PYGMY CHIMPANZEE_G2

ANTHROPOID STEM ★

20 MYA
15
10
5
0

83

FOSSIL MONKEYS AND APES: DISTRIBUTION IN THE EARLY MIOCENE

By the early Miocene (22-16 mya), various primate fossils show distinctive "monkey" and "ape" dental characteristics, with many more apelike than monkeylike species.

Color the time chart and then, using a shade of blue, the Tethys Sea, which still separates Africa from Eurasia. Next color the Miocene volcanoes in the enlarged section of East Africa. The fossil sites are in this area. Now color the North African fossil sites— Moghara in present-day Egypt and Jebel Zelten in Libya, both dated about 19 mya. Color the jaw fragment of the fossil monkey *Prohylobates* and the arrows.

Prohylobates is known from two species, one at each site. In each species, the crested cusp pattern of Old World monkeys is present in an early stage. However, the fifth molar cusp is present, in contrast to living Old World monkeys.

Color the Maboko site, dated about 16 mya, and then *Victoriapithecus*, another fossil monkey found at this site. Notice that *Proconsul* and *Dendropithecus* are also found there.

Victoriapithecus, consisting of two sympatric species, shows well-developed bilophodont molars (Plate 81) and thus resembles living monkeys more than does *Prohylobates*. Of the two sympatric species at *Maboko*, one, whose ulna is shown here, may have been more arboreal.

Two kinds of fossil apes (each with several species) lived in the early African Miocene: the dryopithecid group ("oak ape") represented here by two *Proconsul* species (called *Dryopithecus* by some researchers) and the pliopithecids ("lesser apes"), represented here by *Dendropithecus*.

Color the fossil sites Rusinga, Songhor, and Moroto, where several species of fossil apes have been found. Color the *Proconsul* fossils and the arrows to Moroto, Songhor, Rusinga, and Maboko, where they are found.

The palate belongs to *Proconsul (major)*, the largest species of dryopithecids. Note the parallel tooth rows and relative tooth sizes. (Plate 85 compares it with a modern gorilla.) The lumbar vertebra appears more apelike than monkeylike.

The skull and limb bones are assigned to *Proconsul (africanus)*. The skull is from a subadult individual; its brain size seems comparable to that of living apes, and the teeth are apelike in having five rounded lower molar cusps. The limb bones comprise a mosaic of features. In size, they are similar to some colobines (leaf-eating Old World monkeys), whereas the elbow joint has some apelike features, according to anthropologist Mary Ellen Morbeck. The hand and wrist bones do not show adaptations for apelike brachiation or knuckle walking. Morbeck concludes that *Proconsul (africanus)* was quadrupedal. Bones of the hindlimb also show a mixture of monkey and apelike features, overall indicating a creature unlike any living anthropoid.

Complete the plate by coloring *Dendropithecus*, a pliopithecid, and the arrows from it to Maboko, Rusinga, and Songhor, where it has been found.

The species shown here, *D. macinnesi*, has canine sexual dimorphism and the typical canine honing (sharpening) mechanism, features thought to be present in the common ancestor of both apes and monkeys. The limb bones from this species are more slender than those of *Proconsul*, and for this reason, they have been described by some as gibbonlike. However, other paleontologists believe that gracile limb bones are an ancestral trait for all catarrhines.

In summary, the early Miocene shows a diversity of monkeylike and apelike species in size and proportions of the teeth. In species for which we also have remains of limb bones, there is a mosaic of anatomical features that do not clearly correspond to those of living monkeys or apes. The overall locomotor anatomy of *Proconsul (africanus)*, as well as of the Fayum "dental apes," resembles that of the large New World monkeys (but without the prehensile tail), suggesting that these early Miocene fossil species are much like the ancestral catarrhine that gave rise to both Old World monkeys and apes. None of the dental or limb bone evidence indicates that the pliopithecids or dryopithecids are directly ancestral to any of the living apes.

FOSSIL MONKEYS AND APES: EARLY MIOCENE.★

16★

22★

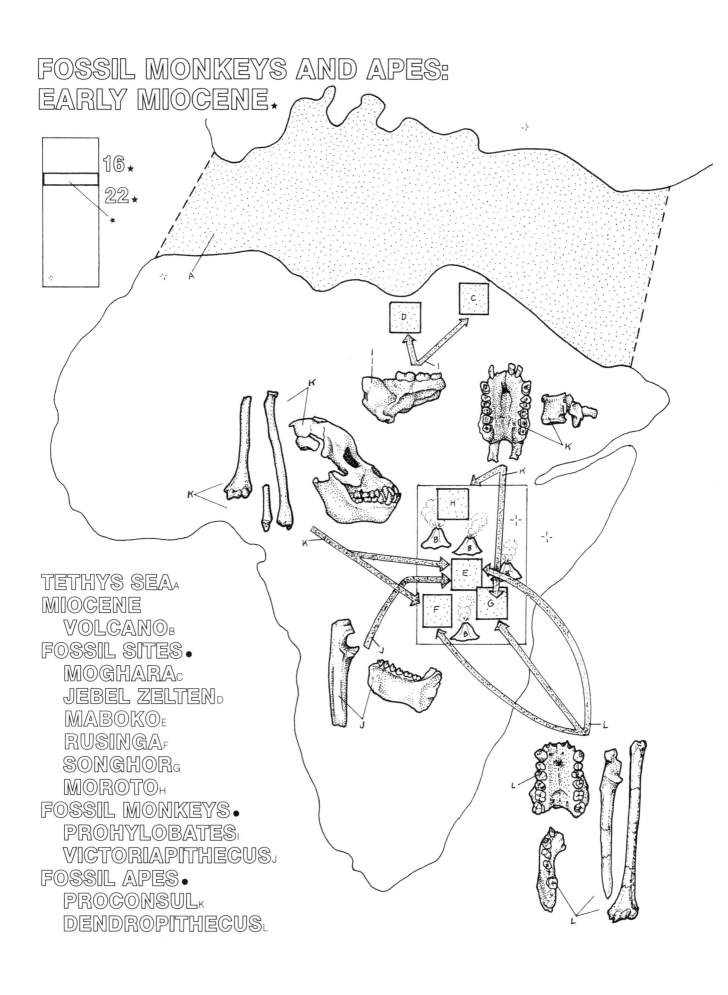

TETHYS SEA_A
MIOCENE
    VOLCANO_B
FOSSIL SITES●
    MOGHARA_C
    JEBEL ZELTEN_D
    MABOKO_E
    RUSINGA_F
    SONGHOR_G
    MOROTO_H
FOSSIL MONKEYS●
    PROHYLOBATES_I
    VICTORIAPITHECUS_J
FOSSIL APES●
    PROCONSUL_K
    DENDROPITHECUS_L

84

THE MIOCENE APES LEAVE AFRICA

Color the time chart on the top left and then the arrows showing the collision of Africa with Eurasia and the formation of the Rift Valley and the Red Sea.

Between 16 and 17 mya *Africa collided with Eurasia;* so the continents encircled the waters that became the Mediterranean Sea. The *Rift Valley* separated the Arabian Peninsula from the rest of Africa, forming the *Red Sea.* This new land connection made possible dispersal and exchange of animals between Africa and Eurasia. During this time, apelike species left Africa and speciated into a number of forms; their fossil remains have been found in Europe and Asia.

In the center of the plate, use light colors for the fossils that represent the three major families of Miocene apes: *Pliopithecus* of the family pliopithecid; *Dryopithecus* and *Proconsul,* both in the family dryopithecid; and *Ramapithecus/Sivapithecus* and *Gigantopithecus* in the ramapithecid family.

Now color the fossil sites, beginning with Spain in western Europe, moving eastward to China. Color the fossils too, noting their geographic distribution.

Pliopithecus and *Dryopithecus* are found only in Europe; *Proconsul,* only in Africa. The ramapithecids are found in Europe, Africa and Asia, but *Gigantopithecus,* only in Asia.

The ramapithecids all have relatively large, thickly enameled cheek teeth, characteristics they share with early hominids and orangutans. In contrast, dryopithecids and the African apes have relatively smaller, thinner enameled molars. Anatomist Richard Kay has shown that thick enamel in primate species correlates with a diet of fruits and nuts enclosed in tough pods.

At *Ft. Ternan,* dated about 14 mya, *Ramapithecus* remains have been found as well as those of *Proconsul.* Even older sediments in *Saudi Arabia* and *Turkey* have yielded specimens similar to the African species, *Ramapithecus wickeri.* Such finds suggest that *Ramapithecus* may have a Eurasian origin. In younger deposits than *Ft. Ternan* in *Turkey* and *Greece* (11 mya), *Sivapithecus* has been found. Fossil sites in *Hungary, China,* and *India/Pakistan*

also yield *Ramapithecus/Sivapithecus.* In general, *Sivapithecus* has larger canines than *Ramapithecus,* but canine sexual dimorphism may cause confusion between a female "Siva" and a male "Rama"; and this uncertainty applies to the Chinese material. Some paleontologists question whether "Rama" and "Siva" are distinguishable entities.

On the *India-Pakistan* border are the important Siwalik deposits. In 1910, Guy Pilgrim first found remains of both *Ramapithecus punjabicus* (which he mistakenly called *Dryopithecus*), and *Sivapithecus indicus.* In the 1930s, G. E. Lewis found more fossils and coined the name *Ramapithecus.* Since 1973, paleontologist David Pilbeam has collected more specimens, of *Ramapithecus, Sivapithecus,* and *Gigantopithecus* in deposits dated 8-12 mya.

Pilbeam originally agreed with Elwyn Simons that *Ramapithecus* was a hominid (directly ancestral to humans), but his recent discoveries in the Siwalik have led him to conclude it was not. On the basis of dental and cranial similarities, he now believes that *Sivapithecus* may be ancestral to the orangutan. Postcranial remains from the Siwalik thought to be ramapithecine are similar to those of apes but do not suggest bipedalism.

In summary, in the Middle Miocene there is still no evidence of the family of living apes (pongids) nor of the human family (hominid, which includes *Australopithecus*). By the middle Miocene, there is a radiation of apelike species out of Africa into Europe and Asia. After 14 mya, there are no fossil hominoids in Africa, until about 4 mya with *Australopithecus.* In Europe there are no fossil apes after 10-11 mya, and none in Asia after 7-8 mya. The reasons for this pattern of extinctions of the Miocene apes are intriguing ones and no doubt relate to changing habitats and climates affected by plate tectonics.

Interestingly, during this same time period, the fossil monkeys began to expand geographically and in numbers of species, a development that may or may not have a causal connection with the decline of Miocene apes. This and other intriguing questions about Miocene primate evolution remain to be answered, as new fossil evidence comes to light and as existing information is reevaluated.

THE MIOCENE APES LEAVE AFRICA ★

MIDDLE MIOCENE ★

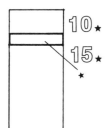

10 ★
15 ★
★

AFRICA COLLIDES WITH EURASIA A
RIFT VALLEY & RED SEA FORMATION B
FOSSIL SITES ●

SPAIN C
FRANCE D
GERMANY E
HUNGARY F
CZECHOSLOVAKIA G

GREECE H
TURKEY I
FT. TERNAN (KENYA) J
SAUDI ARABIA K
INDIA/PAKISTAN L
CHINA M

PLIOPITHECUS N

DRYOPITHECUS O

PROCONSUL O'

RAMAPITHECUS / SIVAPITHECUS P

GIGANTOPITHECUS P'

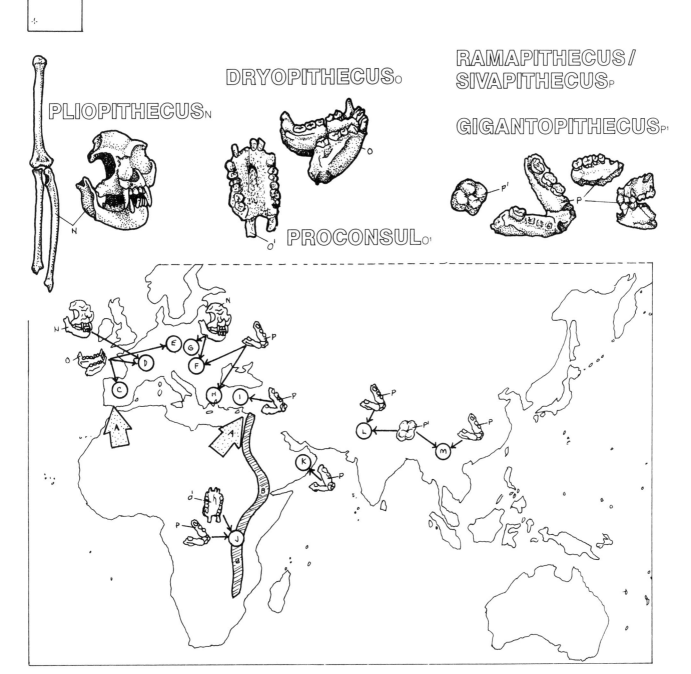

85

MIOCENE AND LIVING APES:
A QUESTION OF ANCESTRY

A radiation of fossil apes of many shapes and sizes into a wide range of habitats occurred throughout Africa, Europe, and Asia during the Miocene. Paleontologists have looked long and hard at these fossil species for potential ancestors to living apes and humans. A number of linkages have been proposed. At one time paleontologists Elwyn Simons and David Pilbeam maintained that chimpanzees and gorillas evolved from two differenct species of *Proconsul* and that *Pliopithecus* was an early gibbon. Such a view gives great time depth to the separation of the different apes, whereas the biochemistry shows a more recent divergence. Is there evidence in the dentition and limb bones of these fossil apes to argue for ancestry to living apes?

This plate examines the anatomy of two fossils: *Proconsul*, a member of the dryopithecids in Africa, and *Pliopithecus*, a European pliopithecid and one of the smallest of the Miocene apes.

Color the dentition of the palate (upper jaw) in *Proconsul* from Moroto and a female gorilla, chosen for comparison because of similar size. Color the incisors and the canines in the skull front views. The skull of Moroto is reconstructed to give an idea of the jaw's orientation relative to the skull.

The *molars* of *Proconsul* have low crowns and relatively thin enamel, as do the other dryopithecids, and are similar to the gorilla in these features as well as in their rounded cusps and in the Y-5 cusp pattern of the lower *molars*. A close look, however, reveals differences in relative size of the teeth. In particular, the *incisors* are smaller in *Proconsul* and differ in orientation. The *molars* and *premolars* also are smaller, but the *canines* are large and conical. According to Richard Kay, the *molars* of *Proconsul major* resemble those of chimpanzees and gibbons, indicating fruit eating rather than foliage eating as in the gorilla. So, although there are dental similarities between *Proconsul* and gorillas, the evidence is not convincing that the former was ancestral to the latter.

Pliopithecus, found in France and Czechoslovakia, and other members of the pliopithecid family from Hungary, Kenya, and Egypt for a long time have been considered directly ancestral to living gibbons. Skeletons from three individuals of *Pliopithecus* were found in Czechoslovakia. Complete forelimb bones of this species are compared to those of a spider monkey from the New World and a gibbon (an ape), the supposed descendent species from *Pliopithecus*.

Color the limb bones of *Pliopithecus*, *Ateles*, and *Hylobates*. Begin with the humerus and color it in all three species, including the side view closeups of the elbow joint. Next color the radius, then the ulna in all views, including the closeups of the wrist joint.

Pliopithecus, like *Ateles*, but unlike *Hylobates*, has a long olecranon process on the *ulna*, a definitely monkeylike feature, and the more prominent lower *ulna* (and anatomy of the carpal bones) indicates more limited movement in its wrist joint compared to the gibbon's. These features indicate an upper limb constructed more for weight bearing (like monkeys) than for suspension (like apes).

Although not shown here, there are several *Pliopithecus* vertebrae, including sacral and caudal ones. Anatomist Friderun Ankel deduced from the sacrum that *Pliopithecus* must have had a tail with about 25 caudal vertebrae, comparable to that of a vervet monkey from Africa. This feature is distinctly not apelike, as apes have no tails at all. What the limb bones and vertebrae tell us is that *Pliopithecus* was more monkeylike than apelike and probably was not ancestral to gibbons.

Paleontologist Peter Andrews points out that a diversity of the pliopithecids (small apes) lived during the Miocene. What distinguishes them from the dryopithecids, the larger apes, is their retention of primitive characters, which are first found in the Oligocene *Propliopithecus*. *Pliopithecus* was considered an ape because it retains a fifth cusp on its lower *molars*, a feature it shares with living apes, but not monkeys. However, the fifth cusp is a trait, found in the ancestors of both groups, that persisted in the apes but was lost in modern monkeys. Therefore, it tells us very little about the relationship of these ancient species to later ones.

Although there were a variety of smaller and larger "apes" in the Miocene, the dentition and limb bones fail to define these fossil species as ancestors of specific living apes.

MIOCENE AND LIVING APES: A QUESTION OF ANCESTRY.★

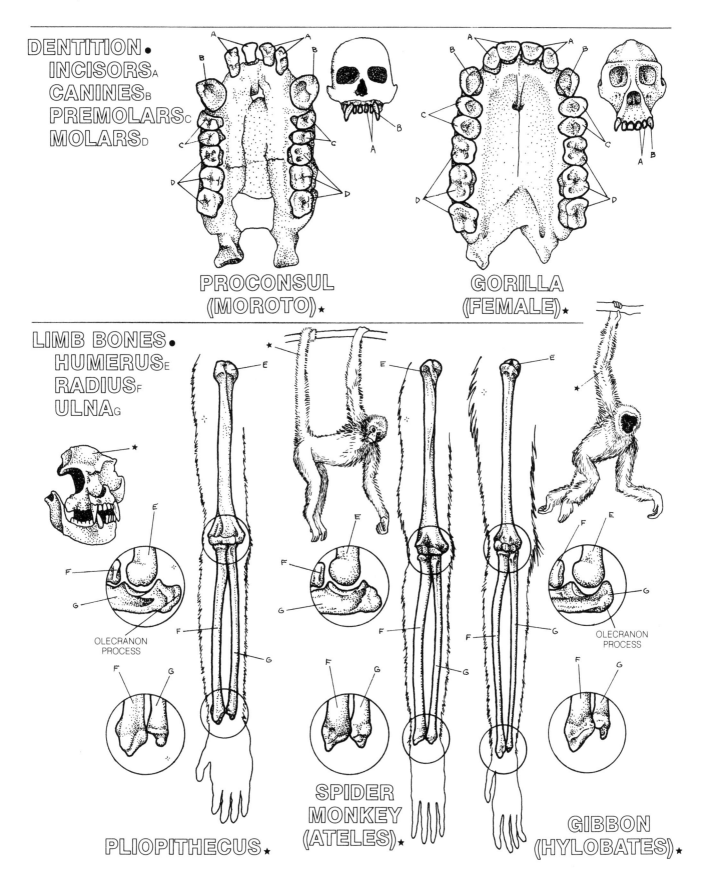

DENTITION.●
INCISORS_A
CANINES_B
PREMOLARS_C
MOLARS_D

PROCONSUL
(MOROTO)★

GORILLA
(FEMALE)★

LIMB BONES.●
HUMERUS_E
RADIUS_F
ULNA_G

OLECRANON
PROCESS

OLECRANON
PROCESS

PLIOPITHECUS★

SPIDER
MONKEY
(ATELES)★

GIBBON
(HYLOBATES)★

86
RAMAPITHECUS: THE APE THAT WOULD BE HUMAN

Ramapithecus, first found in India, was named in the 1930s after the Hindu prince "Rama," hence "Rama's ape." The few teeth and fragmentary jaws were resurrected again in the early 1960s by Yale palentologist Elwyn Simons and put forth as a member of the human family.

Subsequent specimens were found in Pakistan, the Middle East, and one in Kenya by the late Louis Leakey, which was dated at 14 million years old. Many researchers regarded the fossil as a type of *Dryopithecus* until it was nominated and defended as a *human* ancestor.

What was the evidence for this point of view? We here compare the teeth and jaws of *Ramapithecus* with those of a *pygmy chimpanzee* — small in overall jaw size and *canine* size — and with *humans,* to which *Ramapithecus* was supposedly ancestral.

Color the titles for pygmy chimpanzee, *Ramapithecus,* and human. These same colors will be used for upper and lower jaw bones surrounding the teeth. Then color at the top left the two views (from the bottom and side) of the teeth and palate of the pygmy chimpanzee and the upper jaw fragments and teeth from the Kenyan *Ramapithecus.*

Ramapithecus' molar and premolar teeth are larger than those of the *pygmy chimpanzee;* its *canines* are about the same. The fossil *canines* were said to be too small for an ape, yet they are similar to those of a living *pygmy chimpanzee* (an ape). Enamel thickness and cusp patterns of *premolars* and *molars* were also interpreted as being more *human* than apelike. Enamel thickness was further argued to provide evidence of a more ground-living, open country (as opposed to forest) way of life.

Color the teeth and palate of the three species. Leave the lower jaws uncolored for now.

Note that the shape of the tooth row differs in ape and *human.* The middle portion of *Ramapithecus'* palate was missing and consequently no fit was possible. In spite of this lack of evidence, *Ramapithecus* was reconstructed with a humanlike palate shape.

From these dental fragments, an entire *Ramapithecus,* walking upright, has been reconstructed. Color the jaws in the two figures at the top right, one in a bipedal position, the other quadrupedal, like living apes.

The pelvis and foot are the most diagnostic bones of *humans* for locomotor pattern, yet without any pelvic or limb bones at all, *Ramapithecus* was reconstructed as bipedal. It could as easily have been quadrupedal.

Almost 200 years ago, French paleontologist Cuvier boasted that he could reconstruct an animal from its teeth alone. But paleontologists have learned the hard way that teeth do not tell the whole story. A famous example of the unexpected is the chalichothere, a now extinct animal with teeth and skull like a horse, but with claws instead of hooves. For a long time, its teeth and claws were thought to be from two different species.

Paleontologist David Pilbeam's excavations in Pakistan have uncovered a complete lower jaw.

Finish the plate by coloring the three lower jaws at the bottom.

Human and *chimpanzee* jaws are different in shape; the *pygmy chimpanzee's* is somewhat U-shaped, the *human's* more parabolic. The fossil, with its small *incisor* teeth and large *molar* teeth, has a more V-shaped jaw, unlike either the *chimpanzee* or *human.*

Richard Kay, who has studied the *molar* enamel in numerous species of monkeys and apes, suggests that *Ramapithecus* ate hard seeds, nuts, and fruits that could as well have been in the forest as in nonrainforest environments.

Fragmentary skulls and limb bones have been found recently and they are more apelike than humanlike. There is still insufficient cranial, pelvic, and limb bone evidence unequivocally associated with the teeth to show whether *Ramapithecus* had a brain like a hominid, swung through the trees like an ape, or walked upright like a *human.* Until we know more about the rest of the animal, we cannot know its locomotor-feeding pattern or its exact place in ape and *human* evolution.

RAMAPITHECUS:
THE APE THAT WOULD BE HUMAN ★

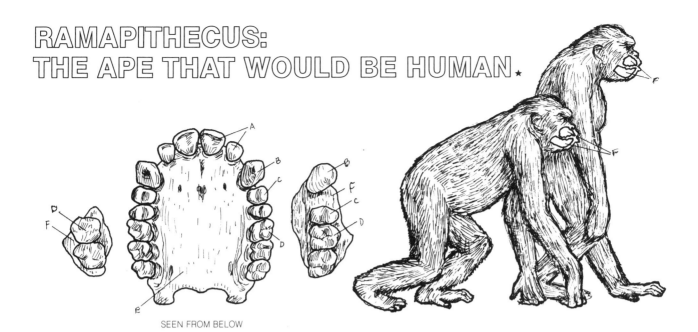

SEEN FROM BELOW

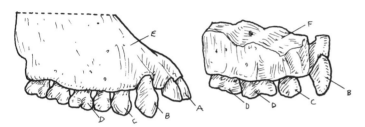

SEEN FROM THE SIDE

DENTITION ●
INCISOR A
CANINE B
PREMOLAR C
MOLAR D

PALATE ●

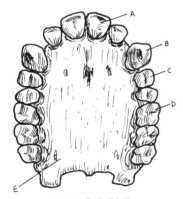

PYGMY
CHIMPANZEE E

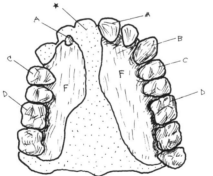

RAMAPITHECUS F

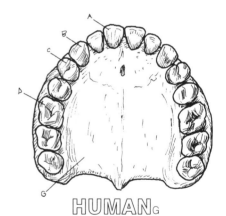

HUMAN G

LOWER JAW ●

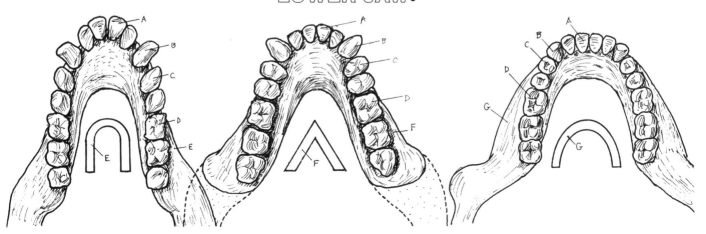

87

The new field of molecular anthropology has provided important new data on primate evolution, but many paleontologists have been slow to accept them. They argue that fossils are the only direct evidence and that molecules are not relevant to the study of extinct species. With regard to human origins, fossil bones and teeth may appear to conflict with molecular data, but, as we have seen with *Ramapithecus,* the fossil evidence is often fragmentary and subject to more than one interpretation.

Using light colors, color the vignettes representing the paleontological and the molecular views of phylogeny.

The two approaches are quite different: paleontologists dig up bones from the earth, study their shapes, and compare them to other living and extinct species in order to establish their place in an evolutionary phylogeny.

Molecular anthropologists study the proteins of living species and deduce how long ago two species diverged from a common ancestor. As we learned on Plates 22-28, proteins are made up of various combinations of the basic twenty amino acids, arranged in definite sequences. A given protein may include hundreds to thousands of amino acids. The proteins of closely related species, such as horse and donkey or dog and fox, are nearly identical; whereas species that diverged more than 100 mya, such as shrew and opossum, have many sequence differences.

These differences can be measured precisely, and their number is approximately proportional to the divergence time. Such "molecular clocks" are particularly valuable for evolutionary study because results can be, and have been, replicated in numerous laboratories, whereas the analysis of fossil bones and teeth is somewhat subjective, and agreement between researchers may be difficult to achieve.

Color the epochs to establish the fossils in time. Color the fossils. Now color the figures in both phylogenies; color the gibbons in each, then the orangutans, gorillas, chimpanzees, and humans. Color the graph lines with each figure.

Paleontological phylogenies have indicated a long,

independent evolution for the five living hominoid groups. They have suggested that *Pliopithecus* was an early *gibbon,* that different species of *Dryopithecus* were ancestral to *gorilla, chimpanzee,* and *orangutan,* and that *Ramapithecus* was ancestral to the *human* line.

In contrast, the *molecular phylogeny* shows a relatively recent divergence for all the apes, with the Asian apes splitting off earlier than the rest. This rules out the possibility that *Pliopithecus* is an ancestral *gibbon* or *Dryopithecus* or *Proconsul* are ancestral apes. The three-way divergence between *human, chimpanzee,* and *gorilla* about 5 mya does not preclude *Australopithecus* as an early representative of the *human* family. So, aside from the equivocal fossil evidence itself (Plates 85, 86), the molecular findings make it even less likely that *Ramapithecus* could have been a *human* ancestor.

To consider *Ramapithecus* as hominid, one would have to assume that primate proteins have evolved at half the rate of shark, fish, frog, snake, kangaroo, mouse, and elephant proteins. Some have argued that proteins evolve more slowly in animals with more time between generations—a presumption refuted by the evidence that mouse and elephant proteins have evolved at the same rate, as have loris and *human* proteins. This external check on the statistical constancy of the molecular clock is further supported by internal evidence: numerous proteins with different rates of change (Plate 28) (cytochrome *c*, albumin, transferrin, hemoglobin, and histones) indicate similar divergence times. That is like timing the same event with an hour hand, minute hand, and second hand and finding out that the times come out the same. The statistical constancy of the molecular clock is not an assumption but an observation based on an enormous amount of data.

From the biochemical evidence, then, there was but one ancestor prior to the ape-*human* split, not three different ones, for *humans* and the African apes. The molecular information has made it clear that we cannot look at *human* origins as an isolated phenomenon, but must consider the event as part of the radiation that included the African apes. If *Ramapithecus* has a place in *human* ancestry, it will have a place in *chimpanzee* and *gorilla* ancestry as well.

TWO VIEWS: ★

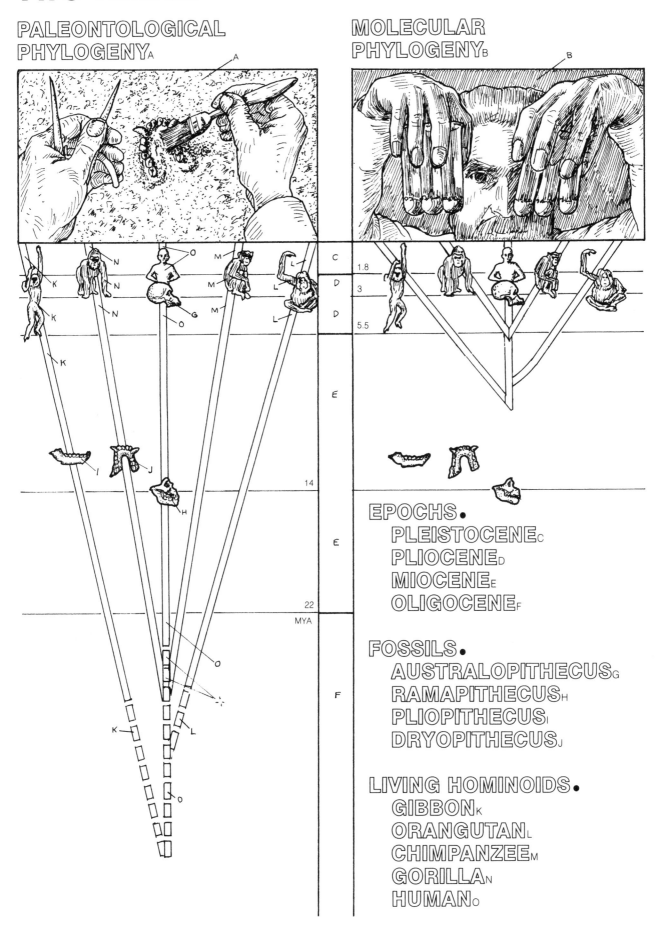

PALEONTOLOGICAL PHYLOGENY A

MOLECULAR PHYLOGENY B

EPOCHS ●
    PLEISTOCENE C
    PLIOCENE D
    MIOCENE E
    OLIGOCENE F

FOSSILS ●
    AUSTRALOPITHECUS G
    RAMAPITHECUS H
    PLIOPITHECUS I
    DRYOPITHECUS J

LIVING HOMINOIDS ●
    GIBBON K
    ORANGUTAN L
    CHIMPANZEE M
    GORILLA N
    HUMAN O

1.8 3 5.5 14 22 MYA

88

APE AND HUMAN: HOW THEY DIFFER

No one would mistake a chimpanzee for a human, but we are captivated by chimpanzees' appearance and behavior, seeing in their actions and expressions an image of what our ancient ancestors might have been.

At the molecular level humans and chimpanzees are very close, but their external appearances, their phenotypes, are distinct. The famous nineteenth-century anatomist Thomas Henry Huxley, in his 1863 book, *Evidence as to Man's Place in Nature,* argued on anatomical grounds that humans were most similar to the African apes, a conclusion confirmed 100 years later by biochemistry.

On this plate the comparison between chimpanzee and human is reviewed. By choosing the chimpanzee for comparison, we can rule out the "noise" of extremes in size, diet, locomotion, or sexual dimorphism.

Color the range in body weight for the ape and for the human at the bottom of the plate.

Small female adult chimpanzees may weigh as little as 25 kilograms, and some populations of humans, such as the !Kung San, who live as hunter-gatherers in Botswana, may weigh as little as 35 kilograms. It is unusual for a chimpanzee to reach 70 kilograms, an average weight for many human populations.

Using dark colors, color brain size and the canine teeth on the large figures and on the skull side views.

Although the modern human *brain* is more than three times as large as the chimpanzees', our earliest ancestors had ape-sized brains (Plate 110).

Human *canines* are small in both sexes; in chimpanzees they are larger in males, but large in females as well.

Using three related colors, color the upper limbs, hands, and thumbs. Begin with the bones, then color the rest and the small figures below. Be sure to color the percentages on the side that indicate the mass of the upper limbs relative to total body weight, as on Plates 51 and 52.

The muscles and bones of the *upper limb* and *hand* are very similar in chimpanzee and human. The main difference is that the chimpanzee has longer, more curved finger bones for hanging and climbing in trees and for walking on its knuckles. The human *thumb* is relatively longer and is well muscled for fine manipulation and control. The lengths of chimpanzee and human arms are similar, but their weights relative to total *body weight* are different. Remember (Plate 53) that chimpanzees also have relatively heavier *hands.*

Color the trunks and related structures.

The chimpanzee's *clavicle* angles upward and gives it the appearance of having no neck; the human shoulder joint is lower. Chimpanzee and human *trunks* are similar in relative size. An important difference is in the shape of the *pelvis,* a hominid adaptation for bipedalism.

Using three related colors, as with the upper limb, color the lower limb, foot, and great toe.

Shorter *lower limbs* give the chimpanzee its squat appearance, despite a similar sized *trunk.* Humans have longer, heavier, *lower limbs.* The human *great toe* lies in line with the others.

The figures at the bottom summarize the story: with similar *trunks,* the chimpanzees' relatively longer *upper limbs* equip them for hanging and climbing in trees, and knuckle walking. The long, muscular *lower limbs* of humans equip them for bipedal locomotion.

Thomas Henry Huxley argued that humans descended from apes through the mechanism of natural selection that had recently been proposed by his colleague, Charles Darwin. Huxley demonstrated the numerous similarities in the skeleton, muscles, teeth, *brain,* and embryos of African apes and humans. He pointed out that the differences were mostly related to differing locomotor patterns, *canine teeth,* and *brain size.* In the mid-nineteenth century, when Huxley was writing, there was no fossil evidence for early humans and therefore no way to test his conclusions. Now we know that the earliest hominids can be described as bipedal chimpanzees.

The DNA and proteins show chimpanzees and humans to be 99 percent similar—as closely related to each other as are horses to zebras and grizzly bears to polar bears. The evidence of the proteins, combined with the fossil record, indicates that relatively few genetic changes have had profound anatomical effects. This supports the hypothesis that a recent divergence between ape and human occurred rapidly in evolutionary time.

APE-HUMAN COMPARISONS.★

BODY WEIGHTA
BRAIN SIZEB
CANINE TEETHC
UPPER LIMBD
HANDE
THUMBF

TRUNKG
CLAVICLEG¹
RIB CAGEG²
PELVISG³

LOWER LIMBH
FOOT
GREAT TOEJ

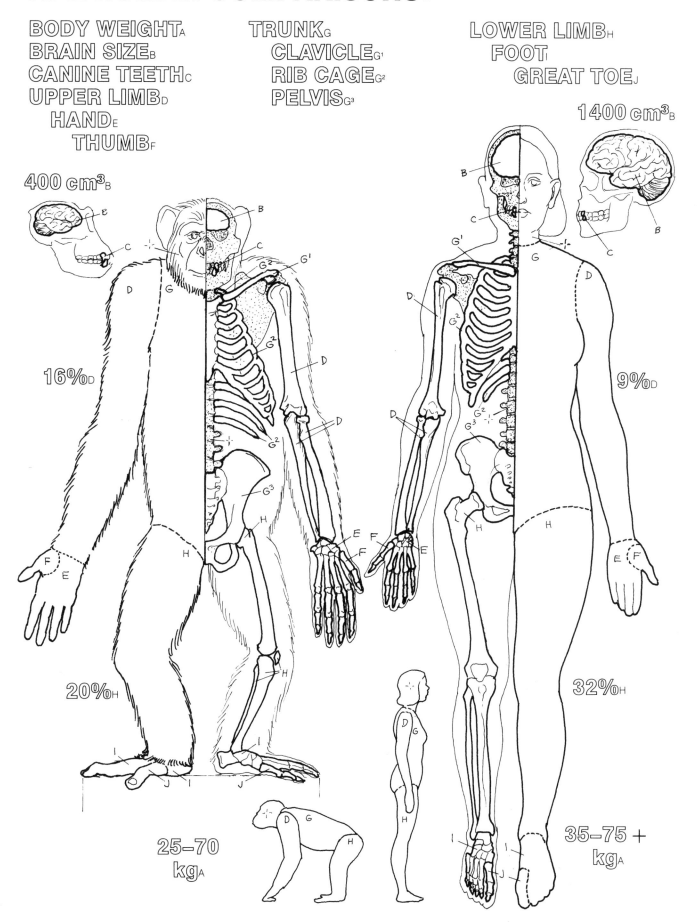

1400 cm³B

400 cm³B

16%D

9%D

20%H

32%H

25–70 kgA

35–75 + kgA

89

PYGMY CHIMPANZEES: A PROTOTYPE FOR THE COMMON ANCESTOR?

If humans and African apes had a common ancestor, what did that ancestor look like? The molecular, fossil, and comparative anatomical data narrow down the choices. Our candidate must have the potential to evolve, within the space of about five million years, into the large dimorphic vegetarian gorilla, the versatile, tool-using *chimpanzee,* and a bipedal human.

Many comparative anatomists have suggested that the common ancestor must have resembled a *chimpanzee*. If it was an ape, the only other choice in terms of genetic closeness would be a gorillalike creature. Something like a *chimpanzee* seems more likely.

The term *chimpanzee* generally implies the *common chimpanzee, Pan troglodytes* (Plate 84). But there is another species, the so-called *pygmy chimpanzee, Pan paniscus,* with a habitat restricted to the Congo River Basin in Zaire. In the 1930s biologist Harold Coolidge noted that *pygmy chimpanzees* are even less specialized than the *common* species, in having shorter arms relative to legs, a narrower chest, and smaller *canine* teeth.

Using two light colors, begin with the pygmy and common chimpanzee figures on the sides of the page. Color the heads on the central figures and the two female skulls, including canines, lower jaws, and cranial capacity at the bottom of the plate.

The *common chimpanzee's cranium* is larger and its lower jaw longer than the *pygmy chimpanzee's. Pygmy chimpanzee* teeth, especially the *canines,* are smaller than those of the *common* species.

Color the male skulls, their canines, lower jaws and cranial capacity.

In contrast with the *common chimpanzee, pygmy chimpanzees* have no dimorphism in *cranial capacity, jaw,* or face size and only a little in *canine* tooth size. In both species, males and females differ in body weight, with females 80–85 percent of male body weight. On average, *common chimpanzee* males have longer limb bones than females, but in *pygmy chimpanzees,* they are the same length.

Using the same two light colors that you chose for the heads, color the trunks of the central

figures. Then color the clavicles, scapulae, and pelvises.

Pygmy chimpanzees have smaller *clavicles* and *scapulae,* narrower and shorter *pelvises.*

Color the bones of the upper limb and those of the lower limb. Leave the mass uncolored for now.

Note that in bone lengths there is little difference between the species but that the upper limb of the *common chimpanzee* is longer relative to its lower one. Thus, the two *chimpanzees* differ more in the size of their teeth, jaws, and skulls than in the rest of the skeleton. *Pygmy chimpanzees* are not really "pygmy."

Bones comprise less than 20 percent of a primate's weight (Plates 51 and 52), so I studied the soft tissues, comparing two female *chimpanzees* of similar body weight, one of each species.

Color the rest of the upper limb, representing its mass, and the 16 percent on each side. Now color the lower limb mass and the percentages.

Pygmy chimpanzees have slightly shorter *upper limb bones*. Their *lower limb bones,* though similar in length to those of *common chimpanzees,* are relatively heavier, almost a third more. Heavy lower limbs are an Old World monkey trait (Plate 51), which emphasizes the more generalized body build of *pygmy chimpanzees*. Long and heavy upper limbs are an ape specialization.

The lessons from this study are several. First, teeth and jaws alone do not predict everything else about an animals's body, thus, the misnomer of *"pygmy" chimpanzee*. Second, the skeleton does not give the full picture of an animal's body build. Third, a common ancestor that resembled a *pygmy chimpanzee,* being the smallest and least dimorphic of African apes, could have evolved into the more specialized relatives—*common chimpanzee,* gorilla, and human. The heavier lower limbs of *pygmy chimpanzees* are more like those of humans (Plate 88), which explains why *pygmy chimpanzees* are frequently upright. A *pygmy chimpanzee* prototype further narrows the gap between a quadrupedal ape ancestor and a small bipedal early hominid and helps us imagine how this transition could have taken place.

A PROTOTYPE FOR THE COMMON ANCESTOR? *

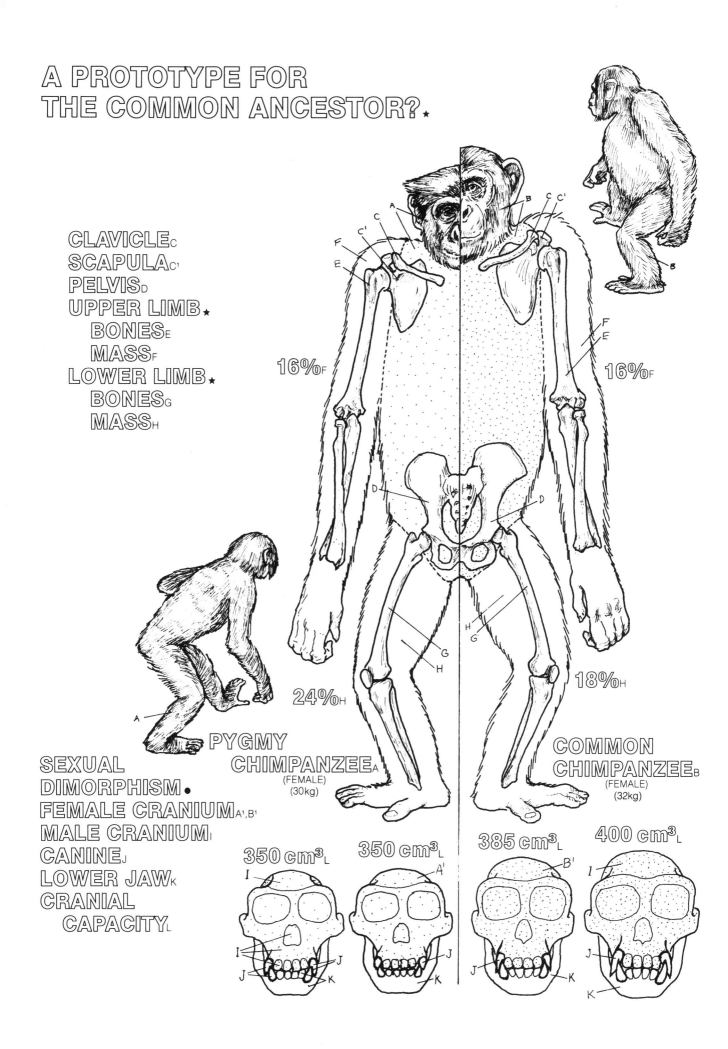

CLAVICLE c
SCAPULA c'
PELVIS d
UPPER LIMB *
   BONES e
   MASS f
LOWER LIMB *
   BONES g
   MASS h

16% f   16% f

24% h   18% h

PYGMY
CHIMPANZEE a
(FEMALE)
(30kg)

COMMON
CHIMPANZEE b
(FEMALE)
(32kg)

SEXUAL
DIMORPHISM •
FEMALE CRANIUM a',b'
MALE CRANIUM i
CANINE j
LOWER JAW k
CRANIAL
   CAPACITY l

350 cm³ l   350 cm³ l   385 cm³ l   400 cm³ l

90

CHIMPANZEE BEHAVIOR: PRECURSORS TO A SAVANNA LIFESTYLE

Evolution is both conservative and experimental. New ways of life are continuous with the old; new patterns emerge by remodeling elements of the old, as the ear evolved from ancestral branchial arches (Plate 12). These principles also apply to human origins.

Previously, we concluded that the human line probably diverged from that of the African apes about 5 mya, that chimpanzees are the least "specialized" of the apes, and that the common ancestor of humans and African apes may have resembled a pygmy chimpanzee in anatomy and body build.

Numerous studies (Plates 24-27) have shown the genes of chimpanzees and humans to be about 99 percent identical. Field observations by Jane Goodall and others and laboratory studies, especially those on *communication* and *tool using,* reveal how much chimpanzee behavior resembles humans. These observations provide a basis for understanding how ancestral human populations may have behaved as they moved out of the forest into the savannas of eastern and southern Africa.

It seems most likely that the divergence of hominids (early humans) from apes derived from a new locomotor feeding pattern involving bipedalism and regular use of tools in gathering and preparing food. The first tools may have been *digging* sticks and containers for collecting and *carrying* small food items dispersed over a large home range. *Bipedalism* made it possible for hominids to carry tools and babies over long distances. *Mothers* gathered food and shared it with their young. Social ties within this *mother and offspring* group, and among individuals in a wider social network, which may have also involved food *sharing,* provided the context for learning and *communicating.* These "elements" of behavior — *bipedalism, tool using, food getting, communication,* and extensive *social ties* — are present in chimpanzees.

Using five light but contrasting colors for this plate, begin by coloring the two examples of bipedalism: food carrying and display. Then color the backgrounds as labeled: food getting (C) for the background of the first figure (A¹) and tool using display (B¹) as background for the second example of bipedalism (A²) to emphasize the interrelatedness of behaviors. Color each be-havior as it is discussed, coloring the backgrounds where designated.

Bipedalism distinguishes humans from apes, but chimpanzees are occasionally *bipedal* when *carrying food* or other objects and when *displaying* through waving arms or brandishing sticks.

Tool using is fundamental to the human adaptation, but again this potential must have existed in the ancestral hominids, for chimpanzees use tools in more versatile ways than any other species except our own. Chimpanzees modify and use twigs to "fish" for *termites* in termite mounds. They also use sticks as *levers,* rocks for *cracking* hard-shelled foods, and crumpled leaves as sponges.

Other than picking fruits or leaves, chimpanzee *food getting* includes *carrying, termiting, cracking* hard-shelled fruits, and *digging* up edible roots. Chimpanzees sometimes *catch* and eat small *animals* with their hands and tear them apart without any kind of tool. All these ways for obtaining foods must have been important for the ancestral hominids.

Chimpanzees communicate through facial expressions (Plate 65), vocalizations, postures, and gestures that often resemble our own. Here a chimpanzee with outstretched hand is peering into a friend's face and *begging* for a piece of meat. Chimpanzees do occasionally *share* food, most often a *mother* with her *offspring. Sharing* is an expression of *social ties* more than it is a significant source of calories.

Social ties between *mother and offspring* and among *siblings* last for life, and these *ties* serve as a basis for *learning.* A *mother* "termites" while her *offspring* looks on; it first *learns by watching,* then by practicing the task alone. By age five, it has mastered *termiting.* Clearly, chimpanzees can learn to use tools without language. *Social ties* that extend beyond this small "kin" group form the basis for flexible groups characteristic of chimpanzees. Such social groups among ancestral hominids, capable of *sharing, communicating,* and *learning* from each other, went out of the forest on to the savanna.

Although other savanna inhabitants, including baboons, lions, and leopards have been studied as potential models for hominid social behavior, chimpanzees help us imagine best how the early hominids might have evolved a savanna way of life through the operation of natural selection on this chimpanzeelike range of behaviors.

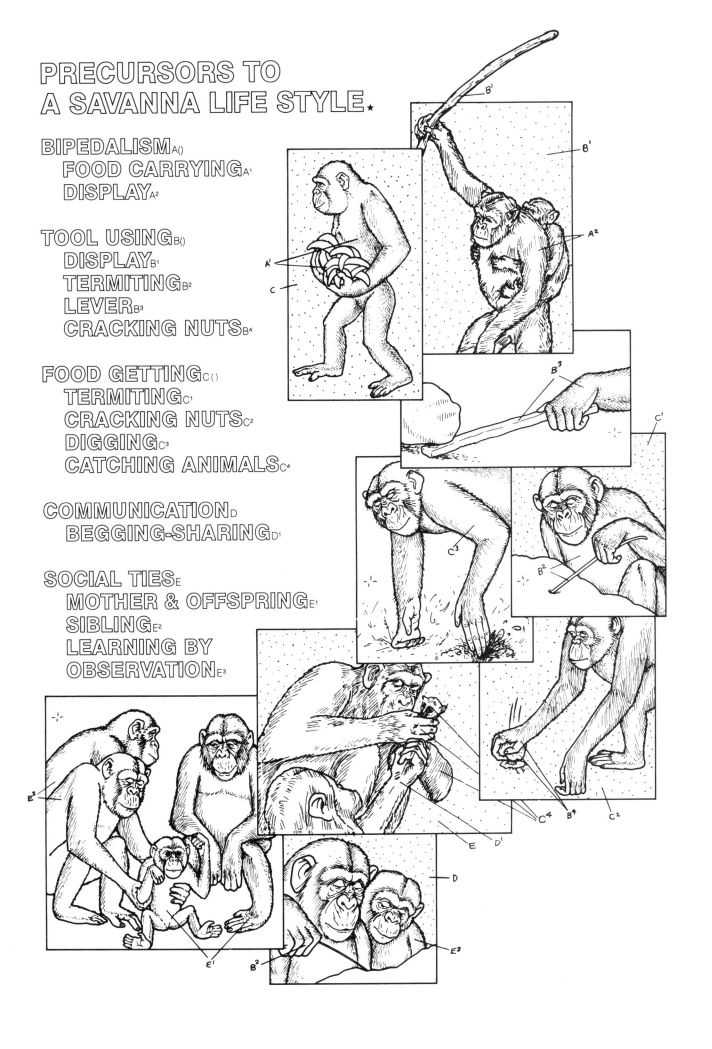

PRECURSORS TO A SAVANNA LIFE STYLE.★

BIPEDALISMA()
    FOOD CARRYINGA1
    DISPLAYA2

TOOL USINGB()
    DISPLAYB1
    TERMITINGB2
    LEVERB3
    CRACKING NUTSB4

FOOD GETTINGC()
    TERMITINGC1
    CRACKING NUTSC2
    DIGGINGC3
    CATCHING ANIMALSC4

COMMUNICATIOND
    BEGGING-SHARINGD1

SOCIAL TIESE
    MOTHER & OFFSPRINGE1
    SIBLINGE2
    LEARNING BY
    OBSERVATIONE3

91

THE ORIGIN OF THE THIRD APE

The earliest fossil evidence of the human line comes from sites in the eastern and southern African *savannas,* removed from forested areas where *chimpanzees* and *gorillas* live. Humans have been called "the third ape," after Darwin's deduction that they originated in Africa and are closely related to *chimpanzees* and *gorillas.*

As ancestral apes presumably moved out of the *forest* into more open country, the gathering of plant food would have been a critical innovation. In contrast to the individual foraging of apes, gathering involves (1) collection of food for more than one individual and for later, rather than on-the-spot consumption; (2) regular tool use for obtaining and preparing food; (3) bipedal carrying; and (4) consistent extensive sharing networks for the distribution of gathered items. The *savanna* was a new and dangerous environment, alive with big cats and other predators, but it also offered new opportunity for obtaining food.

Begin by coloring the major vegetation zones of Africa. Use a bright color like orange for the Mediterranean scrub area, regions of mild climate and seasonal rainfall. Color the desert, which receives less than 32 centimeters of rainfall a year, light yellow or brown. Color the forest green; then color the gorilla, pygmy chimpanzee and half of the common chimpanzee.

In western and central Africa, the tropical *forest* is characterized by thick dense *forest* vegetation and equatorial downpours. The lowland *gorilla* population is found on the west coast, the highland population, on the eastern rift. *Pygmy chimpanzees* are found only in a small enclave in the Zaire (Congo) River Basin.

Color the woodland savanna and the other half of the common chimpanzee in another shade of green.

In the *woodland savanna,* where *forests* are discontinuous, the *common chimpanzee* extends from Senegal in the west to Tanzania in the east.

Complete the plate by coloring the savanna and the early hominid fossils in a bright, contrasting color.

Hominid fossils are found at sites near water sources in *savannas.* The vegetation there today resembles a mosaic, varying from wooded areas along water courses, to low shrub, woodlands, and open grasslands. Plant life is dominated by alternation of wet and dry seasons, probably similar to the Plio-Pleistocene climate.

Some of the human fossils found in ancient *savanna* habitats include "Lucy," almost half a skeleton found at Hadar, Ethiopia, by Donald Johanson; bipedal footprints discovered by Mary Leakey at Laetoli, Tanzania; and the Taung baby, the first *Australopithecus* ("southern ape") named by Raymond Dart in 1925.

The origin of the human line was part of a radiation of new species about 5 mya that included the *chimpanzee* and *gorillas.* How does a new species "radiate" into a new environment? The new geographical niche must be accessible from the old, and the ancestral population must have some "preadaptations" that suit it for living there. Once there, it must not be competing for limited resources with species already established, or else it must have decisive advantages.

Access to the *savanna* obviously existed. The earliest hominids might have established themselves first in *woodland savanna,* areas where *common chimpanzees* sometimes live now. When fossils become numerous about 3.5 mya, *early humans* were established in the *savanna* mosaic. The opportunities for *savanna* living included abundant and varied food sources—fruits, berries, nuts, roots and tubers, seeds, birds, eggs, insects, and many small animals. It was a matter of learning to collect and carry food to safer clusters of trees for consumption. Their diet still included fruit, but tended to contain more seeds, roots, and meat than before, as suggested by the *Australopithecus'* large and heavily worn molars. Then as now there were large herds of hoofed animals like gazelles, gnus, and zebras.

So early hominids came to occupy their new *savanna* niche as a diurnal omnivore, with the unique ability to gather their food. They were not in direct competition with many other species. By walking upright, using tools, and keeping a sharp watch for predators and potential sources of food, *Australopithecus* increased its chances of eating without being eaten.

THE ORIGIN OF THE THIRD APE.★

MEDITERRANEAN SCRUB_A
DESERT_B
FOREST_C
   PYGMY CHIMPANZEE_{C1}
   COMMON CHIMPANZEE_{C2}
   GORILLA_{C3}
WOODLAND SAVANNA_D
   COMMON CHIMPANZEE_{D1}
SAVANNA_E
   EARLY HUMANS_{E1}

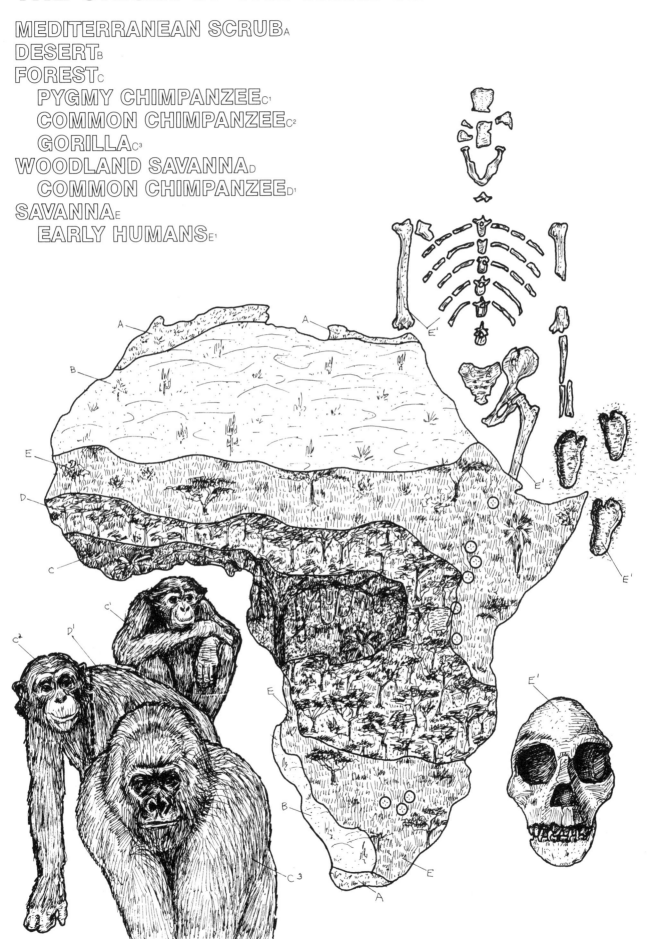

92

FOSSILS IN THE MAKING: FROM DEATH TO DISCOVERY

The making of a fossil—from the *death* of the organism to its *burial* and its final *discovery* in geological deposits—is a story that reveals why the fossil record is biased and incomplete. Through the findings of the relatively new field of "taphonomy" (the study of *death* and *burial* and subsequent preservation of animals), we learn how various kinds of information about past life forms are lost (for example, how one fossil, possibly of an atypical individual, must represent an entire ancient population or species). Eighty percent of an animal, namely, its soft tissues (muscle, skin, brain, and viscera) is almost never preserved, and the skeletal remains may consist of only a few bones or teeth to represent the entire animal. Teeth are preserved more often than any other body part, but teeth comprise only 1 percent of an animal. Errors and uncertainty are inevitable, and caution is highly desirable, in reconstructing the lifeways of extinct species from such scanty evidence.

As an example of the *fossilization* process, we examine an elephant from the moment of its *death* onward. Because they are so large, an elephant's bones are easy to follow as they become *fossilized*. Paleontologist Kay Behrensmeyer reminds us that large body size in fact creates a bias, for large animals are overrepresented and small ones are underrepresented in the fossil record.

Using a light color, begin with the herd of living elephants.

Remember that elephants' locomotor, dietary, and social behavior, as well as their variability due to sex and age differences, must somehow be inferred from the structures that *fossilize.*

Below the herd, color the elephant that has died from disease or starvation, was killed by a predator, or was trapped in a swamp.

If the animal is covered with mud or sand at this point, as apparently happened to one or two found at Olduvai Gorge and others at Torralba in Spain, most of the skeleton will be *fossilized* and recovered.

Color the elephant whose tissues have largely decomposed.

In an arid environment like the savanna in the dry season, the soft tissues, if not eaten, will *decompose,* leaving only the bones. In the case of the elephant, its thick, heavy skin remains intact for a long time; but it too eventually decays or is scavenged by hyenas or vultures.

Color the bones being transported by carnivore action and by stream action.

Scavengers not only eat the flesh and bones of dead animals, they also scatter the remaining bones around, or *transport* them elsewhere, as this hyena is doing. Skeletal parts are of different sizes, shapes, and densities; so these remains may be widely separated by water action. Some, especially teeth, fall to the bottom and are buried in a river at the immediate site of *death.* Others, skulls and vertebrae, for example, may float or roll a great distance downstream.

Color the bones undergoing weathering and burial.

After the predators and scavengers are finished, bones may lie on the surface and undergo *weathering* before final *burial.* Sun, rain, and wind will break, distort, and reshape bones, causing further loss of the original information.

Color the bones undergoing fossilization.

Once completely *buried,* the bones undergo *fossilization,* a process in which the protein material is replaced by minerals from the water in the enclosing soil. This transforms a bone into a fossil while perfectly preserving its form.

Complete the plate by coloring erosion and discovery.

Fossils are usually discovered when the sediments in which they lie become exposed to view, through *erosion* by water and wind action. Timing is essential. As Donald Johanson, the discoverer of "Lucy," remarked of his fossil: five years before, she would have been still buried and unobserved; five years later, disintegrated and lost to science. In the history of a fossil and its recovery, chance, fate, and luck often have the final word.

FROM DEATH TO DISCOVERY.★

HERDᴀ
DEATHʙ
DECOMPOSITIONᴄ
TRANSPORTᴅ
WEATHERING
    AND BURIALᴇ
FOSSILIZATIONꜰ
EROSION AND
    DISCOVERYɢ

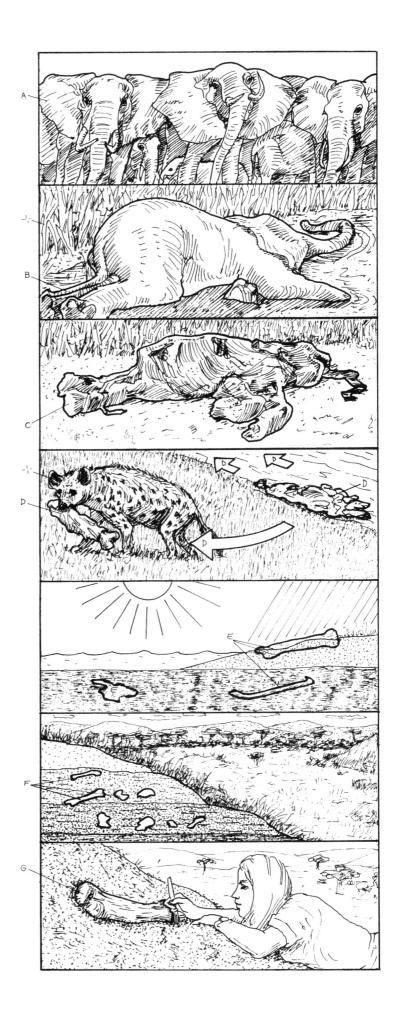

93

MEASURING TIME: THE NUCLEAR CLOCK

For a long time, geologists guessed at the ages of various deposits from their depth or thickness or from the kind of rocks and fossils they contained. And the ages of these same fossils were often estimated from the age of geologic deposits. It is not surprising that until recently many estimates of evolutionary events were inaccurate. Just as the molecular clock revolutionized the study of biological evolution, the discovery of radioactive clocks has revolutionized the earth sciences and dating of evolutionary events.

Radioactive isotopes decay at constant rates, unaffected by temperature, acidity, moisture, or other conditions that may affect rocks, minerals, and fossils and may make them appear older or younger than their actual age. A *parent* isotope decays into one or more *daughter* isotopes. If one knows the decay rate, expressed as a "half-life," and the ratio of *daughter* to *parent* isotopes, one can accurately calculate the time since the process began.

The most commonly used radioactive clocks are carbon 14 (^{14}C) and *potassium 40* (^{40}K). Carbon 14 is being constantly "created" by solar bombardment of the upper atmosphere; it is incorporated into plants as they absorb carbon dioxide. When the plant dies, carbon 14 decays into nitrogen 14, with a half-life of 5,370 years; so wood, for example, can be dated by measuring the ratio of carbon 14 to the normal carbon isotope, carbon 12. Because of the relatively short half-life, this method can date specimens not older than 50,000 years.

Potassium 40 has a much longer half-life (1.3 billion years); so it can be used to date events as old as the formation of the earth 4.5 billion years ago, as well as lava flows only a few thousand years old.

Color the top picture, noting that potassium 40 decays into two daughter isotopes, calcium 40 (^{40}Ca) and argon 40 (^{40}Ar). The proportions of represented atoms is diagrammatic only.

Because *calcium 40* is widely distributed in rock and soil, its presence is not informative. *Argon 40*, a gas, is rarely found in rocks unless it was trapped in lava by the decay of *potassium 40*. In liquid lava, inside the earth, the gaseous *argon 40* boils off; so the clock of

the *argon 40* to *potassium 40* ratio begins when the lava flows out and solidifies. At that time, the ratio is zero, as there is no *argon 40* present. One half-life (1.3 billion years) later, 50 percent of the *potassium 40* has decayed; 89 percent of the *daughters* are *calcium 40*, and 11 percent *argon 40*, giving an *argon* to *potassium* (Ar:K) ratio in the lava of 0.11; this ratio will continue to increase with time, and when measured can be used to date the lava.

Color the chart in the center of the plate. Notice how the amount of potassium 40 decreases with time as it decays into its daughter isotopes.

This chart shows the fate of *potassium 40* in rocks formed 3.8 billion years ago, when the earth's crust first formed. Notice that for each time period of earth history, there is a characteristic ratio of ^{40}Ar to ^{40}K.

Color the picture on the bottom of the plate, noticing how volcanic tuffs (lava flows) laid down 20 mya contain a higher ratio of argon to potassium than those in more recent deposits.

Luckily, the East African Rift Valley, where so many human fossils have been found, has been a volcanically active region for millions of years, so that many fossils can be dated accurately by measuring the Ar:K ratio in the lava found with them or in layers above or below them. By comparing the Ar:K ratio with a chart similar to the one here, the age of the fossils can be determined.

Other radioactive methods for dating rocks, fossils, and artifacts include uranium and fission-track dating. Uranium 238 (half-life 4.5 million years) decays in a complex manner, eventually ending up as stable lead isotopes. As with *potassium-argon* dating, the ratio of *daughter* to *parent* isotopes measures the time elapsed since the uranium clock was "set" when the minerals in which it is found first crystallized. Uranium isotopes also undergo spontaneous fission. When they do, heavy energetic fragments of the uranium atom fly apart and leave "tracks" in certain minerals that can be seen with a microscope and counted. The number of tracks indicates the elapsed time since the mineral was formed.

THE NUCLEAR CLOCK.★

POTASSIUM 40 (K) (PARENT)A
CALCIUM 40 (Ca) (DAUGHTER)B
ARGON 40 (Ar) (DAUGHTER)C

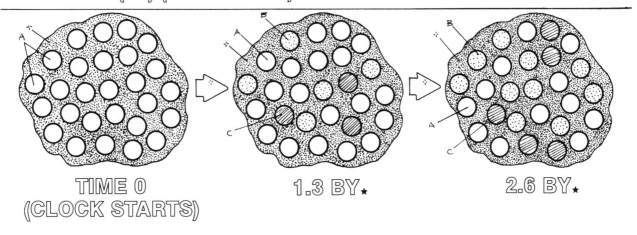

TIME 0
(CLOCK STARTS)

1.3 BY ★

2.6 BY ★

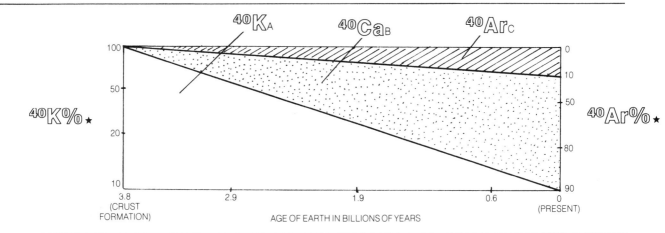

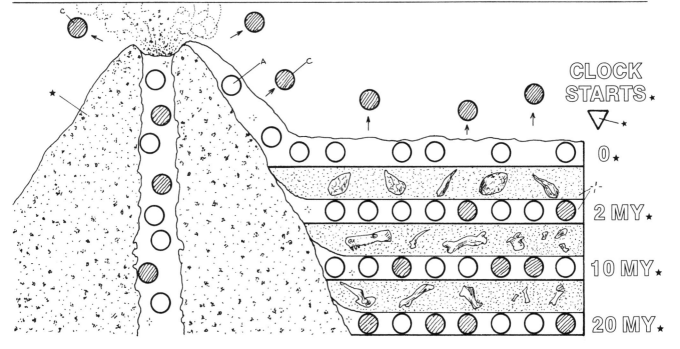

Mary Leakey's 1959 discovery of the nearly complete hominid skull "Zinjanthropus" focused world attention on Olduvai Gorge. Louis Leakey made his first trip there in 1931 and immediately discovered primitive stone *tools*. Over the next 30 years, Mary and Louis Leakey uncovered numerous species of extinct animals, and many stone *tools* before discovering the hominids who made them.

Color the plant, top right, green.

This plant (in Masai "ol" = place, "dupai" = wild sisal) accounts for the Gorge's name "the place of the wild sisal."

Color the titles of the Olduvai beds on the left and in the larger cross section in the middle. Color the corresponding layers in the diagram at top left. Notice the four major beds, and the more recent ones referred to by specific place names: Masek, Ndutu, and Naisuisiu.

The Gorge, 25 miles long and over 300 feet deep, is abruptly cut into the Serengeti Plain by recent rivers, as the Grand Canyon has been cut by the Colorado River, exposing deposits formed in an ancient lake. Its sediments covered animals, hominids, and their *tools*.

Color the potassium-argon dates on the right.

Lava from prehistoric volcanic activity makes it possible to date the Olduvai sediments by the potassium-argon method (Plate 93). When, in 1961, geologists Jack Evernden and Garniss Curtis announced that these sediments were nearly 2 million years old, the scientific community reacted with surprise and disbelief, for this was far older than these early humans had been thought to be. But many subsequent measurements, and the independent fission-track method, have confirmed the dates.

Bones of more than 40 individual hominids and more than 20 "occupation sites" with stone *tools* have been found at Olduvai. There are two distinct kinds of hominids, *Australopithecus* and *Homo,* and two *tool* traditions, known as Oldowan (for example, choppers and crude scrapers) and Acheulian (for example, hand axes sharpened on both sides).

Color *Australopithecus* in Bed I. ("OH" stands for Olduvai hominid, the number is the specimen number.)

Louis Leakey named it *"Zinjanthropus boisei"* (Zinj = Persian for East Africa; anthropus = man; Boise had provided financial support to the Leakeys). However, anatomist Phillip Tobias recognized its resemblance to some South African hominids *(Australopithecus robustus)* and "Zinj" was renamed *Australopithecus boisei.* Its enormous teeth earned it the nickname "nutcracker man"; its cranial capacity was a mere 530 cm³. Other skeletal parts of this species include the femoral fragment shown here.

Color the *Homo* fossils in Bed I.

Within two years of the discovery of "Zinj," the Leakeys found another species of the same age with smaller teeth, a bigger brain (650 cm³), and a smaller body. They found a complete foot (Plate 103), hand bones (Plate 104), a tibia and fibula, and skulls and teeth. Leakey, Tobias, and Napier named this new species *Homo habilis* ("handy").

Color the hominids in Beds II, IV, Masek, and Ndutu.

The fragmentary skull OH 13 (Olduvai Hominid 13) is classified as *Homo habilis,* but the partial skull OH 9 (capacity 1000 cm³) at the top of *Bed II* and those above it are considered *Homo erectus,* suggesting that *habilis* had evolved into *erectus* during this time interval. Femoral and innominate fragments (OH 28) suggest a larger body size.

Complete the plate by coloring the tools in all beds.

Olduvai was the first hominid site with an excellent association of *tools,* hominids, and animal bones. It put to rest the debate whether these small-brained creatures could have made and used *tools*.

The Leakeys' discoveries at Olduvai Gorge had a profound impact on our understanding of human evolution. Human origins were pushed back to 2 million years, twice what was previously known, and the co-existence of *Australopithecus* and *Homo,* suspected from the South African finds, was convincingly demonstrated.

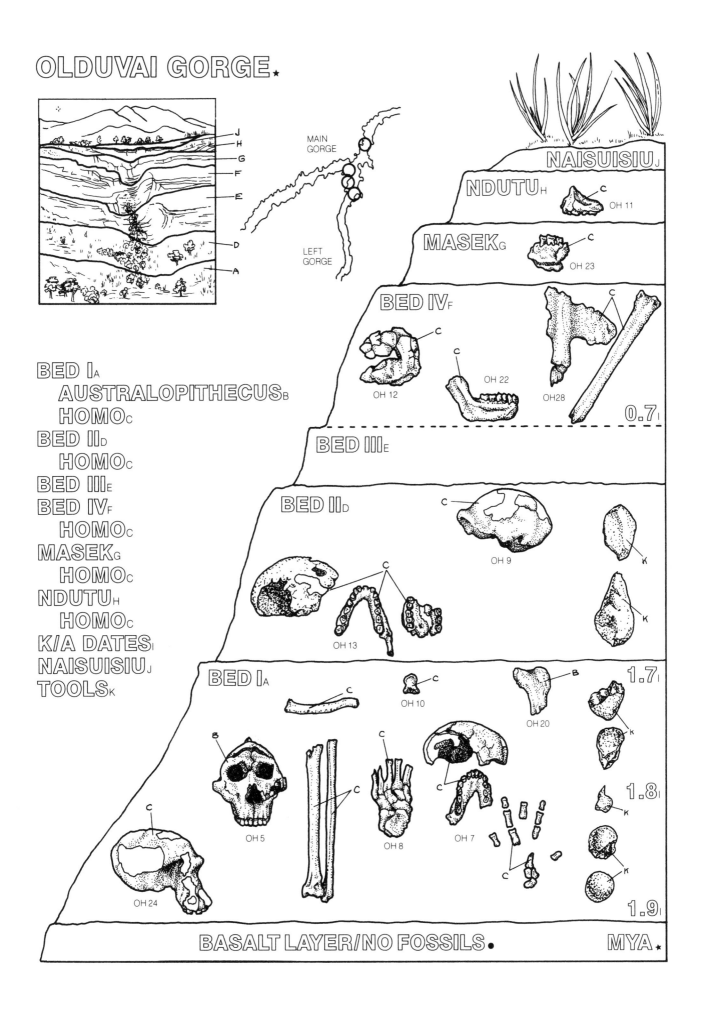

95

MAJOR HOMINID SITES: KOOBI FORA, LAKE TURKANA

In 1967, young Richard Leakey, son of fossil hunters Louis and Mary Leakey, was flying his plane over *Lake Turkana* and noticed stratified sediments. Later he flew there by helicopter and found fossil fragments of the *Plio-Pleistocene*. The following year he organized an expedition and began excavating one of the richest early hominid sites in Africa.

The Koobi Fora Research Project, as it is called, is composed of an interdisciplinary team that includes co-organizer/archeologist Glynn Isaac, palentologist Kay Behrensmeyer, for whom the famous *KBS tuff* is named, anatomist Bernard Wood, and many others. These specialists work together not only to find and study the fossils but to discover evidence of human activity.

Color Lake Turkana on the upper left map, noting its proximity to Ethiopia and Sudan. The lake is part of the Rift Valley system that also includes Olduvai Gorge and Hadar. Color the Plio-Pleistocene deposits on the right, then the present-day fossil sites, which extend for 800 square kilometers, noting the changed landscape. Color the rest of the top right diagram except H, I, J, and K.

This schematic diagram shows that hominids and other animals living in the *Plio-Pleistocene* became fossilized when they died near the lake's edge and their bodies were quickly covered by sediments washed over them. Subsequent layers of ash from the periodic volcanic eruptions make accurate dating by the potassium-argon method possible. The sites that contain fossils and *tools* have been exposed through erosion. The fossil sites are 13 kilometers from the shore of the present-day lake.

Color the volcanic tuffs, which contain potassium and argon for dating the sediments, in the upper right diagram and on the main illustration. The fossils may lie between the tuffs and can be "bracketed" by two dates or may lie within the tuff. Now color the hominids and tools in each layer, noting that they span a time range similar to Olduvai Gorge. Placement of

fossils between the KBS and Tuli Bor tuffs is schematic only, as the fossils and tools are either in the KBS tuff or directly below it, and there is a large non-fossil-bearing hiatus above Tuli Bor.

The now famous ER 1470 skull was sighted by Kenyan Bernard Ngeneo in 1972, and the many fragments were reconstructed by paleontologist Meave Leakey. Its estimated cranial capacity is 780 cm³. The leg bones from a single individual show evidence of bipedalism approaching that of modern humans.

Much controversy has surrounded the *KBS tuff* and the probable age of these specimens. The *KBS tuff* was dated at 2.6 mya, but the fauna suggested a younger date of 1.6–1.8 mya, which G. Curtis later confirmed. The older date put *Homo habilis* (represented by ER 1470) at *Lake Turkana* almost a million years earlier than anywhere else. With the younger date, all the African sites are concordant. The first evidence of both stone *tools* and of hominids with a brain size greater than 700 cm³ occurs less than 2 mya.

The fossils spanning the *KBS* and *Okote tuffs* include at least two species of hominids. *Australopithecus* (ER 406) is very similar to the "Zinj" skull (OH 5) and the Swartkrans skull (SK 48) from South Africa; they have cranial capacities of 530 cm³. More than 40 *Australopithecus* specimens have been unearthed from the time range of 1 to 2 mya.

The other hominid is represented by *Homo erectus* skull (ER 3733), with a volume of 850 cm³, similar to Java specimens of the species. An innominate bone (ER 3228) resembles that of modern humans. *Homo* and at least one species of *Australopithecus* seem to have coexisted for at least 500,000 years. ER 3883 pictured at the top is believed to be of similar age to skull ER 3733. The similarity of these two skulls to the Asian specimens suggests that *Homo erectus* ranged over large areas of the Old World (Plate 106).

The *Lake Turkana* discoveries of more than 150 hominid fossils have confirmed several important features of human evolution: the coexistence of *Australopithecus* (robust species) and *Homo*, evidence of a larger-brained *Homo habilis* (ER 1470), and the early appearance of *Homo erectus,* possibly from a *Homo habilis* ancestor.

KOOBI FORA, LAKE TURKANA ★

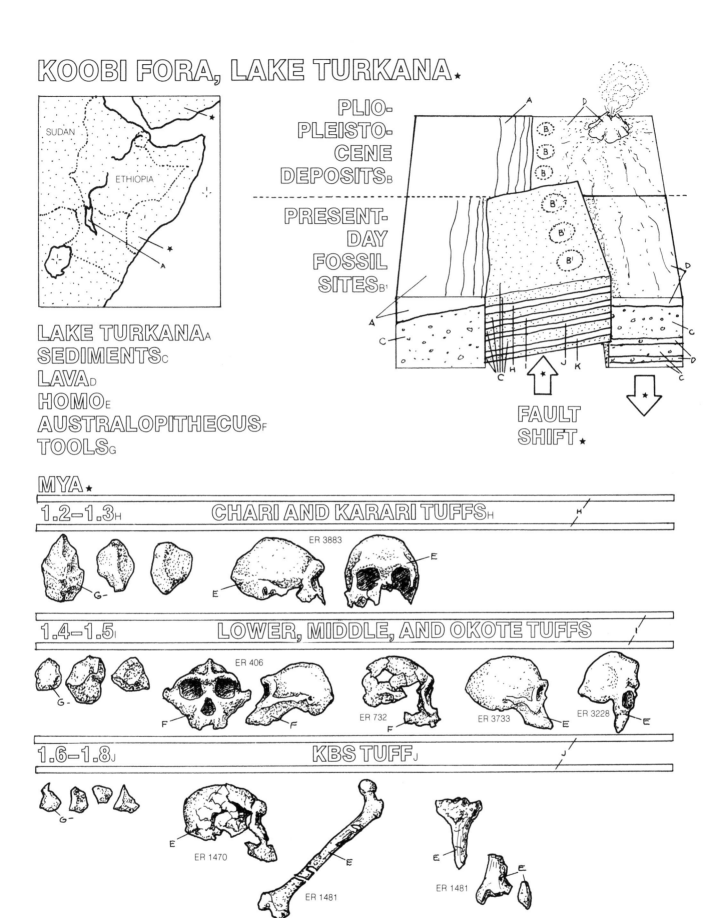

SUDAN

ETHIOPIA

PLIO-PLEISTO-CENE DEPOSITS B

PRESENT-DAY FOSSIL SITES B¹

LAKE TURKANA A
SEDIMENTS C
LAVA D
HOMO E
AUSTRALOPITHECUS F
TOOLS G

FAULT SHIFT ★

MYA ★

1.2–1.3 H	CHARI AND KARARI TUFFS H
1.4–1.5 I	LOWER, MIDDLE, AND OKOTE TUFFS
1.6–1.8 J	KBS TUFF J
3.2 K	TULI BOR TUFF K

ER 3883

ER 406

ER 732

ER 3733

ER 3228

ER 1470

ER 1481

ER 1481

96

MAJOR HOMINID SITES: STERKFONTEIN AND SWARTKRANS

Although the most accurately dated hominid sites are in East Africa, South Africa yielded the first evidence for human origins more than 50 years ago. Young Raymond Dart, arriving from London to teach anatomy at the University of Wiwatersrand in Johannesburg, had a keen interest in human evolution. When a lime miner brought him the face, endocast, and jaws of an immature skull, Dart recognized that this was no ordinary ape. The Taung baby had a full set of milk teeth, first permanent molars erupting, small canines, high forehead, and centered foramen magnum (where the spinal cord leaves the braincase—Plate 100). Dart could see from its unapelike, protohuman traits, that it was a sort of "missing link," even though its projected adult brain size of 440 cm³ was in the ape range. He named it *Australopithecus africanus* (southern ape of Africa).

Despite the meager evidence, Robert Broom, a Scottish physician and fossil hunter, was convinced that Dart was right. Beginning in 1936, he undertook explorations in South Africa, first at Sterkfontein, later (1948) at Swartkrans. In 1947, Dart began excavating another site, Makapansgat. All three were ancient dolomitic limestone caves opening to the surface. The bones and other debris that had accumulated inside the caves became cemented in the *breccia* (a hard calcium carbonate) that filled them. (There is no evidence that hominids ever lived in these caves.) Dozens of fossils and *tools* have been recovered from these caves, but they are not arranged in easily dated layers, as are the fossils at Olduvai Gorge and Koobi Fora.

Color Member 1 through 6 in this schematic section through Sterkfontein cave deposits, based on work by Phillip Tobias. Color the associated fossils. The different members represent different time periods. The hominid fossils are found only in Members 4 (estimated age 2.5–3 mya) and 5 (1.5–2 mya).

In 1936, Sterkfontein yielded an adult *Australopithecus* skull, with small canine teeth and small brain (about 500 cm³). An important find in 1947 was a nearly complete humanlike pelvis (Sts 14) and a com-

plete skull (Sts 5), confirming that these small-brained creatures were bipedal. In 1956, *stone tools* similar to those from Olduvai were found in *Member 5.*

In 1966, new excavations were undertaken by Alun Hughes and Phillip Tobias—Dart's successor in the anatomy department at the University of Wiwatersrand. These clarified the sequence and relations of the sedimentary layers in the cave. In 1976, *Member 5,* where the *stone tools* had been found, also yielded a partial cranium (StW 53), thought to be the Transvaal equivalent of *Homo habilis,* known from Olduvai and Koobi Fora.

Color the illustration of the Swartkrans cave and the associated fossils.

Here Robert Broom and his assistant John Robinson found teeth, jaws, skulls, endocasts, and postcranial bones from at least 80 individuals, mostly robust australopithecines (Plate 101), first named *"Paranthropus."* As early as 1949, Robinson discerned that a lower jaw (Sk 15) was unlike the others and called it *"Telanthropus."* It is now assigned to the genus *Homo,* and it has been determined that it comes from the younger *brown breccia* (estimated age 0.5–1 mya) rather than the older *pink breccia,* which is the source of the *Australopithecus* fossils.

In 1969, while examining fragments of a palate and a skull, supposedly from two different individuals, anthropologist Ron Clarke found that the pieces fit perfectly together. The composite cranium (SK 847) is a *Homo* from the *pink breccia*—similar to StW 53 from Sterkfontein but coexistent with *Australopithecus robustus*—another proof that these two early humans lived at the same time. The *Homo* fossils have a larger braincase, higher forehead, and smaller face than the *Australopithecus* fossils.

The material from the South African caves confirms the widespread savanna adaptation and apparent success of early hominids in Africa. Found there were both gracile and robust *Australopithecus* (see Plate 101), early *Homo* (comparable to *habilis* or possibly *erectus* in East Africa) and Oldowan and Acheulian *stone tools.*

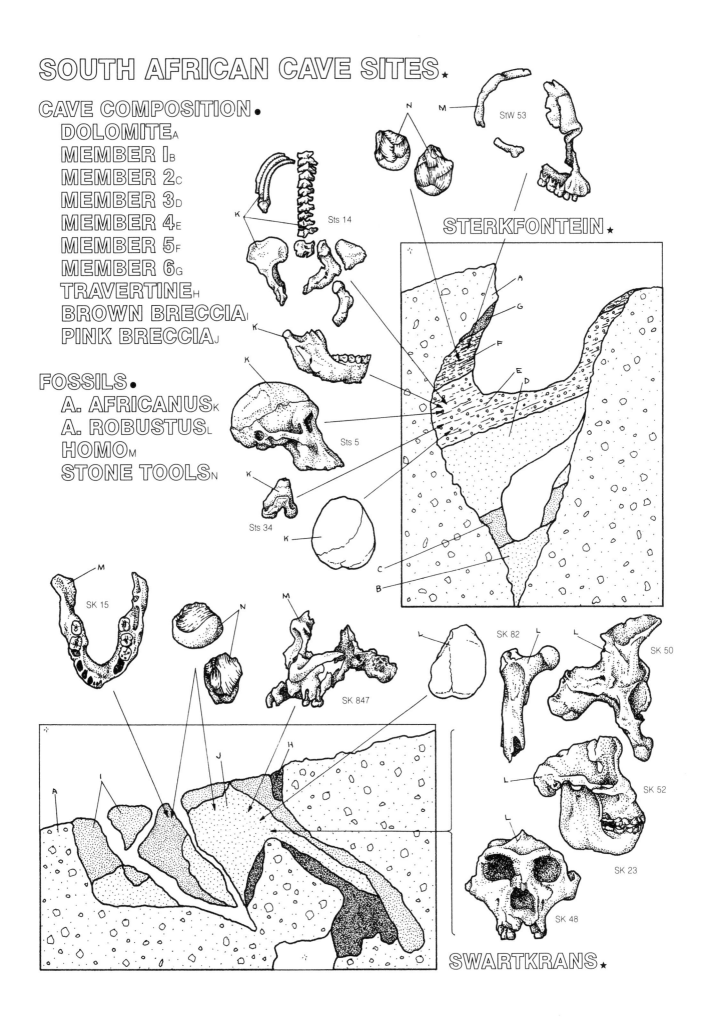

SOUTH AFRICAN CAVE SITES. ★

CAVE COMPOSITION. ●

DOLOMITE A
MEMBER 1 B
MEMBER 2 C
MEMBER 3 D
MEMBER 4 E
MEMBER 5 F
MEMBER 6 G
TRAVERTINE H
BROWN BRECCIA I
PINK BRECCIA J

FOSSILS. ●

A. AFRICANUS K
A. ROBUSTUS L
HOMO M
STONE TOOLS N

StW 53

Sts 14

STERKFONTEIN ★

Sts 5

Sts 34

SK 15

SK 847

SK 82

SK 50

SK 52

SK 23

SK 48

SWARTKRANS ★

97

BONES IN THE SOUTH AFRICAN CAVES: THE LEOPARD HYPOTHESIS

Numerous *hominid* and other *fossil bones* have been found in limestone caves in South Africa. How did they get there?

We know (Plate 92) that fossil bones have often had a complex history. The case study of a fossil *hominid skull* at Swartkrans is like a detective story.

Color the illustration of Swartkrans as it is today, the small square on the right.

The *fossil bones* are imbedded in *breccia. Trees* take root, now as in ancient times, in the damp, protected area near the opening of the cave.

Color the leopard and antelope in the three remaining small squares.

Rather than the *bone* accumulations being due to hominid activity, paleontologist Bob Brain suspected that carnivores must have had a role at Swartkrans. Raymond Dart had previously argued that the *fossil animal bones* in the caves were the remains of prey that early hominids had killed with weapons and eaten in the caves. This theory was popularized by the late Robert Ardrey in *African Genesis,* in which he wrote of early humans as "killer-apes" and suggested that the key to human evolution was that early man (male) became a hunter to provide meat for his mate and offspring.

To the contrary, Brain suspected that the hominids might have been the hunted rather than the hunters. Brain observed that *leopards* often take their prey up into *trees,* where they make several meals of it. High in the *trees,* the *leopard* safeguards its prey from the more powerful lions or packs of hyenas. As the *leopard* munches its prey, *bones* fall from the branches to the ground or into a cave below, if the *tree* is so situated.

Color the large picture on the right showing a reconstructed scene at Swartkrans over a million years ago, when bones were accumulating in the cave.

A *leopard* in the *tree* is eating its meal, and the *bones* are falling into the cave.

Complete the plate by coloring the hominid skull and leopard jaw on the bottom right.

At Swartkrans, an australopithecine *skull* of a child had been found in 1950. The *skull* had two perfect puncture wounds with the bone flaps inside still in place, revealing that this young hominid had died before the healing process began. There are *leopard* and hyena *fossils* in the cave, too, and the canine teeth of a *fossil leopard jaw* fit exactly into the pair of holes. This does not mean, of course, that that particular *leopard* killed that particular child, but it provides strong circumstantial evidence that some *leopard* did. The particular parts of the skeleton left provided further clues: *leopards* eat soft bones like ribs and vertebrae and leave the rest, including the hard teeth and jaws, which fits with the *hominid* and other animal *bones* most common in the caves.

The evidence from Swartkrans and other South African caves suggests that early hominids may have been among the *leopard's* favorite prey, as baboons, a little smaller than hominids, are today. Fierce predators like *leopards,* hyenas, and lions must have been a constant danger to early hominids, especially for their more vulnerable youngsters. This emphasizes the importance of early hominid social behavior and communication for minimizing the dangers of predation.

Was early man, then, the hunter? In light of taphonomic studies like those of Bob Brain, the evidence points to *leopards* and not hominids as hunters. More likely, early hominids gathered plant foods (tough ones, judging by their large, thick-enameled teeth). Females probably played a major role, as they do in human gathering-hunting societies today. Meat sources, such as small animals caught by hand or large dead ones scavenged and butchered, were no doubt important dietary items, but hunting with tools appears much later.

The denouement of this mystery carries the lesson that when *fossil* human and animal *bones* are associated, you cannot leap to conclusions about which was killing the other. Most important, the *leopard* hypothesis provides a beautiful example of the scientific method: gathering data, carrying out experiments, thus ruling out some hypothesis and putting forth others that fit best with available evidence.

THE LEOPARD HYPOTHESIS.

SWARTKRANS
CAVE SITE.
TREE A
DOLOMITE B
TRAVERTINE C
BONES D
    FOSSILS D1
      HOMINID SKULL D2
BRECCIA E
LEOPARD F
    JAW BONE F1
ANTELOPE G

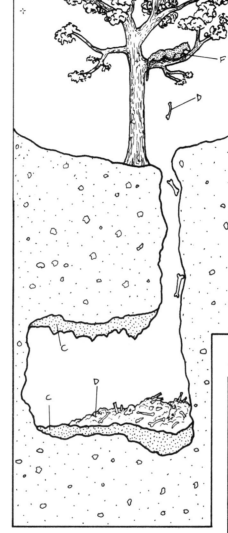

SITE TODAY.

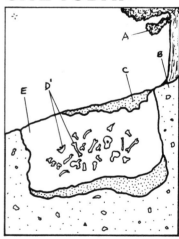

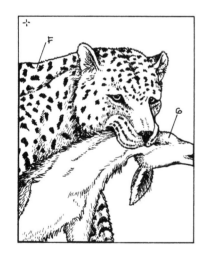

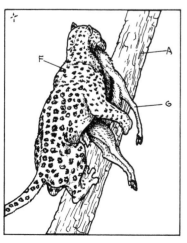

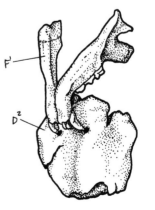

98

EARLY HOMINID SITES: DISTRIBUTION IN TIME AND SPACE

During the 1960s and 1970s, new Plio-Pleistocene hominid sites were discovered in eastern Africa, and new excavations of previous sites in southern Africa yielded numerous hominid specimens and stone *tools*. A multidisciplinary approach has been brought to bear on the study of these sites. The result has been a more complete profile of dates, past environments, and speciation events.

The abundance of potassium-rich volcanic tuffs, lavas, and basalts in eastern Africa has made it possible to establish accurate dates for the sediments in which hominid remains have been found and to establish time correlations between the different sites.

Color the well-known sites in East Africa, first on the map, including the enlargement of the small boxed map area, and then on the chart below. Use related colors (darker for small areas, lighter for larger ones) for East African sites. Use other related colors for the South African sites. As you color each site, notice the time range it spans. Color the associated tools and hominids found at each site, using contrasting colors.

The oldest East African site is *Laetoli*, in *Tanzania*, which dates to 3.7 mya. Just south of *Olduvai Gorge*, *Laetoli* has yielded 3.5 million-year-old human footprints.

Several hundred miles to the north lies the next oldest site, *Hadar*, in the Afar triangle of *Ethiopia*, excavated by Donald Johanson and Maurice Taieb. Numerous hominid remains dated between 3.4 and 2.8 mya have been found and a new species, *Australopithecus afarensis*, proposed by Johanson, Tim White, and Yves Coppens.

The longest and best dated fossil sequence in East Africa comes from the *Omo* beds of the *Omo River Basin*, which drain into *Lake Turkana*, investigated by an interdisciplinary team assembled by Clark Howell and Yves Coppens. This site has yielded more than 240 hominid fossils, mostly teeth, dated between 3.4 and 1.0 mya. The many other fossil species present in the *Omo* beds, such as pigs, gazelles, and elephants, have enlarged our understanding of the evolution of these groups in Africa, provided "faunal dating" to supplement the potassium-argon dates, and helped to correlate similar time levels at different sites. *Omo* is the only East African site that documents the time period between 2.5 and 2 mya which preceded the appearance of *Homo* and the robust australopithecines. Stone *tools* here have been dated back to 2.1 mya.

Olduvai Gorge (Plate 94) and *Lake Turkana* (Plate 95) are similar in age, with hominid fossils dated to almost 2 million years. Robust australopithecines, *Homo* (*habilis* and *erectus*), and stone *tools* have been found at both these sites.

The South African cave sites have produced a rich record of hominid fossils and stone *tools,* but their dating has posed a serious problem. There was no volcanic activity; so the potassium-argon method cannot be applied. The primary method has been "faunal dating," supplemented by paleomagnetic information, neither of which is as precise as the radioactive techniques. Particularly useful for faunal dating are the many fossil bovids (antelopes, buffalos, gazelles, and so forth) found at these sites, and studied by paleontologist Elizabeth Vrba. They indicate that *Sterkfontein* cave sediments may be as old as 3 million years, with *Makapansgat* (gat = Afrikaans for cave) a bit older and *Swartkrans* somewhat younger. *Sterkfontein* and *Swartkrans* contain both gracile and robust australopithecines, *Homo,* and stone *tools* (Plates 96 and 101). Not shown here is the Taung site, which has been destroyed; its fauna remain virtually unknown and its date, uncertain.

Early hominids have inhabited eastern and southern Africa for at least 3.7 million years. Until 1924, we modern humans were unaware of where and how long ago the human line originated. In the past 20 years, new information about our origins has been coming from Africa at an ever-increasing tempo. The record remains fragmentary and incomplete, some of the dates and species relations uncertain, but the very ancient history of emerging humanity is now coming to light.

DISTRIBUTION IN TIME AND SPACE ★

TOOLS_A
AUSTRALOPITHECUS_B
HOMO_C

EAST AFRICA ●
HADAR (ETHIOPIA)_D
OMO RIVER BASIN (ETHIOPIA)_E
EAST LAKE TURKANA (KENYA)_F
OLDUVAI GORGE (TANZANIA)_G
LAETOLI (TANZANIA)_H

SOUTH AFRICA ●
MAKAPANSGAT_I
STERKFONTEIN_J
SWARTKRANS_K

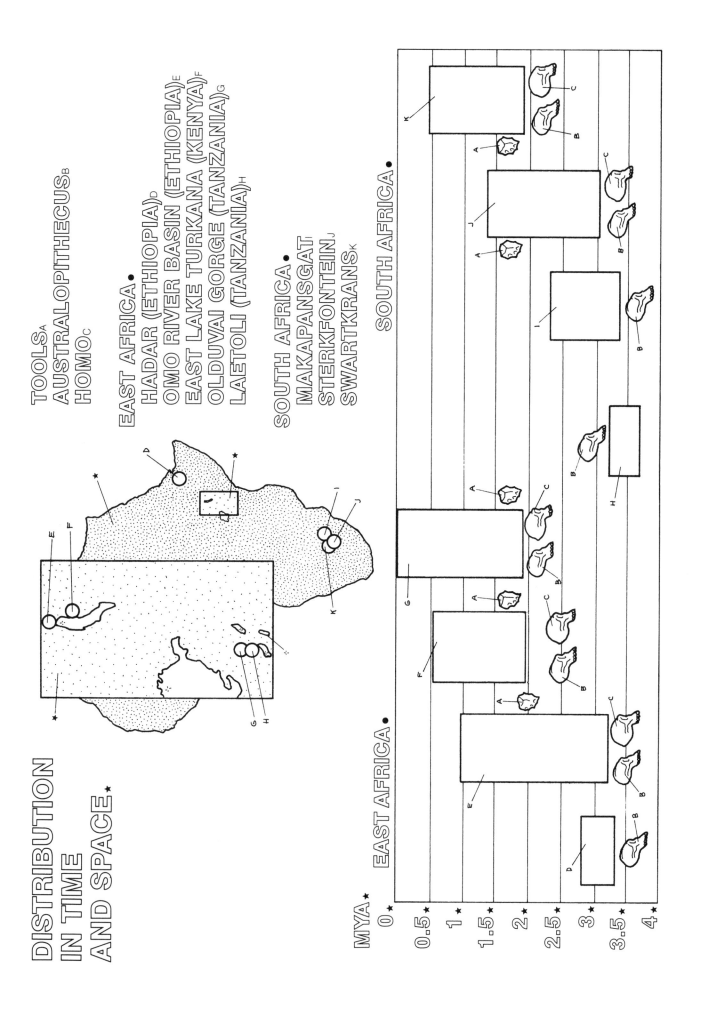

Some of the earliest and latest trends in human evolution can be discerned by comparing the skull of a chimpanzee with those of the earliest hominid, *Australopithecus,* and the latest, *Homo sapiens. A. africanus* from Sterkfontein was chosen for this comparison. The skull of *A. afarensis* is even older and in some respects more chimpanzeelike.

Color cranial capacity, a measure of brain size. Next color the temporal regions in all three views.

The *cranial capacity* of *Australopithecus* is only marginally larger than the chimpanzee. The chimpanzee's entire braincase is covered by the temporal muscle; the hominid's slightly larger skull shows around the muscular attachment; and in the human the *temporal area,* though really the same size, appears much smaller because of the greatly enlarged cranium and can hardly be seen in the top view.

Color the nuchal (neck) area in the side and rear views.

The *nuchal area* lies toward the rear in the ape but more inferiorly in the early hominid and human as a result of different skull proportions. The chimpanzee has more massive neck muscles (to use its canines effectively) attaching to this area, and there is a nuchal crest and a protuberance where these muscles and the temporals meet. They do not meet in *Australopithecus;* the neck muscles are not as powerful, due to reduction of the *canines;* so the back of the skull is more rounded. These muscles are even less confluent in living humans, but occasionally a nuchal crest is seen in very muscular males.

Color the zygomatic arches in the side and top views.

Two slender bony processes form the *zygomatic arch,* one extending from the temporal bone, the other, from the maxilla. The masseter, a main chewing muscle, attaches to this *arch* and the temporal muscle passes under it, attaches to the mandible, and is also important in chewing. Therefore, the strength and size of the *zygomatic arch* provide clues, along with the dentition, as to an animal's diet—at least as to the need for stronger or weaker chewing muscles. The *arch* flares out from the skull more in *Australopithecus* than in the chimpanzee, meaning a bigger masseter and temporal muscle to go with its big grinding teeth. The teeth are more nearly underneath the *arch,* which is mechanically more efficient for chewing. Modern humans, because of tools for preparing food, do not have the big chewing muscles and big teeth, and the rather gracile *zygomatic arch* is dwarfed by the large braincase.

Color the brow ridges and notice the amount of face in front of them and braincase above them.

Even though *Australopithecus* has prominent *brow ridges,* rather like the chimpanzee's, there is considerably less face in front of them and considerably more cranium above them. Unlike the chimpanzee's, the early hominid's *brow ridge* cannot be seen from the rear. The *brow ridges* are much smaller in the modern human and can be seen from neither the top nor the rear. There is very little face forward of them and a great deal of brain above them.

Color the canine teeth.

In comparison with the chimpanzee's, the *canines* of *Australopithecus* have been reduced nearly to human size, a change that has both dietary and social significance. The reduction implies less need for anterior teeth to procure food; when there is heavy chewing, big *canines* tend to get in the way. Male great apes have larger *canines* than females. Yet they eat the same diet; so the big *canines* must serve some social function, such as threat displays against other males. The small early hominid *canines* suggest two possibilities: they were using their hands more, and possibly tools as well, rather than their front teeth to procure food or the male–male aggression requiring large dimorphic *canines* was no longer such an important social interaction among them as it had been in their ape forebears.

Skulls, even long dead ones, tell many tales about the living ways of their vanished possessors.

SKULL COMPARISON.★

CRANIAL CAPACITY_A
TEMPORAL AREA_B
NUCHAL AREA_C
ZYGOMATIC ARCH_D
BROW RIDGE_E
CANINE_F

CHIMPANZEE.★	AUSTRALOPITHECUS.★	HUMAN.★
400 cm³_A	450 cm³_A	1400 cm³_A

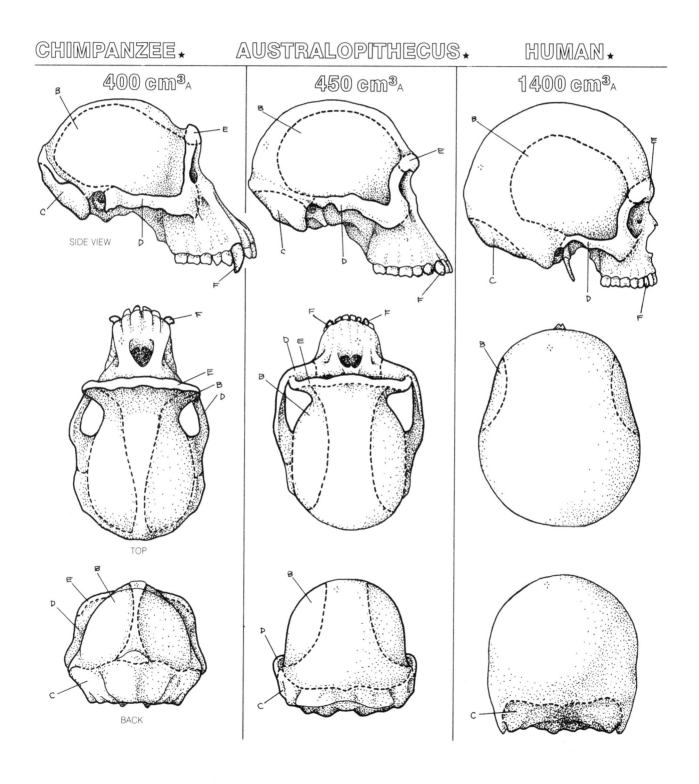

Much of the cranial, facial, and dental anatomy of *Australopithecus* is influenced by an adaptation for grinding hard foods with big teeth.

Color the incisors of upper and lower jaws of chimpanzee, *Australopithecus*, and modern human.

The chimpanzee's *incisors* are larger and protrude more.

Color the canines.

Note the diastema between the chimpanzee's *canines* and second *incisors* on the top jaw that allows space for interlocking of the long *canines*. This feature is absent in the hominids, whose *incisor*like *canines* occlude rather than interlock.

Color the premolars.

The chimpanzee's first lower "sectorial" (cutting) *premolars* differ in shape and orientation from that of the others. *Australopithecus' premolars* are larger to provide more grinding surface; the modern human's are small and bicuspid.

Color the molars.

The *molars* are large and thickly enameled in *Australopithecus* as compared to chimpanzee and human, though the dental formula is the same in all three. Note the differences in palate length and shape of the dental arcade—rather rectangular in the chimpanzee, more U-shaped in *Australopithecus*, and parabolic in humans. These features have sparked many prolonged debates as to which fossil hominoids are or are not on the human lineage.

Color the zygomatic arch and the foramen magnum.

In the previous plate we looked at the *zygomatic arch*. Now we take a new view of it, from below, to see its relations to the palate and to the skull. Going from chimpanzee to early hominid to modern human, it appears that the palate is progressively "tucked in" under the expanding cranium; the *foramen magnum*, where the spinal cord enters the skull, appears to move progressively forward as the back of the skull becomes more rounded.

Color the ascending ramus (branch) of the mandible.

The *ramus* goes higher in *Australopithecus*, providing a longer lever arm for forceful chewing. The angle of the jaw is more perpendicular in both hominids than in the chimpanzee, reflecting the more efficient placement of the grinding teeth beneath, rather than in front of, the chewing muscles.

Color the temporal muscles.

As in the previous plate, the temporal attachment to the skull is about the same size in the three species, but it appears progressively smaller because of increasing hominid cranial capacity. However, the mass of the chimpanzee's *temporal muscle* is as much as four times as great as that of modern humans, due mostly to the chimpanzee's use of its large *canines*. *Australopithecus' temporal muscles* were also quite large, as shown by the large space between the skull and *zygomatic arches*, where the *temporal muscles* passed. But *Australopithecus* had small *canines* and the muscle was part of its heavy grinding apparatus.

Color the masseter muscles, also used for chewing and biting.

The chimpanzee's *masseter muscles* weigh two and a half times as much as the modern human's. *Australopithecus* had even larger *masseters* than modern-day chimpanzees, as indicated by the marks of muscle attachment on the *zygomatic arch* and mandibular angle.

In summary, though *Australopithecus* was unquestionably a bipedal hominid, some features of its skull are similar to those of modern chimpanzees; others show definitely human tendencies; and still others, especially the big grinding teeth and craniofacial modifications that went with them, are uniquely its own. The evolution of the human face and dentition, like that of the pelvis (Plate 103), exemplifies the mosaic nature of the evolutionary process—that is, everything doesn't evolve at the same time and the same rate. The big teeth, jaws, and chewing muscles of early hominids were an adaptation to their life and diet on the African savanna. Later, in response to different ways of life and a different diet, the human brain increased in size and the teeth became smaller. The overall trend was a marked increase in the ratio of the braincase to the face and teeth beneath it.

DENTITION. *

INCISOR_A   ZYGOMATIC ARCH_E
CANINE_B   FORAMEN MAGNUM_F
PREMOLAR_C   ASCENDING RAMUS OF MANDIBLE_G
MOLAR_D   TEMPORAL MUSCLE_H
                    MASSETER MUSCLE_I

CHIMPANZEE *   AUSTRALOPITHECUS *   HUMAN *

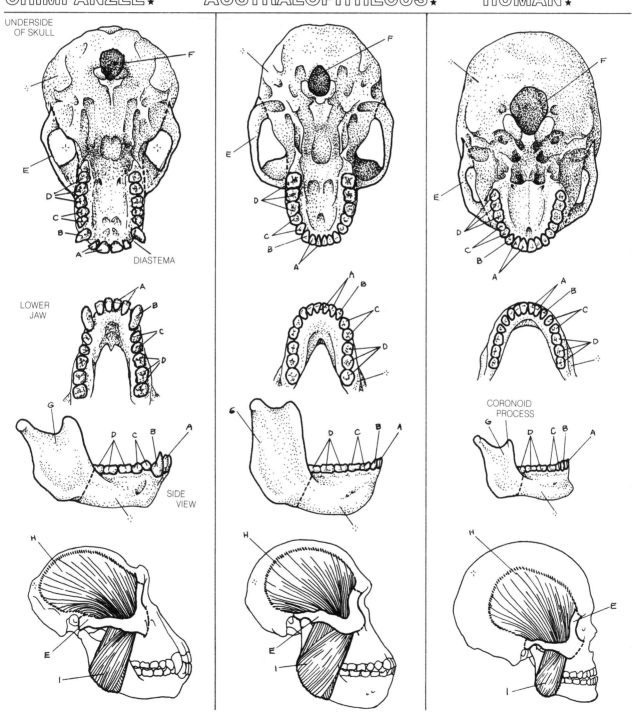

UNDERSIDE OF SKULL

DIASTEMA

LOWER JAW

SIDE VIEW

CORONOID PROCESS

AUSTRALOPITHECUS: GRACILE AND ROBUST

The two species of *Australopithecus* known prior to 1975, *africanus* and *robustus* (including *boisei*), also known simply as gracile and robust, are sufficiently similar in their extraordinarily large *molars* to have suggested an "australopithecine adaptation" for feeding on tough, hard-to-chew vegetation. The older and more recently named species, *afarensis,* shares this tooth structure, so presumably was similarly adapted. What does this adaptation entail, anatomically and behaviorally, and what has it to do with the eventual extinction of this genus?

Here an *A. robustus* skull from Olduvai ("Zinj" OH 5) dated at 1.8 mya and *A. africanus* skull dated about 2.5 mya from Sterkfontein are compared.

Color cranial capacity.

The slightly larger *cranial capacity* of the robust species may simply reflect a somewhat larger body size. The *Homo* species contemporary with *A. robustus* already had a much larger brain (850 cm³).

Color the temporal regions and sagittal crest.

The *sagittal crest,* which anchors the temporal muscles on both sides, is present only in *robustus,* showing that this chewing muscle was much larger in the robust species. In the gracile species, the attachment of the temporal muscle does not even go to the top of the skull.

Color the zygomatic arch.

The *zygomatic arches,* which leave space for the temporal muscle, flare away from the skull more in *africanus* than in either chimpanzee or human, and still more in *robustus,* giving further evidence of temporal muscle enlargement. The root of the *arch* attaches farther forward on the face of *robustus* than of *africanus,* which gives the robust australopithecines their characteristic "dish-faced" appearance.

Color the incisors and canines.

Both have the hominid character of small *canine* teeth; the anterior teeth (*incisors* plus *canines*) are relatively smaller in the robust than in the gracile hominids.

Color the premolars.

The *premolars* are much larger in *robustus* and have extra cuspules that make them more like *molars,* showing a further "grinding adaptation" in this species.

Finally, color the molars.

Robustus' cheek teeth (*premolars* plus *molars*) have a greater area than *africanus.* The anterior teeth are smaller. The same trend is seen when *africanus* is compared to *Homo,* that is, now *africanus* has the larger cheek teeth and smaller anterior teeth. Therefore, compared to *africanus, robustus* and *Homo* are dentally moving in opposite directions.

A pygmy versus common chimpanzee comparison is somewhat analogous to that between the gracile and robust australopithecines.

Complete the plate by coloring in gray the skulls of the two chimpanzee species at the bottom of the plate.

The common chimpanzee is on average about 20 percent larger than the pygmy chimpanzee and has a slightly larger *cranial capacity.* Its skull and dentition are larger and more robust, with more bony knobs on the skull and jaw; and it is more likely than the pygmy chimpanzee to have a *sagittal crest.* With the chimpanzees, as with the hominids, the gracile species is probably more like the common ancestor.

Understanding these species holds a key to understanding human origins and possibly the evolution of tool use. The gracile species is earlier than the robust, which may have developed from it. The "australopithecine adaptation" was part of the savanna way of life, with tough, gritty food requiring much chewing. *A. robustus* may have carried this adaptation to such an extreme that it became extinct about 1 mya, a fate that often overtakes highly specialized species when more successful competitors arrive on the scene or there is a change in climate. In this case, the more successful competitor may have been *Homo,* who used tools.

If Don Johanson and Tim White are correct in their hypothesis that *afarensis* gave rise both to *africanus* and *Homo, Homo habilis* and its descendent species are the sole survivors of the human experiment (see Plate 109). Interestingly, stone tools appear on the scene at the same time as does *Homo*—about 2 mya, and may indicate their importance in food preparation. Anthropologist Yoel Rak, who has studied the facial architecture of all apes and hominids, also believes that *africanus* was already too far along in the "australopithecine adaptation" to have been directly ancestral to modern humans.

AUSTRALOPITHECUS. ★

GRACILE. 450 cm³ A

A. AFRICANUS

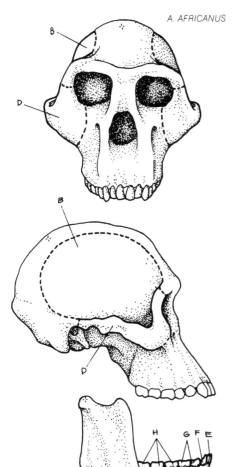

ROBUST. 530 cm³ A

A. ROBUSTUS

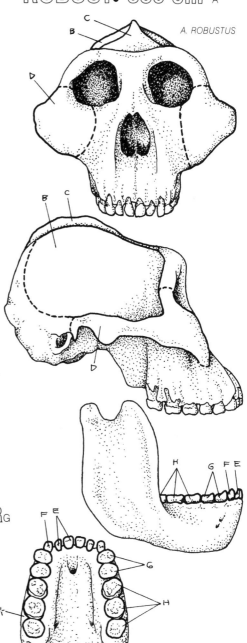

CRANIAL
CAPACITY A
TEMPORAL
REGION B
SAGITTAL
CREST C
ZYGOMATIC
ARCH D

INCISOR E
CANINE F
PREMOLAR G
MOLAR H

PYGMY CHIMPANZEE. ●

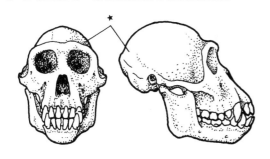

COMMON CHIMPANZEE. ●

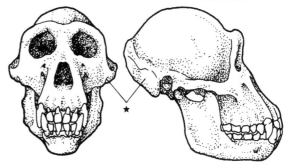

Humans have long been thought unique among animals in their ability to make and use *tools*. The gap between ourselves and our closest relatives, the apes, was believed so large that it was hard to imagine how it was crossed in the evolutionary process. New evidence from archeology and from field studies of *chimpanzees* has narrowed the gap to a crack.

Ancient *tools* of copper, bronze, and iron had long been known, but stone *tools* were first recognized as human artifacts in 1790 by Englishman John Frere (Mary Leakey's great great great grandfather), who found handaxes associated with extinct animal bones and described them as "fabricated and used by people who had not the use of metals," from a very ancient period.

In Europe and Africa, especially at Olduvai, stone *tools* are found in much greater abundance than *hominid* remains and were discovered much earlier.

Color the titles: chimpanzee, early hominid, and human, using light colors. Then color the corresponding hands.

Wrist and hand *bones* of *Australopithecus* and *Homo* are often not associated with teeth when discovered and so are difficult to identify as to species. Here a composite *early hominid* hand is compared with that of a *chimpanzee* and a modern *human*.

Color the thumb, finger, and wrist bones. Color the long flexor muscle tendons in the chimpanzee and the human.

The *early hominid thumb phalanges* and *finger phalanges* (in side view) are from Olduvai Bed I, analyzed by anatomist John Napier. These *finger bones* are more curved than in modern *humans* but less than in *chimpanzees*. The terminal *phalanx* of the thumb is intermediate in size between the *chimpanzee's* and the modern *human's*. It's breadth suggests strong muscles and a sturdy *tendon* of the *long flexor* going to the *thumb*, unlike the tiny one in the *chimpanzee*—again, an intermediate condition between *chimpanzee* and modern *human* that indicates *early hominid* hands were well equipped to use *tools*.

The skull provides more evidence, for in *Australopithecus* it appears that the cerebellum, important for skilled and coordinated hand movements, was larger than a *chimpanzee's*.

Complete the plate by coloring the hand grips (power and precision) and the objects in them.

Chimpanzees, like other apes and monkeys, have good hand control, but their *power grip* barely involves the *thumb* and their *precision grip* opposes the *thumb* to the proximal rather than the distal *phalanx* of the index *finger,* making these *grips* less flexible and effective than the *human* ones. The first stone *tools* were "choppers," made from pebbles, only a few flakes removed, which were also used as *tools* and must have required a well-muscled *thumb* for *precision* use (possibly to butcher animal carcasses or cut up plants). But these only appeared 2 mya, and we have *hominid* remains as old as 3.5 million years. How can we explain this seeming jump from no *tool* use to stone *tools*?

The answer lies in the bias of the fossil record. *Tools* made of wood or other organic materials, like those that *chimpanzees* make or the sticks, gourds, skins, bark, and vegetation used by contemporary gathering-hunting peoples, would not have survived in the fossil record. Yet it is very likely that our earliest ancestors depended heavily on such *tools* long before they systematically made ones of stone. It is also likely that the early pebble choppers were used to make organic ones.

Stone *tools* do mark a kind of technological "revolution," based on millions of years of perfecting organic *tools*. It has often been assumed that stone *tools* and evidence of butchering large animals indicate that *early hominids* were hunters, but weapons for hunting, such as spearheads suitable for hafting, have not been found; such unequivocal evidence appears only 150–200,000 years ago. Stone *tools* became increasingly varied and sophisticated within 0.5 million years of their first appearance, but metal *tools* emerged only several thousand years ago.

HAND AND TOOLS.★

THUMB PHALANXA
FINGER PHALANXB
WRIST BONESC
**LONG FLEXOR
  MUSCLE TENDON**D

HAND GRIPS.●
  POWERH
  PRECISIONI

**OBJECT/
  TOOL**J

CHIMPANZEE E

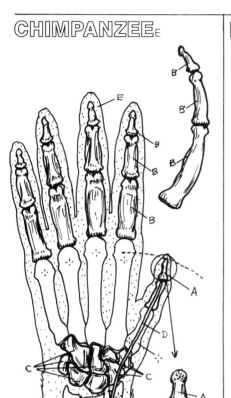

TERMINAL
PHALANX

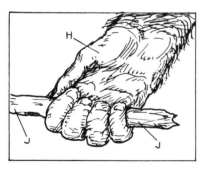

EARLY HOMINID F

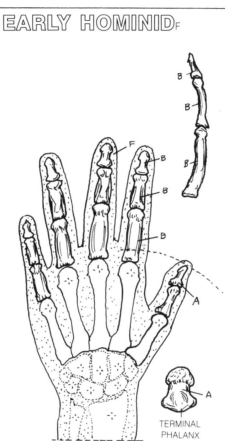

TERMINAL
PHALANX

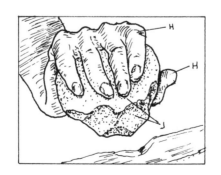

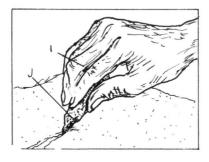

HUMAN G

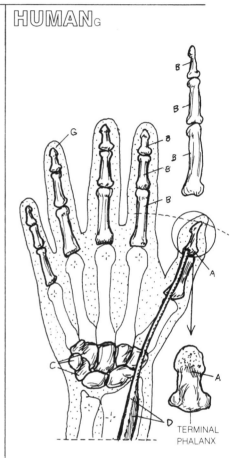

TERMINAL
PHALANX

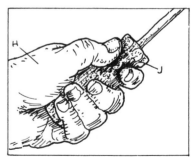

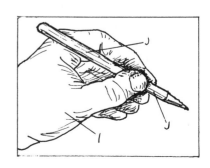

103

EVIDENCE FOR BIPEDALISM: PELVIS AND LEG

Bipedal walking emerged late in the structural history of the hominoids. Because evolution is conservative, muscles and bones already present in apes were remodeled in *humans* for new purposes. This resulted in functional compromises. The pelvis, for example, accommodates many different functions—locomotor, postural, visceral, and obstetrical. The wide female pelvis, necessary for giving birth to large-brained infants, is less efficient in locomotion than the narrower male pelvis. The imperfections in *human* bipedal structures are reflected in such common ailments as low back pain, herniated discs, arthritis (especially of the lumbar region), fragile knee and ankle joints, and flat feet.

The first australopithecine pelvis and thigh bones were found in South Africa in the 1940s: two adolescent ilia at Makapansgat, and a pelvis, *sacrum*, vertebrae, and fragmentary *femur* at Sterkfontein. Anthropologist Sherwood Washburn pointed out over 30 years ago that in a mosaic fashion, the hominid locomotor system evolved long before the large brain and small teeth. The new fossil hominid pelvis and leg bones from Hadar and the Laetoli footprints confirm this.

Evidence from the fossil record compared to a quadrupedal ape and a bipedal *human* gives clues about the transformation from quadruped to biped.

Color the titles, chimpanzee, *Australopithecus*, and human, using light colors. Then color the corresponding thighs indicating the body outline. Next color the innominate bones, including the acetabulum (hip socket), in side view at the top of the plate and in the central figures.

In the side view, you can see that the fossil ilium is shorter and broader than the *chimpanzee's*, even though the *acetabula* are the same size. The *human* ilium is curved more than that of the fossil and the *acetabulum* is notably larger.

Now color the sacrum.

The fossil *sacrum*, like the *human* one, is broad and gives the pelvis a bowl shape. The short, broad fossil ilium indicates that the gluteal muscles (Plate 55) were already reoriented horizontally, as in *humans,* rather than still vertical, like apes going bipedally. The gluteus maximus, important for upright posture and walking, was oriented at the hip as in *humans.*

Complete the plate by coloring the femur and the weight-bearing axis.

Australopithecus' femur, like its pelvis, is unique. Its length and head are chimpanzee-sized, but its long neck and robust shaft, and the oblique angle to the ground, are unlike either *chimpanzee* or *human.* The oblique shaft of the *femur* suggests that *Australopithecus* had its knees and feet well planted under its body and that the *weight-bearing axis* approached that of living *humans,* in whom the weight of the upper body is borne largely on the outside of the *femur.* The *chimpanzee's axis* is on the inside of the *femur.*

The *weight-bearing axis* is reflected in the shape of the *femur.* In the *chimpanzee,* the medial condyle is higher and larger; in the *human,* the lateral condyle is larger; in *Australopithecus,* the morphology (form) suggests a function not exactly like either of the other two.

Although not quite like those of any living species, the australopithecine pelvis and *femur,* like its foot, are diagnostic of bipedal locomotion. The apelike features—the less recurved iliac crest, the small *acetabulum,* the longer ischium—tell us that the transition to bipedalism had only recently occurred. The small size of the hip joint, like that of a *chimpanzee's,* is not due just to small body size, but indicates that the weight-bearing capacity of *Australopithecus* was not yet as great as that of modern *humans.*

The "refinements" in the locomotor system occurred subsequent to *Australopithecus,* sometime in the last two million years. Unfortunately, our fossil record of the pelvis, leg bones, and foot is scanty for *Homo erectus;* so we cannot be sure when the locomotor system as we know it in *Homo sapiens* was achieved.

PELVIS AND LEG ★

PELVIS ●
INNOMINATEA
ACETABULUMB
SACRUMC

FEMURD
WEIGHT-
BEARING
AXISE

CHIMPANZEE F   AUSTRALOPITHECUS G   HUMAN H

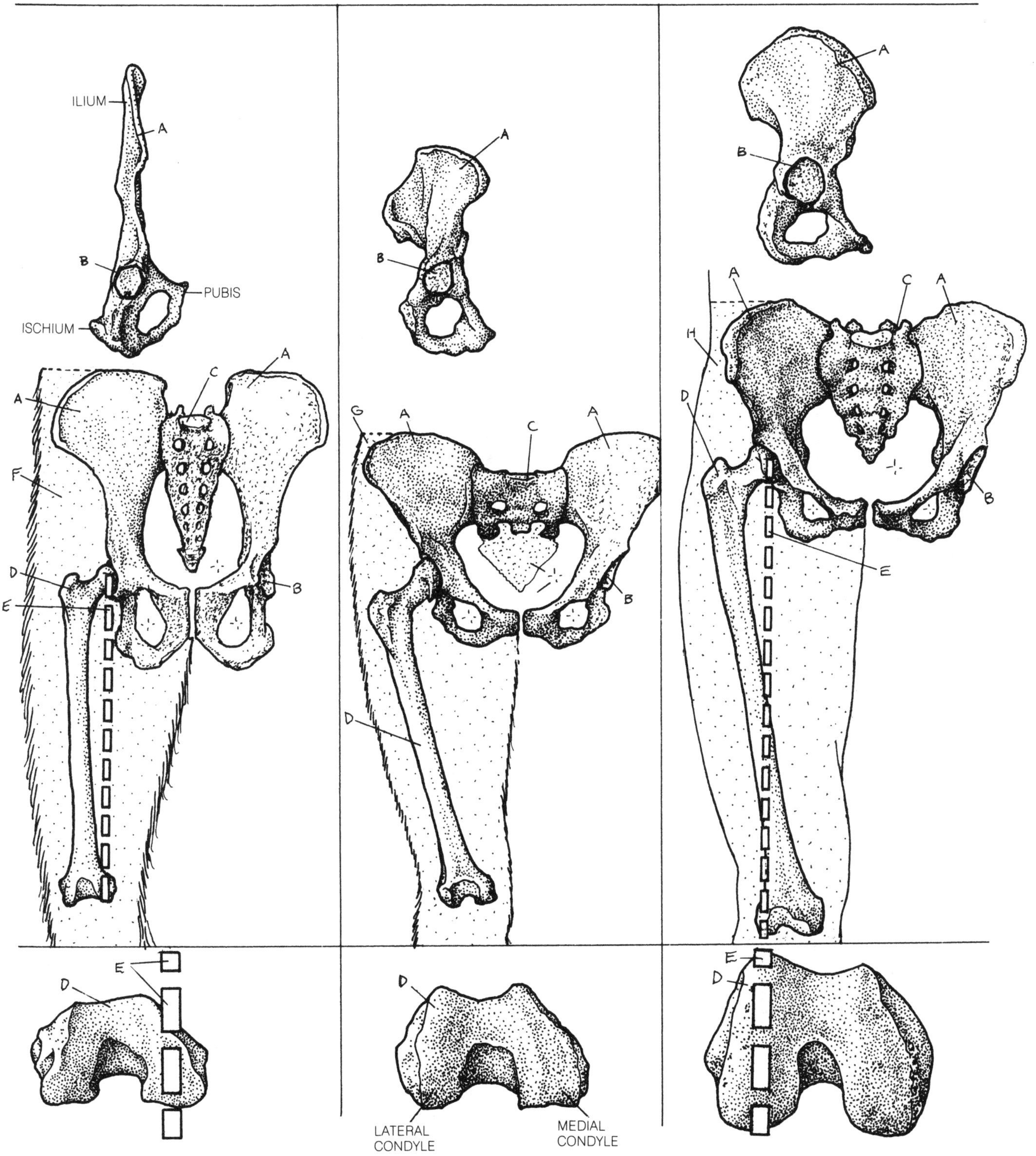

ILIUM
A
B
PUBIS
ISCHIUM
A
C
A
F
D
B
E

A
B
G
A
C
A
B
D

A
B
C
A
H
A
D
E
B

E
D

D

E
D

LATERAL
CONDYLE
MEDIAL
CONDYLE

104

EARLIEST EVIDENCE FOR BIPEDALISM: FOOT AND FOOTPRINTS

One day 3.5 million years ago hominids walked on ground softened by a recent rainshower; soon after, their footprints were covered, preserved by volcanic ash, and only recently discovered by Mary Leakey. Many other animals left footprints there at about the same time; so it is possible to visualize the comings and goings of the monkeys, gazelles, guinea fowl, and hominids of that ancient world.

The footprints and foot bones of *chimpanzee, Australopithecus,* and *human* are compared here, in search of clues to the origin of *human* bipedalism.

Color the chimpanzee, *Australopithecus,* and human titles, using light colors, their respective footprints at the top, and the outline of the foot around the bones.

Chimpanzee footprints show long, curved toes with the great toe divergent. The *human* footprints show a large convergent great toe, longitudinal arch, and relatively narrow heel. The fossil footprints, less well delineated because of the soft surface on which they were made, show a great toe more convergent than the *chimpanzee's,* but perhaps less so than the *human's.*

Color the sole of the chimpanzee and human left feet, noting the surface that makes prints. Color the human and chimpanzee foot bones in the top and side views.

Note the short toes, large *calcaneus* (heel bone), and arch of the *human* foot bones.

A *talus* (ankle bone) from the South African cave site, Kromdraai, provided the first look at the australopithecine foot. Subsequently, a tibia and fibula, 12 foot bones, and a terminal great toe bone (pictured here) were uncovered at Olduvai (Plate 94).

Color the foot bones from Olduvai Gorge, starting with the talus and calcaneus. The metatarsals and calcaneus are broken, and the great toe bone comes from a different individual.

The Kromdraai *talus* and articulated Olduvai foot possess characteristics of the *human* foot: plantar arch, adducted great toe, broad robust first *metatarsal.* Anatomist Owen Lewis concluded that many of its features resemble those of the African apes. The main difference is a more convergent great toe. The fossil evidence suggests that one of the earliest adaptations for bipedalism was a supportive foot for bearing the body's weight. But the early hominid foot and ankle differed from those of modern humans.

Complete the plate by coloring the baby chimpanzee and baby human feet.

Baby *chimpanzees* cling to their mother's hair with grasping toes, as all primate babies except *humans* do. *Human* babies lack grasping toes and mothers have no hair for babies to cling to; so they must be carried. Among early hominids, the burden of infant transportation literally shifted to the mother. Babies without grasping toes had to be carried continually for two and occasionally for another three or four years. Even before birth, the baby's weight would add stresses on the now-bipedal mother's hip, knee, and ankle joints. It is sensible to imagine, then, that a sling to support and carry an infant may well have been one of the earliest hominid "tools," invented to free the mother's arms for gathering and carrying food. The survival of an early hominid child depended upon its mother's ability to carry it long distances while she gathered food to share with it.

How did *Australopithecus* actually walk? Various authors have imagined it waddling, shambling, shuffling, or strolling along. Judging by the Laetoli footprints, its gait may have appeared remarkably similar to our own, though the underlying mechanism of the muscles and joints was certainly different.

The fossil pelvis, femur, and foot bones support the interpretation that bipedalism emerged rapidly and that perhaps *Australopithecus* had not been bipedal for much more than a million years. We estimate that these early changes included (1) relatively minor alterations of the foot, which had important behavioral consequences; (2) shortened and broadened sacrum and ilium for reorienting hip muscles, creating a lumbar curve for supporting a vertical trunk; (3) decrease in mobility of hip, knee, and ankle joints; and (4) decrease in weight and length of the upper limbs (Plate 105). These modifications through natural selection permitted our earliest ancestors to take their first permanently bipedal strides.

FOOT AND FOOTPRINTS.★

FOOT BONES.●
TALUS(A)
CALCANEUS(B)
OTHER TARSALS(C)
METATARSAL(D)
PHALANGES(E)

CHIMPANZEE(F)	AUSTRALOPITHECUS(G)	HUMAN(H)

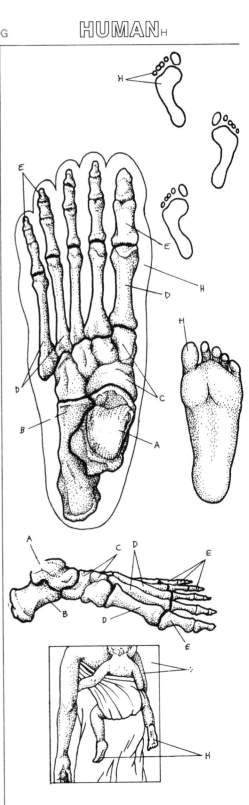

105

WHAT DID LUCY LOOK LIKE?

Though most anthropologists would agree that *Homo sapiens* is closely related to living apes and, like them, had an ape ancestor, they disagree on which of the living apes is most like that long extinct ancestor.

Most recently, a partial skeleton from Hadar over three million years old known as "Lucy" provides the oldest evidence of small bipedal hominids. Some of the bones are complete, and together with bones from Sterkfontein they provide a basis for reconstructing the body build of *Australopithecus*. In my opinion, the living species most like the hypothetical ancestor is the *pygmy chimpanzee* (Plate 89). This hypothesis can be critically examined by comparing the two and so "reconstructing" the fossil.

There is no way to be sure that "Lucy" is not "Louis." In modern humans, males and females can be distinguished 95 percent of the time on the basis of pelvic features. But in *Australopithecus,* with a different pelvis and no sample through which to assess variation, sexing is merely guesswork, though there is a 50 percent chance that AL 288 ("Lucy") is female!

Color the pygmy chimpanzee and *Australopithecus* titles, cranial capacities, and head. The shadow in the background represents a modern human, giving an indication of scale in height. Color height estimated from femur length for the fossil, and weight, estimated from bones of the skeleton.

The fossil's *cranial capacity* is somewhat larger than the average *chimpanzee's,* but within its range, and much smaller than that of modern humans (Plate 88).

Color the upper jaw dentition.

Australopithecus' canines are only slightly smaller than the *chimpanzee's,* but the molars and premolars are much larger and the shape of the jaw, very apelike. These changes may relate to a new savanna diet that includes tough, chewy plants.

Color the overall body form. Color the bones of the upper limb.

Lucy's *humerus* is *235 millimeters* long, much shorter than the average *pygmy chimpanzee's.* I estimate *radial* length as *205 millimeters,* based on the ratio of *radius* to *humerus* in *chimpanzees* and humans. Notice the overall difference in appearance. "Lucy's" shoulder probably hung lower than the *chimpanzee's,* like ours; her arm bones were more slender; and her hand was shorter. The upper limbs of both *pygmy* and common *chimpanzees* are about 16 percent of body *weight*; I estimate "Lucy's" to have been only 12 percent.

Color the innominate, sacrum, and vertebrae.

The *chimpanzee's innominate* is much longer, the *sacrum,* much narrower and longer. The shorter, broader pelvis and longer lumbar region of *Australopithecus* are a bipedal hominid characteristic, and the lumbar region is longer.

Color the bones of the lower limb. Color the feet. Color the height and weight figures.

"Lucy's" *femur,* both its length and femoral head size, is within the range of the *pygmy chimpanzee's,* indicating similarity in *height.* The *tibia,* though fragmentary, is estimated to be similar in length to the *pygmy chimpanzee's.*

Despite the similarity in lengths of the bones, notice the very different orientation of the lower limbs in the two species. The long-necked *femur* in *Australopithecus* is brought in (adducted) under the body, a bipedal trait. Each lower limb is estimated at 14 percent of body *weight,* compared to the *pygmy chimpanzee's* 12 percent.

"Lucy's" pelvic and limb bones could have evolved with only minimal changes from those of a small ape with a gracile upper body and equal upper and lower limb lengths—an ape like a *pygmy chimpanzee.* In the shift to upright locomotion, reduction in upper limb length and weight—as we see in "Lucy"—lowered the center of gravity, making upright balance easier.

Using the *pygmy chimpanzee* for comparison with the early hominids, we can pinpoint in greater detail changes in the transition from ape to human.

WHAT DID LUCY LOOK LIKE?★

CRANIAL
   CAPACITYc
HUMERUSd
RADIUSe
ULNAf
PELVIS★
   INNOMINATEg
SACRUMh
VERTEBRAEi
FEMURj
TIBIAk
FIBULAl
FOOTm
HEIGHT/
   WEIGHTn

DENTITION★

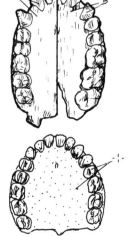

MODERN
HUMAN

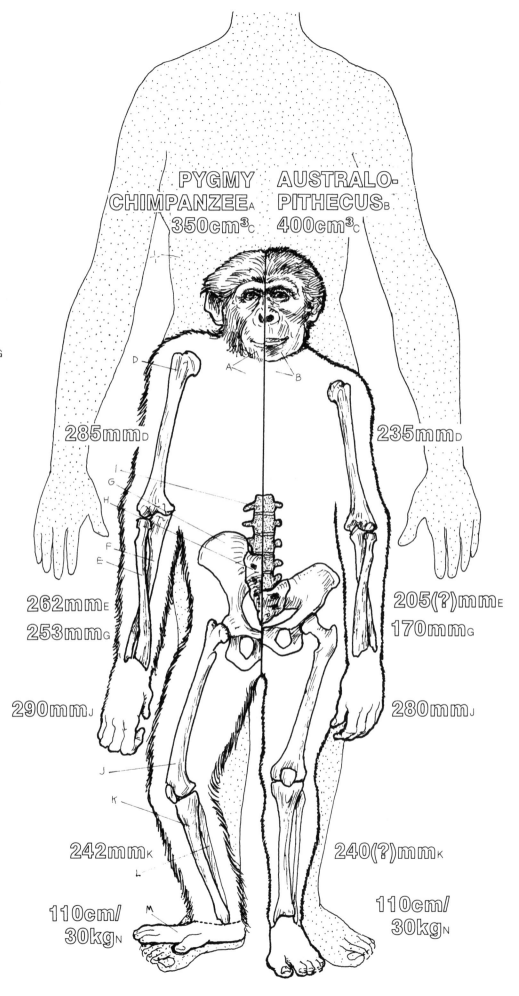

PYGMY
CHIMPANZEEa
350cm³c

AUSTRALO-
PITHECUSb
400cm³c

285mmd

235mmd

262mme
253mmg

205(?)mme
170mmg

290mmj

280mmj

242mmk

240(?)mmk

110cm/
30kgn

110cm/
30kgn

106

HOMINIDS LEAVE AFRICA: MIGRATION ROUTES OF HOMO ERECTUS

Remains of *Australopithecus,* the earliest known hominid, have been found only in Africa, often in association with members of the genus *Homo.* African fossils of the species *Homo erectus* have been dated as early as 1.4 mya. After 1 mya, and possibly earlier, *Homo erectus* inhabited Asia; and by half a million years ago, the species had apparently spread throughout Asia and Europe.

The earliest evidence of what we now know as *Homo erectus* was discovered in *Java* in the 1890s by Eugene DuBois, a Dutchman. He was convinced that the "missing link" between humans and apes had lived in tropical Asia; so he went to look for it there. He called his earliest finds *Pithecanthropus erectus* (erect apeman) or "Java man." The widespread distribution of this kind of early human was revealed by subsequent discoveries in China (*Sinanthropus* or "Peking Man"), Europe ("Heidelberg Man"), and North Africa *(Atlanthropus).* All these variants are often known by the same name, *Homo erectus.*

Color the time scale, the sites in Africa and the fossils associated with them.

Homo remains have been found at *Olduvai Gorge,* Lake Turkana, *Swartkrans,* and *Sterkfontein.* It is difficult to tell whether the fossils from the latter two sites belong to the species *H. habilis* or *H. erectus,* but at *Olduvai* and Lake Turkana, there were unquestionably *H. erectus,* identifiable by their larger cranial capacity and other facial and dental features. Body size was approaching that of modern peoples.

Color the arrow indicating expansion of the species into northern Africa.

Lower jaws have been found at *Ternifine* in Algeria, in Morocco, and at other sites. The age of the *Ternifine* site is not firmly established, but might be as much as a million years old.

Color the arrows leading to Asia, first to Java, then north into China. Color the associated fossils.

In all, there are about a dozen skulls from several sites; they range in cranial capacity from 850 to 1000 cm³. Some upper and lower jaws and teeth are similar to those of *H. habilis* at *Olduvai Gorge,* according to

studies by anatomists G. von Koenigswald and P. Tobias. Some Javanese fossils are probably more than a million years old; others are younger according to ages of tectites found in the same sediments.

H. erectus fossils were found in the 1930s at Choukoutien caves near Beijing *(Peking).* The remains are primarily skulls and jaws, over a dozen individuals with cranial capacities more than 1,100 cm³, dated about 3–400,000 years old. The skull from *Lantian,* a site in southwestern China, is perhaps some 750,000 years old.

Complete the plate by coloring the migration route into Europe, the European sites, and their fossils.

H. erectus fossils in Europe are generally younger than those from Asia and have cranial capacities of at least 1,000 cm³. Historically, the *Heidelberg* jaw was the first evidence of *Homo* other than *sapiens* in Europe. Its age is unclear, as the stratigraphic relations are confused and no potassium-argon dates are available. *Steinheim* and *Swanscombe* skulls and an occipital bone from *Hungary* may be intermediate between *Homo erectus* and *sapiens.* Clark Howell, Christopher Stringer, and others find that they resemble *sapiens* more than *erectus.*

Europe was covered by ice sheets on and off during the Pleistocene; so it is unlikely that humans could have survived there without a fairly advanced culture that included the use of fire. A number of European sites (Torralba and Ambrona in Spain, Terra Amata near Nice—not shown) give evidence of fire and tools and the butchering of large animals for meat.

The best evidence for hominid activities has been found in the East African sites where "living floors" undisturbed for a million years give evidence of crude rock-based shelters, and many kinds of well-manufactured tools. Fewer tools have been found in Asia and not in the "living" context established in East Africa.

It appears then that *Homo erectus* left Africa by at least a million years ago, by way of North Africa, the Middle East, Southeast Asia, then to northern Asia, and finally to Europe. Much of this scenario is still speculative, and we have much to learn about this stage of human evolution.

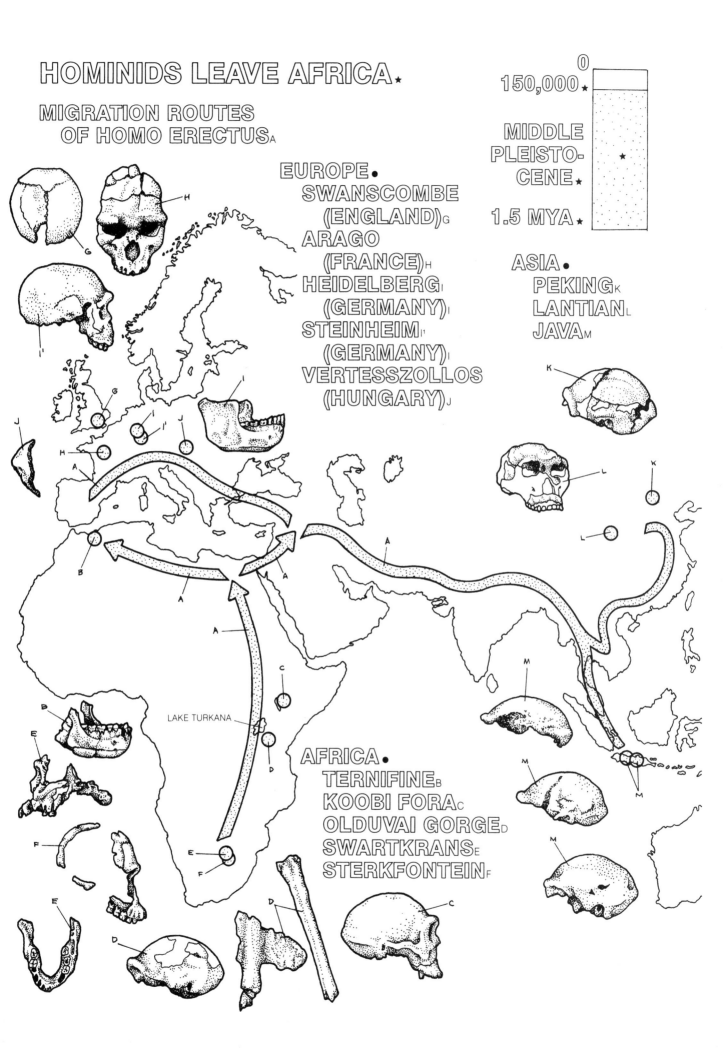

HOMINIDS LEAVE AFRICA ★

MIGRATION ROUTES OF HOMO ERECTUS ᴀ

EUROPE ●
SWANSCOMBE (ENGLAND)ɢ
ARAGO (FRANCE)ʜ
HEIDELBERG (GERMANY)ɪ
STEINHEIM (GERMANY)ɪ'
VERTESSZOLLOS (HUNGARY)ᴊ

ASIA ●
PEKINGᴋ
LANTIANʟ
JAVAᴍ

0
150,000 ★
MIDDLE PLEISTO-CENE ★
1.5 MYA ★

LAKE TURKANA

AFRICA ●
TERNIFINEʙ
KOOBI FORAᴄ
OLDUVAI GORGEᴅ
SWARTKRANSᴇ
STERKFONTEINꜰ

107

THE TRANSITION TO HOMO SAPIENS

We come now to the origin of our own species, *Homo sapiens*. It seems to have come into being between 130,000 and 75,000 years ago. Was there a gradual transition from *Homo erectus,* or did the species emerge relatively suddenly and spread rapidly throughout the world? Dating this transition is a problem, for the time period is too old for carbon 14, and most sites did not have volcanic activity so potassium-argon dating is inapplicable.

Some scientists consider the Steinheim and Swanscombe skulls (previous plate) transitional between *erectus* and *sapiens*. There is rarely more than a single good fossil specimen from any one site; so we don't have statistical data to establish male-female differences or variability within and between different populations. Cranial capacity is in the range of modern *H. sapiens,* but overall they are sufficiently different from us and from Neanderthals to be labeled "archaic *H. sapiens*." Many European specimens have been known for decades; recent Ethiopian and Tanzanian finds give a broader geographical perspective.

Color the time scale. Then color Fontechevade and the arrow showing its location. Proceed to color each fossil named for the site where it was found—its title, skull, and arrow—as it is discussed in the text.

Fontechevade, a partial cranium (about 1,400 cm³), is thick like Swanscombe and was found in a cave in *France. Saccopastore* skulls, from a tributary of the Tiber, have a flattened frontal region, a cranial capacity of 1,200–1,300 cm³, a massive face and an interglacial date, to judge by associated elephant bones. Some consider *Saccopastore* a forerunner of Neanderthal populations (Plate 108). Another interglacial skull, associated with extinct elephant and rhino, was found at *Ehringsdorf* near Weimar, and has an estimated cranial capacity of 1,400 cm³.

Petralona has one of the best preserved faces and a cranial capacity of 1,200 cm³ with similarities to *Broken Hill.* At *Krapina* portions of several skulls, many teeth, and parts of a skeleton, in association with an interglacial rhino, were discovered in the

floor of a rock shelter. The skulls have strong brow ridges, sloping forehead, and powerful jaws—some features like Neanderthal and some like modern *sapiens.*

Jebel Ighoud has yielded two skulls (capacity 1,430 cm³) and a juvenile mandible similar to *Saccopastore. Bodo,* recently discovered, has a broad, thick nasal region, robust zygomatic arches, and is similar in appearance to *Broken Hill* and *Petralona.*

From *Omo* we have a partial skeleton and two skulls about 130,000 years old. Omo II has a cranial capacity of 1,435 cm³ and other skull dimensions within the range of modern *Homo.* The newly reported *Laetoli* skull (capacity 1,200 cm³), described by anatomist Michael Day, may be 120,000 years old. It resembles *Omo, Bodo,* and *Saldanha.*

Broken Hill, a mine in what is now *Zambia,* produced one nearly complete skull (cranial capacity 1,280 cm³) and fragments of others, in association with tools and animal bones dated by archeologist Richard Klein at a minimum of 125,000 years. It resembles *Bodo* and *Saldanha.* The latter, a skull cap (capacity 1,280 cm³) found near Cape Town, is dated by fauna and artifacts at 120,000 years.

The largest single collection for this time period comes from *Solo* (also called Ngandong): parts of nine adult skulls (capacities 1,035–1,255 cm³). Found in the Solo River valley in Java in association with other mammal bones, the best guess for a date is about 130,000 years. Skull dimensions are distinct from *Homo erectus* but "archaic" with respect to modern *sapiens.*

The long persistence of *Homo erectus* ended with the appearance of individuals having expanded, less angular skulls and thinner bone structure, but without any dramatic changes in their stone tools. These archaic *Homo sapiens* preceded the Neanderthals by only a short time, and their relationship with the subsequent human populations of Europe, Africa, and Asia remains problematic. As yet, we know little of the cultural transitions that took place as the species came to rely less on physical adaptation and more on technological and social innovations.

TRANSITION TO HOMO SAPIENS ⋆

FONTECHEVADE
  (FRANCE)ₐ
SACCOPASTORE
  (ITALY)ʙ
EHRINGSDORF
  (DDR)c
PETRALONA
  (GREECE)ᴅ
KRAPINA
  (YUGOSLAVIA)ᴇ
JEBEL IGHOUD
  (MOROCCO)ꜰ

BODO
  (ETHIOPIA)ɢ
OMO
  (ETHIOPIA)ʜ
LAETOLI
  (TANZANIA)ı
BROKEN HILL
  (ZAMBIA)ᴊ
SALDANHA
  (SOUTH AFRICA)ᴋ
SOLO
  (INDONESIA)ʟ

0
(THOUSANDS
OF YEARS)

75,000 ⋆
UPPER PLEISTOCENE ⋆
150,000 ⋆

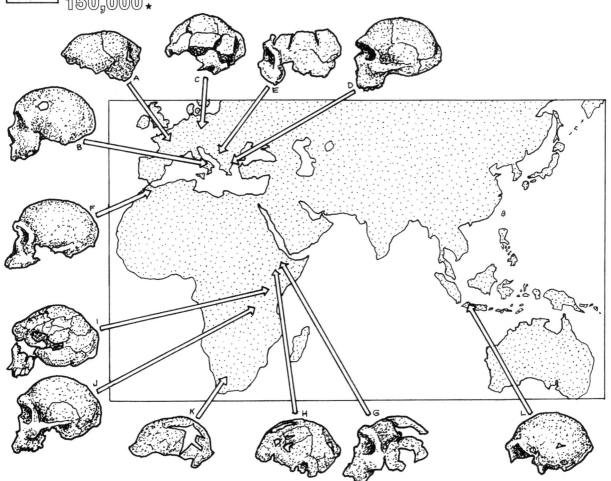

108

NEANDERTHAL: BRUTISH BUT BRIGHT

The word *Neanderthal* has become undeservedly synonymous with the image of a subhuman ancestor, stooped, shambling, beetle-browed, a not-very-smart caveman. This image has been so perpetuated in cartoons, movies, and popular writings that it will surprise most people to realize that *Neanderthals* were in fact humans of our own species with brains on average slightly larger than our own.

The misunderstanding began with the discovery of the first nearly complete *Neanderthal* skeleton in France. This individual happened to have severe arthritis so was, in fact, stooped, bent, and would have had a shuffling gait like victims of this disease. The survival of such an individual to an advanced age was in itself a hallmark of humanity: an individual who could clearly not take care of itself had been taken care of.

Color the time chart and the Neanderthal discovery sites, many concentrated in Europe, from Russia in the east to the Channel Isles in the west. Shanidar in Iraq and the Dordogne area of France are among the richest sites.

A few *sites,* like Shanidar, have yielded carbon 14 dates, but the time of the *Neanderthals,* the last glacial time period, 35,000 to 70,000 years ago, is generally too old for good carbon 14 dates, so exact dating of many *sites* remains problematic.

Neanderthals lived mostly in areas with cold climates. Even the Mediterranean region was much cooler then than now. The first fossil specimen was found in 1856 in the Neander Valley (*Neanderthal,* in German) near Dusseldorf. The name came to be applied to fossil hominids of this general time period, particularly in Europe. Their place in human evolution has been and remains clouded.

The skull form is striking. Color in gray the modern *Homo sapiens* skull and its average cranial capacity; note the high rounded forehead. Use a bright color for the skull of the typical Neanderthal and its relatively large cranial capacity. Color the side view of the head.

Note *Neanderthal's* more prominent face, brow ridges, and lower forehead and the bunlike occipital swelling. The jaw is stout; the large molar teeth are well worn, with characteristic enlarged pulp cavities (taurodontism).

Neanderthals provide the best hominid fossil record since the Plio-Pleistocene. The earlier hominids lived near lakes and rivers. The *Neanderthals* lived in cave *shelters* and *buried* their dead. Both ways of life favored fossilization. Unlike early hominids, except for "Lucy" (Plate 105), the *Neanderthals* left many complete or nearly complete skeletons. From the studies of anthropologists Eric Trinkaus and William Howells, *Neanderthals* were generally less than five feet (150 centimeters) tall, stockily built, with barrel chest, and short heavy arm and leg bones curiously curved with distinct muscle markings. Many bones had been broken and healed; one injury may have been a spear wound. Their life, it seems, was a tough one. Women were as muscular as men, but slightly smaller.

Complete the plate by coloring the prey, tools, shelter, and burial.

These people were hunters. Plant food would have been seasonal and unavailable during the long, cold winters of the glacial period. We have evidence that the *Neanderthals* hunted woolly mammoth, woolly rhinoceros, cave bear, ibex, and other game. They doubtless made use of the warm animal skins for clothing, blankets, and *shelter.*

Tools from *Neanderthal sites* include fine points, sharp knives, and scrapers, a great advance on the handaxes and choppers of the prior "Acheulian" culture. These Middle Paleolithic *tools* are called Mousterian, after a site in the Dordogne of France. Some of the stone points were hafted onto spears apparently used for thrusting and jabbing rather than throwing.

They lived in caves and in some areas apparently constructed wood *shelters,* supported by mammoth tusks, covered by skins, with hearths for heat. Individuals were *buried* on their sides, with knees up near the chest, covered with ochre or flowers, and surrounded by bear skulls or ibex horns. We thus have evidence for abstract thought (about death) and the beginning of religious ritual.

What happened to the *Neanderthals?* Whether this distinct *Homo sapiens* population was absorbed into the more modern Cro-Magnons or whether it died out altogether is as yet unknown.

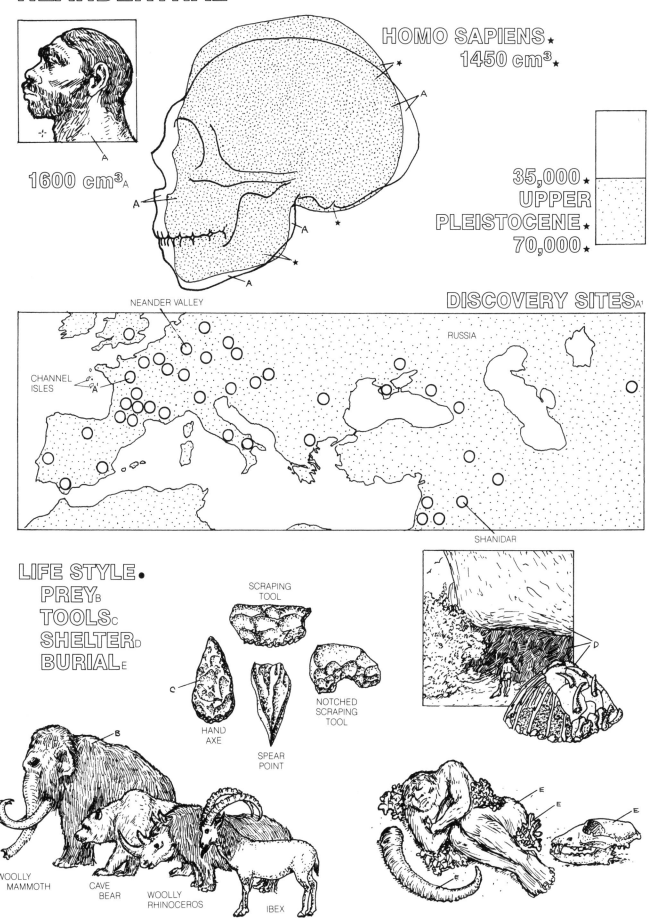

NEANDERTHAL A

HOMO SAPIENS ★
1450 cm³ ★

1600 cm³ A

35,000 ★
UPPER
PLEISTOCENE ★
70,000 ★

DISCOVERY SITES A¹

NEANDER VALLEY

RUSSIA

CHANNEL
ISLES

SHANIDAR

LIFE STYLE ●
PREY B
TOOLS C
SHELTER D
BURIAL E

SCRAPING
TOOL

HAND
AXE

SPEAR
POINT

NOTCHED
SCRAPING
TOOL

WOOLLY
MAMMOTH

CAVE
BEAR

WOOLLY
RHINOCEROS

IBEX

109

PATHWAYS OF HUMAN EVOLUTION: CHANGING HYPOTHESES

There was but a single human evolutionary history, but as new evidence accumulates, our ideas about this history undergo their own rapid evolution.

When Charles Darwin and Thomas Henry Huxley wrote about human evolution in the 1860s and 1870s, the only fossils of possible human ancestry known were a *Dryopithecus* jaw and a skull top from the Neander Valley. *Homo erectus* was not discovered in Europe, China, and Java until the late nineteenth and early twentieth centuries. Although Dart, and later Broom and Robinson, began finding fossils from two species of small-brained hominids in the 1930s, *Australopithecus* was not widely accepted as a human ancestor until about 1950. Anatomist W.E. Le Gros Clark studied the original specimens in 1947 and became convinced that they were those of a very early human species having a mosaic of apelike and humanlike features. His views held sway with the European scientific community. Not for another decade were the "experts" convinced that this small-brained biped could have made tools. Sherwood Washburn integrated all this information within the framework of mosaic evolution, and stressed the role of tools in shaping the course of human evolution.

Color the fossils and their place in time as interpreted before 1960.

By 1960, based on the findings from South Africa, *Australopithecus africanus* was considered by many ancestral to *Homo erectus*, with *A. robustus* a side branch. *Neanderthal* people were considered a much later side branch with the same fate as *A. robustus*.

Color the family tree as some saw it in 1965.

The robust australopithecine, *A. boisei*, and the newly named *Homo habilis*, both from Olduvai, had been shown to coexist and to date back to nearly two million years. One way of putting all this together was the "single-species hypothesis" of anthropologist Loring Brace. He believed that all australopithecines belonged to the same species, with a wide range of geographic and individual variation and sexual dimorphism. The australopiths then evolved into *Homo habilis*, transitional to *Homo erectus*, followed by the *Neanderthals*, which were ancestral to modern *Homo sapiens*.

Color the pathway as viewed in 1970.

By 1970, new hominid finds were pouring in from many sites, especially from the Omo and Lake Turkana. Potassium-argon dates at Omo pushed hominid evolution back to almost three million years, and the fragmentary jaws and isolated teeth indicated that a gracilelike australopithecine existed earlier than a robust species which appeared later, about 2.1 mya. Additional evidence from Lake Turkana for the coexistence of robust australopithecines with *Homo* was inconsistent with the single-species hypothesis. *A. robustus* and *A. boisei*, perhaps merely geographical variants of the same species rather than two distinct species, both disappear from the fossil record by 1 mya.

Complete the plate by coloring the tree for 1980, noting that the period between 2.1 and 2.5 mya is to be colored gray.

During the 1970s, hominid fossils as old as 3.5 million years from Laetoli and Hadar were proposed as a new species, *A. afarensis*, somewhat older than *A. africanus* in South Africa. Some taxomonists maintain that the new species designation is not justified due to the number of anatomical features it shares with *A. africanus*.

More evidence and detailed comparisons will resolve this controversy, as well as confirm or disprove whether *A. africanus* or *A. afarensis* was ancestral to *Homo* and whether these two australopithecine species coexisted. New fossil discoveries also will clarify the origin of the robust australopithecines. Fossils from the time range older than 3.7 mya will provide evidence to test the hypothesis that the pygmy chimpanzee is the hominid prototype.

Homo erectus is now dated from Lake Turkana at 1.4 mya, the earliest known for this species, and *Homo sapiens* from Laetoli and Omo, at 120,000 years. From France new discoveries may support the interpretation that *Neanderthal* was contemporary with modern *sapiens*, raising again the possibility that *Neanderthal* people were a side branch rather than ancestral.

PATHWAYS OF EVOLUTION: CHANGING HYPOTHESES ★

MYA ★

0.1 ★
0.5 ★
1 ★
1.5 ★
2 ★
2.5 ★
3 ★
3.5 ★
4 ★
4.5 ★
5 ★

BEFORE 1960 ★   1965 ★   1970 ★   1980 ★

TO THE LIVING
AFRICAN APES

A. AFARENSIS[A]   HOMO HABILIS[E]
A. AFRICANUS[B]   H. ERECTUS[F]
A. ROBUSTUS[C]   H. SAPIENS NEANDERTHAL[G]
A. BOISEI[D]   H. SAPIENS SAPIENS[H]

110

EVOLUTION OF THE BRAIN: THE INCREASE IN SIZE

The expansion of the brain during human evolution was of greater magnitude and occurred more rapidly than in any other animal group in the history of life. Given the dangers that big brains impose on females and their babies during childbirth and the compromises to efficient bipedalism with a wide pelvis that must be made in females with large birth canals, we can infer that there must have been strong selection pressures for large brains that outweighed these significant disadvantages.

Color the top chart, using light colors.

How can we compare human brain size with that of other animals? Clearly, an important consideration is relative brain size, for large animals need larger brains to control their larger muscle mass. The graphs here are adopted from the work of Harold Jerison and plot brain versus body weight for many species. The upper polygon comprises the relatively high brain/body ratios of living *mammals* and *birds;* the lower one includes living and extinct fish, amphibians, and *reptiles.* The third group, falling between the other two, consists of values estimated from the fossil record for *extinct* Mesozoic *mammals,* which lived 150–200 mya.

The second chart depicts the relationship between brain size and body weight for anthropoids and fossil hominids. Color the rectangles for living monkeys and apes (D and E).

Though they overlap, the size of *ape* brains on average is definitely larger than that of *monkey* brains, implying that larger brain size has been evolving throughout primate evolution, not just since the origin of the human lineage.

Color the rectangles for the extinct and living hominids. Color the braincases and cranial capacities for the fossils on the time line at the bottom of the plate.

Fossil brain weights can be accurately estimated from skull endocasts and volume measurements; body weights must be estimated from skeletal reconstructions (Plate 104). These charts show how relative brain size in hominids has been increasing for at least four million years. The hominoid ancestor 5 mya probably had a brain of 400 cm³, like that of *living apes.* Brain size increased slowly during the first three million years of hominid evolution but seems to have accelerated about 2 mya, paralleling the appearance of the earliest stone tools. *Homo habilis* had a brain size of 600–700 cm³ on a body not much larger than that of a gracile australopithecine. By 0.5 mya, after *Homo erectus* had left Africa and expanded into Eurasia, another 300 cm³ had been added to the average cranial capacity.

Modern brain size averages about 1,400 cm³, ranging from 1,000–2,000 in all human groups. Individual brain size does not correlate well with intelligence, because it is not so much the sheer size of our brains that makes us distinctively human as it is the neural reorganization. On average, the human brain is more than three times larger than a chimpanzee's, but it is not simply an enlarged version. The human neocortex and association areas (Plate 47) have increased in volume more than other parts have. The cerebellum, coordinator of learned motor activity patterns, has also expanded disproportionately, especially those tracts involved with the hands and face. The limbic system, which lies beneath the neocortex and is involved with motivation, emotion, and social communication in nonhuman primates has also become reorganized as humans evolved. But we are not certain how limbic functions in humans have become integrated with cortical functions.

The evolution of brain size is a crude but important index of human evolutionary change. In computers, the best (also crude) analog we have to the brain, size is roughly correlated with "storage capacity," "memory," or computational power. But in computers, as well as in the brain, we need to know much more about the programs, hardware and software, before we can give a good description of how the device works. This kind of research on the brain (and "artificial intelligence," using computers) is an exciting scientific frontier which may eventually result in deeper insights into ways in which the human brain has evolved its "programs" as well as its size.

EVOLUTION OF THE BRAIN: THE INCREASE IN SIZE. ★

MAMMALS AND BIRDS_A
EXTINCT MAMMALS_B
REPTILES_C

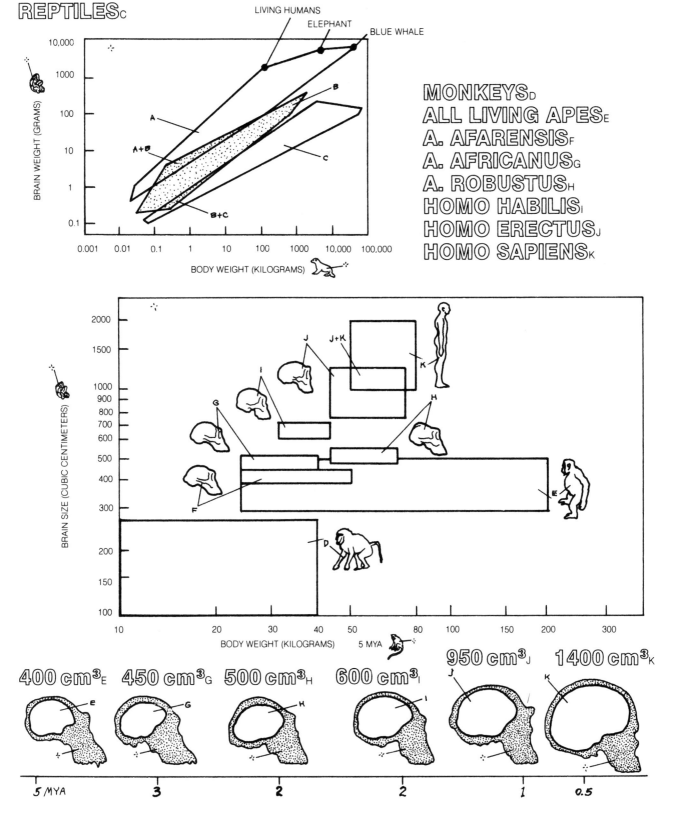

MONKEYS_D
ALL LIVING APES_E
A. AFARENSIS_F
A. AFRICANUS_G
A. ROBUSTUS_H
HOMO HABILIS_I
HOMO ERECTUS_J
HOMO SAPIENS_K

111

EVOLUTION OF THE BRAIN: LANGUAGE AND THE BRAIN

All human groups studied are mostly right-handed, by a ratio of at least three to one, whereas chimpanzees and other primate groups have equal numbers of right and left handers. So the brain centers that control skilled human hand movements lie mostly in the left hemisphere. The brain centers that control human speech are even more often located in the left hemisphere—in nearly all right handers and many left handers—whereas vocalization centers in nonhuman primates are not lateralized (Plate 66).

Research on the nondominant (mostly right) side of the brain suggests that it also has specialized for such functions as facial recognition, *musical* and *artistic* abilities, and *intuitive* responses. That these special right and left brain abilities are not present in our close relatives, the chimpanzees, reveals how much the reliance of our ancestors on tools and language during the past five million years of evolution has brought about not only a tremendous increase in brain size but also a major reorganization of its functional capacities.

Color the corpus callosum in the left sagittal view of a human brain sliced in half and in the top view in the center of the page.

This tract evolved in placental mammals more than 100 mya when the neocortex began to expand. It has two million nerve fibers, more than any other primate's, due to the need in our species for exchange of information between the two hemispheres.

Color Broca's area and Wernicke's area in the left lateral view.

The left lateral view of the human brain shows *Broca's* and *Wernicke's areas* within the neocortical association areas. They are found only in the language-dominant hemisphere and have no counterpart in nonhuman primates. *Broca's area* controls muscles of the face, tongue, palate, and larynx used in speech. When this area is injured, speech becomes slow and labored. *Wernicke's area* contains circuits for speech comprehension. When it is damaged, speech may still seem fluent but the message is garbled and confused.

Color the motor and somatosensory strips and the visual sensory cortex on the left lateral view (for review see Plate 47). Now color the rest of the plate, which depicts how the two hemispheres are specialized. Begin with the right visual field and the left visual cortex, which analyzes its input. Color the left visual field and the right visual cortex, then the hands and the sensory and motor strips where they are represented in the neocortex.

Experimentally, one can test the different abilities of the two hemispheres by flashing images that will be processed only by the *left* or by the *right visual cortex*. It has been found that one side (usually the left) is better at *reading* and comprehending words and phrases, while the other is superior in recognizing faces and geometric designs. Auditory stimuli elicit similar patterns of lateralization: The dominant hemisphere is usually better at analyzing speech, the other at perceiving music and nonverbal acoustic communication.

Brain specialist Jerre Levy has suggested that this hemispheric "division of labor" evolved out of a basic antagonism between the two types of cognitive processes—the sequential analytic type needed for speech versus the Gestalt or holistic type, in which reality is perceived nonverbally, intuitively, as a total pattern. Both hemispheres contribute to an individual's well-roundedness, to the capacity for rational thought and speech, as well as social and emotional sensitivity and responsiveness.

To fully understand the story of human evolution, we need to unravel the mysteries of human brain evolution. But the questions are difficult and multiplex. How do both genes and environment affect brain development? What are the relationships between brain structure and function, emotion, language, thought, behavior, culture? These are lines of anthropological inquiry which lie between scientific explanation and ways in which the linguist, artist, and poet can help guide the scientist to a more complete understanding of human nature.

LANGUAGE AND THE BRAIN ★

CORPUS CALLOSUM A
BROCA'S AREA B
WERNICKE'S AREA C
MOTOR STRIP D
SENSORY STRIP E
VISUAL AREA F

RIGHT
HEMISPHERE ●
LEFT VISUAL
FIELD H
RIGHT VISUAL
CORTEX H'
ARTISTIC I
INTUITIVE J
SPATIAL K
MUSICAL L

LEFT
SAGITTAL
VIEW

LEFT
LATERAL
VIEW

LEFT
HEMISPHERE ●
RIGHT VISUAL
FIELD G
LEFT VISUAL
CORTEX G'
CATEGORICAL M
VERBAL N
WRITING O
READING P
LINEAR Q

SENSORY/
MOTOR
STRIPS R

SENSORY/
MOTOR
STRIPS S

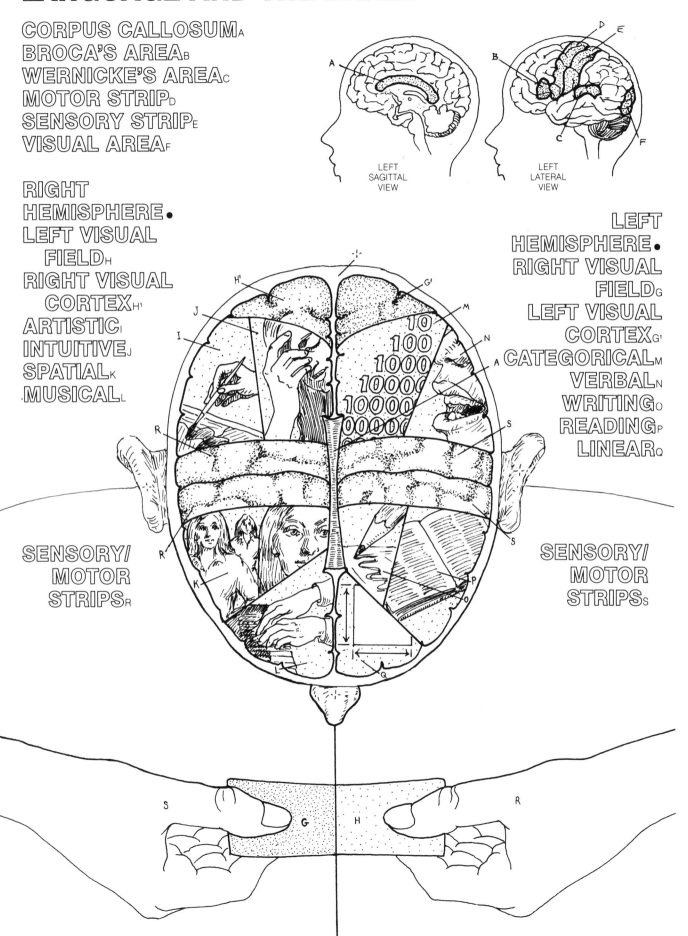

APPENDIX

CLASSIFICATION OF LIVING PRIMATES
Order: Primate

SUBORDER/ INFRAORDER	SUPERFAMILY	FAMILY	GENUS	COMMON NAME	DISTRIBUTION
	Tupaioidea		*Tupaia*	tree shrew	Philippines, India, Southwest Asia
			Tana		
			Dendrogale		
			Urogale		
			Anathana		
			Ptilocercus		
PROSIMIANS (lower primates)					
	Tarsioidea	Tarsiidae	*Tarsius*	tarsier	Philippines, Southeast Asia
	Lorisoidea	Galagidae	*Galago*	galago, bush baby	Africa
		Lorisidae	*Loris*	slender loris	India, Sri Lanka
			Nycticebus	slow loris	Africa (sub-saharan)
			Perodicticus	potto	Southeast Asia
			Arctocebus	angwantibo, golden potto	West Africa
	Lemuroidea	Lemuridae	*Lemur*	lemur	Madagascar
			Hapalemur	gentle lemur	Madagascar
			Lepilemur	sportive lemur	Madagascar
			Microcebus	mouse lemur	Madagascar
			Cheirogaleus	dwarf lemur	Madagascar
		Indriidae	*Indri*	indri	Madagascar
			Propithecus	sifaka	Madagascar
			Lichanotus	avahi	Madagascar
		Daubentoniidae	*Daubentonia*	aye-aye	Madagascar
ANTHROPOIDEA (higher primates)					
Platyrrhine	Ceboidea (New World Monkeys)	Cebidae	*Cebus*	capuchin	Central, South America
			Saimiri	squirrel monkey	Central, South America
			Ateles	spider monkey	Mexico to Amazon Basin
			Lagothrix	woolly monkey	South America
			Brachyteles	woolly spider monkey	Brazil
			Alouatta	howler monkey	Tropical America (to Mexico)

SUBORDER/ INFRAORDER	SUPERFAMILY	FAMILY	GENUS	COMMON NAME	DISTRIBUTION
			Callicebus	titi monkey	Tropical South America
			Aotus	night monkey	Tropical Southern Mexico
			Pithecia	saki	Amazon Basin
			Cacajao	uakari	South America
		Callithricidae (marmosets)	*Callithrix*	marmoset	Tropical South America
			Cebuella	pygmy marmoset	South America
			Callimico	Goeldi's marmoset	Amazon
			Leontocebus	golden or lion marmoset	Tropical South America
			Saguinus	marmoset	Tropical South America
Catarrhine (Old World Monkeys, Apes and Humans)	Cercopithecoidea (Old World Monkeys)	Cercopithecidae (cheek pouch monkeys)	*Cercopithecus* (Erythrocebus)	guenons, vervet, patas	West, Central, East, South Africa
			Cercocebus	mangabey	West, Central, East Africa
			Papio (Theropithecus)	baboon, gelada, hamadryas	Sub-saharan Africa East Africa
			Mandrillus	drill, mandrill	West, Central Africa
			Macaca	macaques, rhesus monkey	Europe, North Africa, all of Asia
			(Cynopithecus)	celebes black ape	Suluwasi (Celebes) Southeast Asia
		Colobidae (leaf monkeys)	*Colobus*	guereza, colobus	Sub-saharan Africa
			Presbytis	Langur, Leaf monkey	India, South Asia, Southeast Asia
			Nasalis	proboscis	Borneo
			(Pygathrix)	douc langur	Vietnam
			(Rhinopithecus)	snub-nose langur	China, Tibet
	Hominoidea (apes and humans)	Pongidae (apes)	*Pongo*	orangutan	Borneo, Sumatra
			Pan	pygmy and common chimpanzees, gorilla	Equatorial Africa
			Hylobates (Symphalangus)	gibbon siamang	Southeast Asia
		Hominidae (humans)	*Australopithecus* (extinct)	ape-people	Eastern and Southern Africa
			Homo	women, men	World-wide

Primary Source: Simpson, G.G. *The principles of classification and a classification of mammals.* Bulletin, American Museum of Natural History, vol. 85, 1945.

Secondary Source: Napier, J. and P. Napier. *Handbook of Living Primates.* Academic Press, New York. 1967.

BIBLIOGRAPHY

Part I

Cloud, Preston. *Cosmos, Earth, and Man.* Yale University Press, New Haven, 1978. Includes the evolution of everything!

Colbert. Edwin H. *Evolution of the Vertebrates.* 2nd edition, John Wiley and Sons, New York, 1966.

Crompton, A.W., and P. Parker. "Evolution of the mammalian apparatus," In: *American Scientist,* Vol. 66: 1978; pages 192-201.

Hildebrand, Milton. *Analysis of Vertebrate Structure.* John Wiley and Sons, New York, 1974.

Irvine, William. *Apes, Angels and Victorians.* McGraw-Hill, New York, 1955. The story of Darwin, Huxley, and evolutionary theory.

Lack, David L. *Darwin's Finches: an Essay on the General Biological Theory of Evolution.* Harper and Row, New York, 1961.

McAlester, A. Lee. *The History of Life.* Prentice-Hall, Inc., Englewood Cliffs, New Jersey, 1968.

Medawar, P.B. *Advice to a Young Scientist.* Harper and Row, New York, 1979. Practical advice on how to become a scientist.

Moore, Ruth. *Evolution.* Time-Life Books, New York, 1964.

Romer, Alfred S. *The Vertebrate Story.* University of Chicago Press, Chicago, 1959.

Scientific American. *Evolution.* W.H. Freeman and Company, San Francisco, 1978.

Scientific American. *Evolution and The Fossil Record.* Introduction by Léo Laporte. W.H. Freeman and Company, San Francisco, 1982.

Simpson, George Gaylord. *The Meaning of Evolution: A Study of the History of Life and its Significance for Man.* Yale University Press, New Haven, 1949.

Stone, Irving. *The Origin.* Doubleday and Company, Garden City, New York, 1980. A novelized biography of Darwin.

Thompson, D'Arcy. *On Growth and Form.* Abridged edition, J.T. Bonner, ed. Cambridge University Press, Cambridge, 1977. First published in 1917.

Torrey, Theodore W. *Morphogenesis of the Vertebrates.* John Wiley and Sons, New York, 1971.

Part II

Ayala, Francisco J., and J.A. Kiger, Jr. *Modern Genetics,* Benjamin/Cummings Publishing Company, Menlo Park, California, 1980.

Brues, Alice M. *People and Races.* Macmillan, New York, 1977.

Damon, Albert. (ed.) *Physiological Anthropology.* Oxford University Press, New York, 1975.

Kretchmer, Norman. "Lactose and Lactase." In: *Scientific American,* Vol. 227(4), Reprint #1259. W.H. Freeman and Company, San Francisco, 1972.

Lerner, I.M. and W.J. Libby. *Heredity, Evolution and Society.* 2nd edition, W.H. Freeman and Company, San Francisco, 1976.

Moore, Ruth. *The Coil of Life: The Story of Great Discoveries in the Life Sciences.* Alfred Knopf, New York, 1961.

Vayda, Andrew P. *Environment and Cultural Behavior.* Natural History Press, Garden City, New York, 1969. Particularly, see Stephan L. Weisenfeld, "Sickle-cell trait in human biological and cultural evolution."

Watson, James D. *The Double Helix.* Atheneum, New York, 1968.

Wilson, A.C., S.S. Carlson, and T.J. White. "Biochemical Evolution." *Annual Review of Biochemistry,* Vol. 46: 1977; pages 573-639.

Wilson, E.O. and T. Eisner, (eds). *Life on Earth.* Sinauer Associates, Inc., Sunderland, Massachusetts, 1975. Particularly, see R.E. Dickerson, "Molecular evolution."

Part III

Altmann, Jeanne. *Baboon Mothers and Infants.* Harvard University Press, Cambridge, Massachusetts, 1980.

Bateson, P.P.G., and R.A. Hinde (eds.). *Growing Points in Ethology.* Cambridge University Press, Cambridge, Massachusetts, 1976. Particularly, see N.K. Humphrey, "The social function of the intellect."

Chalmers, Neil. *Social Behavior in Primates.* Edward Arnold, London, 1979.

Chevalier-Skolnikoff, Suzanne, and Frank E. Poirier, (eds.). *Primate Bio-Social Development: Biological, Social and Ecological Determinants.* Garland Publishing, Inc., New York, 1977.

DeVore, Irven, (ed.) *Primate Behavior: Field Studies of Monkeys and Apes.* Holt, Rinehart and Winston, New York, 1965.

Eimerl, Sarel and Irven DeVore. *The Primates.* Time-Life Books, New York, 1974.

Goodall, Jane van Lawick. *In the Shadow of Man.* Houghton Mifflin Company, Boston, 1971.

Grand, Theodore I. "Body weight: its relation to tissue composition, segment distribution, and motor function." *American Journal of Physical Anthropology.* Vol. 47(2): 1977; pages 211-239, 241-248.

Jay, Phyllis, C. (ed.). *Primates: Studies in Adaptation and Variability.* Holt, Rinehart, and Winston, New York, 1968.

Jolly, Alison. *The Evolution of Primate Behavior.* Macmillan, New York, 1972.

Lancaster, Jane B. *Primate Behavior and the Emergence of Human Culture.* Holt, Rinehart, and Winston, New York, 1975.

Lindburg, Donald G., (ed.). *The Macaques: Studies in Ecology, Behavior and Evolution.* Van Nostrand Reinhold, New York, 1980.

Mitchell, G. *Behavioral Sex Differences in Nonhuman Primates.* Van Nostrand Reinhold, New York, 1979.

Napier, J.R., and P.H. Napier. *Handbook of Living Primates; Morphology, Ecology, and Behavior of Nonhuman Primates.* Academic Press, New York, 1967.

Napier, J.R. and P.H. Napier. *Old World Monkeys: Evolution, Systematics, and Behavior.* Academic Press, New York, 1970. Particularly, see Suzanne Ripley, "Leaves and leaf-monkeys: the social organization of foraging gray langurs."

Schultz, Adolph H. *The Life of Primates.* Weidenfeld and Nicolson, London, 1969.

Steklis, Horst D. and Michael J. Raleigh. *Neurobiology of Social Communication in Primates. An Evolutionary Perspective.* Academic Press, New York, 1979.

Sussman, Robert W. (ed.). *Primate Ecology: Problem-Oriented Field Studies.* John Wiley and Sons, New York, 1979.

Teleki, Geza. *The Predatory Behavior of Wild Chimpanzees.* Bucknell University Press, Lewisberg, 1973.

Part IV

Beck, Benjamin B. *Animal Tool Behavior: The Use and Manufacture of Tools by Animals.* Garland STPM Press, New York, 1980.

Charles-Dominique, Pierre. *Ecology and Behavior of Nocturnal Primates.* Columbia University Press, New York, 1977.

Ciochon, Russell, and A. Brunetto Chiarelli, (eds.). *Evolutionary Biology of the New World Monkeys and Continental Drift.* Plenum Press, New York, 1981. See especially articles by D. Tarling, R. Hoffstetter, R. Lavocat, and V. Sarich and J. Cronin.

Clutton-Brock, T.H. (ed.) *Primate Ecology.* Academic Press, New York, 1977.

Doyle, G.A., and R.D. Martin, (eds.). *The Study of Prosimian Behavior.* Academic Press, New York, 1979.

Goodman, Morris and R.E. Tashian, (eds.). *Molecular Anthropology: Genes and Proteins in the Evolutionary Ascent of the Primates.* Plenum Press, New York, 1976.

Grand, Theodore I. "A mechanical interpretation of terminal branch feeding." *Journal of Mammalogy.* Vol. 53(1): 1972; pages 198-201.

Hamburg, David A., and E.R. McCown, (eds.). *The Great Apes.* Benjamin/Cummings, Menlo Park, California, 1979.

Hershkovitz, Philip. *Living New World Monkeys,* Volume I. University of Chicago Press, Chicago, 1977.

Kleiman, Devra G., (ed.). *The Biology and Conservation of the Callitrichidae.* Smithsonian Institution Press, Washington, D.C., 1977.

LeGros Clark, W.E. *The Antecedents of Man.* Harper and Row, New York, 1959.

Luckett, W.P., and F.S. Szalay, (eds.). *Phylogeny of the Primates.* Plenum Press, New York, 1975.

Montgomery, G. Gene. *The Ecology of Arboreal Folivores.* Smithsonian Institution Press, Washington, D.C., 1978. In particular, see articles by Kay and Hylander, and Grand.

Morbeck, Mary Ellen, H. Preuschoft, and N. Gomberg, (eds.). *Environment, Behavior, and Morphology: Dynamic Interaction in Primates.* Gustav Fischer New York, Inc., New York, 1979.

Press, Frank and Raymond Siever. *Earth.* 2nd edition, W.H. Freeman and Company, San Francisco, 1978.

Simons, Elwyn L. *Primate Evolution: An Introduction to Man's Place in Nature.* Macmillan Company, New York, 1972.

Szalay, Frederick S., and E. Delson. *Evolutionary History of the Primates:* Academic Press, New York, 1979.

Zihlman, A.L., J.E. Cronin, D.L. Cramer, and V.M. Sarich. "Pygmy chimpanzee as a possible proto-type for the common ancestor of humans, chimpanzees, and gorillas. " In: *Nature,* Vol. 275(5682): 1978; pages 744-746.

Zihlman, A.L., and J.M. Lowenstein. "False start of the human parade." *Natural History,* Vol. 88(7): 1979; pages 86-91.

Part V

Behrensmeyer, Anna K., and A.P. Hill, (eds.). *Fossils in the Making: Vertebrate Taphonomy and Paleoecology.* University of Chicago Press, Chicago, 1980.

Bishop, W.W., and J.A. Miller, (eds.). *Calibration of Hominoid Evolution; Recent Advances in Isotopic and Other Dating Methods as Applicable to the Origin of Man.* Scottish Academic Press, Edinburgh, 1972.

Butzer, Karl W. *Environment and Archeology: An Ecological Approach to Prehistory.* 2nd edition, Methuen and Company, Ltd., London, 1971.

Cole, Sonia. *Leakey's Luck: The Life of Louis Seymour Bazett Leakey.* Harcourt Brace Jovanovich, Inc., New York, 1975.

Coppens, Yves, F.C. Howell, G.L. Isaac, R.E.F. Leakey (eds.). *Earliest Man and Environments in the Lake Rudolf Basin.* University of Chicago Press, Chicago, 1976.

Dahlberg, Frances, (ed.). *Woman the Gatherer.* Yale University Press, New Haven, 1981.

Day, Michael H. *Guide to Fossil Man.* University of Chicago Press, Chicago, 1977.

Eicher, Don L. *Geologic Time.* Prentice-Hall, Inc., Englewood Cliffs, New Jersey, 1968.

Hay, Richard L. *Geology of the Olduvai Gorge.* University of California Press, Berkeley, 1976.

Howell, Clark F. *Early Man.* Revised edition, Time-Life Books, New York, 1971.

Howells, William (ed.). *Ideas on Human Evolution: Selected Essays, 1949–1961.* Atheneum, New York, 1967.

Jerison, Harold. *Evolution of the Brain and Intelligence.* Academic Press, New York, 1973.

Johanson, Donald C., and Maitland A. Edey. *Lucy: The Beginnings of Humankind.* Simon & Schuster, New York, 1981.

Jolly, Clifford, (ed.) *Early Hominids of Africa.* Gerald Duckworth and Company, London, 1978.

Klein, Richard G. *Ice-Age Hunters of the Ukraine.* University of Chicago Press, Chicago, 1973.

Laporte, Léo F. *Ancient Environments.* 2nd edition, Prentice Hall, Inc., Englewood, New Jersey, 1979.

Leakey, Louis S.B. *By the Evidence: Memoirs: 1932–1951.* Harcourt Brace Jovanovich, Inc., New York, 1974.

Leakey, Meave G., and Richard E. Leakey, (eds.). *Koobi Fora Research Project, Volume 1: The Fossil Hominids and an Introduction to Their Context, 1968–1974.* Clarendon Press, Oxford, 1978.

Leakey, Richard E., and Roger Lewin. *Origins.* E.P. Dutton, New York, 1977.

Lee, Richard Borshay. *The !Kung San: Men, Women, and Work in a Foraging Society.* Cambridge University Press, Cambridge, 1979.

Le Gros Clark, W.E. *Man-Apes or Ape-Men; The Story of Discoveries in Africa.* Holt, Rinehart, and Winston, Inc., New York, 1967.

Le Gros Clark, W.E. *The Fossil Evidence for Human Evolution.* 3rd edition. University of Chicago Press, Chicago, 1978.

Maglio, Vincent J., and H.B.S. Cooke, (eds.). *Evolution of African Mammals.* Harvard University Press, Cambridge, Massachusetts, 1978. Particularly, see H.B.S. Cooke, "Africa: the physical setting" and F. Clark Howell, "Hominidae."

Marshall, Lorna J. *The !Kung of Nyae Nyae.* Harvard University Press, Cambridge, Massachusetts, 1976.

Moore, Ruth. *Man, Time, and Fossils.* Alfred Knopf, New York, 1961.

Time-Life Editors. *Emergence of Man Series.* Time-Life Books, New York, 1973.

Trinkaus, Erik and William W. Howells. "The Neanderthals." *Scientific American,* Vol. 241(6): 1970; pages 118-133.

Washburn, S.L., and Ruth Moore. *Ape into Human.* 2nd edition, Little Brown and Company, Boston, 1980.

Zihlman, A.L., and L. Brunker. "Hominid bipedalism: then and now." *Yearbook of Physical Anthropology.* Vol. 22: 1979; pages 132-162.

INDEX